T0263083

CONTROL AND DYNAMIC SYSTEMS

Advances in Theory and Applications

Volume 70

CONTRIBUTORS TO THIS VOLUME

DIRK AEYELS
HAMED M. AL-RAHMANI
YUSUF ALTINTAS
F. BERNELLI-ZAZZERA
STANOJE BINGULAC
SABRI CETINKUNT
GENE F. FRANKLIN
ZORAN GAJIC
TADASHI ISHIHARA
MICHEL KINNAERT
MYOTAEG LIM
P. MANTEGAZZA
YOUBIN PENG
XUEMIN SHEN
ALLAN D. SPENCE
HUGH F. VANLANDINGHAM
JACQUES L. WILLEMS
SIJUN WU

CONTROL AND DYNAMIC SYSTEMS

ADVANCES IN THEORY AND APPLICATIONS

Edited by

C. T. LEONDES

School of Engineering and Applied Science
University of California, Los Angeles
Los Angeles, California

VOLUME 70:   DIGITAL CONTROL SYSTEMS
IMPLEMENTATION TECHNIQUES

ACADEMIC PRESS

San Diego   New York   Boston   London   Sydney   Tokyo   Toronto

Academic Press, Inc.
A Division of Harcourt Brace & Company
525 B Street, Suite 1900, San Diego, California 92101-4495

United Kingdom Edition published by
Academic Press Limited
24-28 Oval Road, London NW1 7DX

International Standard Serial Number: 0090-5267

International Standard Book Number: 0-12-012770-9

Printed and bound in the United Kingdom
Transferred to Digital Printing, 2011

CONTENTS

CONTRIBUTORS

Numbers in parentheses indicate the pages on which the authors' contributions begin.

Dirk Aeyels (353), *Faculty of Engineering, Universiteit Gent, 9052 Gent (Zwijnaarde), Belgium*

Hamed M. Al-Rahmani (1), *Department of Electrical and Computer Engineering, Kuwait University, 13060 Safat, Kuwait*

Yusuf Altintas (243), *Department of Mechanical Engineering, The University of British Columbia, Vancouver, British Columbia, Canada V6T 1Z4*

F. Bernelli-Zazzera (67), *Dipartimento di Ingegneria Aerospaziale, Politecnico di Milano, 40-20133 Milano, Italy*

Stanoje Bingulac (113), *Electrical and Computer Engineering Department, Kuwait University, 13060 Safat, Kuwait*

Sabri Cetinkunt (291), *Department of Mechanical Engineering, University of Illinois at Chicago, Chicago, Illinois 60680*

Gene F. Franklin (1), *Department of Electrical Engineering, Stanford University, Stanford, California 94305*

Zoran Gajic (199), *Department of Electrical and Computer Engineering, Rutgers University, Piscataway, New Jersey 08855*

Tadashi Ishihara (163), *Graduate School of Information Sciences, Tohoku University, Aoba-ku, Sendai 980-77, Japan*

Michel Kinnaert (25), *Laboratoire d'Automatique, Universitè libre de Bruxelles, 1050-Brussels, Belgium*

Myotaeg Lim (199), *Department of Electrical and Computer Engineering, Rutgers University, Piscataway, New Jersey 08855*

P. Mantegazza (67), *Dipartimento di Ingegneria Aerospaziale, Politecnico di Milano, 40-20133 Milano, Italy*

Youbin Peng (25), *Laboratoire d'Automatique, Universitè libre de Bruxelles, 1050-Brussels, Belgium*

Xuemin Shen (199), *Department of Electrical Engineering, University of Alberta, Edmonton, Alberta, Canada T6G 2G7*

Allan D. Spence (243), *Department of Mechanical Engineering, The University of British Columbia, Vancouver, British Columbia, Canada V6T 1Z4*

Hugh F. VanLandingham (113), *The Bradley Department of Electrical Engineering, Virginia Polytechnic Institute and State University, Blacksburg, Virginia 24061*

Jacques L. Willems (353), *Faculty of Engineering, Universiteit Gent, 9052 Gent (Zwijnaarde), Belgium*

Sijun Wu (291), *MagneTek, Drives and Systems, New Berlin, Wisconsin 53151*

PREFACE

Effective control concepts and applications date back over millennia. One very familiar example of this is the windmill. It was designed to derive maximum benefit from windflow, a simple but highly effective optimization technique. Harold Hazen's 1932 paper in the *Journal of the Franklin Institute* was one of the earlier reference points wherein an analytical framework for modern control theory was established. There were many other notable landmarks along the way, including the MIT Radiation Laboratory Series volume on servomechanisms, the Brown and Campbell book, *Principles of Servomechanisms*, and Bode's book entitled *Network Analysis and Synthesis Techniques*, all published shortly after mid-1945. However, it remained for Kalman's papers of the late 1950s (wherein a foundation for modern state space techniques was established) and the tremendous evolution of digital computer technology (which was underpinned by the continuous giant advances in integrated electronics) for truly powerful control systems techniques for increasingly complex systems to be developed. Today we can look forward to a future that is rich in possibilities in many areas of major significance, including manufacturing systems, electric power systems, robotics, and aerospace systems, as well as many other systems with significant economic, safety, cost, and reliability implications. Thus, this volume is devoted to the most timely theme of "Digital Control Systems Implementation Techniques."

The first contribution to this volume is "Techniques in Multivariate Digital Control," by Hamed M. Al-Rahmani and Gene F. Franklin. One of the most significant areas of implementation techniques in digital control systems is that of multivariate digital control system techniques. This contribution presents and illustrates techniques in this broad area, and, as such, this is a most appropriate contribution with which to begin this volume.

The next contribution is "The Design of Digital Pole Placement Controllers," by Michel Kinnaert and Youbin Peng. The basic idea of pole placement is to assign the closed-loop poles at desired locations with respect to achieving desired design specifications. This contribution is an in-depth treatment of a comprehensive set of design rules to develop an assisted pole placement control with a number of essential design features. A number of illustrative examples clearly demonstrate the effectiveness of the techniques presented in this contribution.

The next contribution is "Linearization Techniques for Pulse Width Control of Linear Systems," by F. Bernelli-Zazzera and P. Mantegazza. Pulse width control techniques are an important means of implementing digital control systems. However, such types of control implementations are intrinsically nonlinear and discrete in time and fall into the broader class of pulse modulated systems. This contribution presents linearization design techniques for pulse width modulated control of linear systems and demonstrates their viability and substantive effectiveness through a number of illustrative examples.

The next contribution is "Algorithms for Discretization and Continualization of MIMO State Space Representations," by Stanoje Bingulac and Hugh F. VanLandingham. The modeling of digital control system implementations requires algorithms for the problems of discretization of continuous-time models as well as the inverse problem of recreating a continuous-time model from a given discrete-time model. This contribution presents and illustrates seven robust algorithms for dealing with these problems. With these algorithms, the system design engineer has complete flexibility to move between the continuous and discrete model domains. As such, this contribution is an essential element of this volume.

The next contribution is "Discrete-Time Control Systems Design via Loop Transfer Recovery," by Tadashi Ishihara. Most theoretical results on Loop Transfer Recovery (LTR) techniques, i.e., the system open-loop transfer function which produces the required closed-loop transfer function, have been given in the case of continuous-time system formulations. This contribution presents LTR techniques for designing discrete-time controls, taking into account the inherent limitations of digital control systems. A number of illustrative examples exemplify the effectiveness of the techniques presented in this contribution.

The next contribution is "The Study of Discrete Singularly Perturbed Linear-Quadratic Control Systems," by Zoran Gajic, Myotaeg Lim, and Xuemin Shen. The main goal in the control theory of singular perturbations is to deal with the system design problem by decomposition into slow and fast system time scales. This has been a fruitful control engineering research and design area for the past 25 years, and has been investigated for digital control systems since the early 1980s. This contribution is an in-depth treatment of this area of discrete-time systems with numerous illustrative examples which demonstrate the effectiveness of these techniques.

The next contribution is "Modeling Techniques and Control Architectures for Machining Intelligence," by Allan D. Spence and Yusuf Altintas. Certainly among the very many implementations of digital control systems in practice is that of Digital Numerical Control (DNC) of machine tools, which dates back to the early 1960s. This contribution is an in-depth treatment of the modern techniques and technology of DNC of machine tools and is a most appropriate element of this volume.

The next contribution is "Techniques in Discrete-Time Position Control of Flexible One Arm Robots," by Sijun Wu and Sabri Cetinkunt. Whereas DNC of machine tools was among the first implementations of digital control systems, robot systems are among the most recent implementations of digital control systems, particularly over the past decade. This contribution is an in-depth treatment of the techniques for a broad class of robot systems, and is also a most appropriate contribution for this volume.

The final contribution to this volume is "Pole Assignment by Memoryless Output Feedback," by Dirk Aeyels and Jacques L. Willems. The design requirement to achieve desired or specialized discrete-time dynamic systems is inherently related to the position of the closed-loop system poles. This contribution is an in-depth treatment of general techniques for achieving the required closed-loop pole positions for a digital control system implementation.

The contributors to this volume are all to be highly commended for their contributions to this unique and rather comprehensive treatment of implementation techniques in digital control systems, which should provide a unique reference on the international scene of this broad subject for many individuals in diverse areas of activity for many years to come.

Techniques In Multirate Digital Control

Hamed M. Al-Rahmani

Department of Electrical and Computer Engineering
Kuwait University

Gene F. Franklin

Department of Electrical Engineering
Stanford University

I. INTRODUCTION

When a digital or a sampled-data system possesses different sampling rates at different locations, it is called a multirate system. Due to their importance, multirate systems have received a great deal of attention in the past years [1-16]. Unlike single-rate systems, multirate sampling provides more freedom in selecting the sampling rates than does single-rate sampling. For example, in a single-rate system, the sampling rate can be limited by the sensor speed or the A/D converter speed. A slow output sampling results in slow input updates which, in turn, may result in poor performance. On the other hand, with multirate sampling, a higher performance might be achieved by a higher control update rate. A typical example can be seen in robotics systems that utilize video cameras as sensors. Here, the accurate tracking requires a high control update rate as compared to the relatively slow rate at which the pictures are taken. Notice that a fast video camera together with the required large buffers and high-speed A/D's can prove very costly.

A novel multirate scheme for the control design of linear periodic systems was introduced in [4]. It has been shown that such a scheme produces the near high performance of complex multirate methods while maintaining the design simplicity of single-rate systems. In what follows, we study that multirate design for the case when only the output is available for measurements. In particular, a multirate LQG design algorithm is formulated in Section 4, while a complete design example for the control of resonance in disk-drive systems is considered in Section 5. But first, a description of the multirate scheme together with a theoretical background on its controllability and stabilizability properties are provided in Sections 2 and 3.

II. SYSTEM FORMULATION

The following notation will be used throughout: Given an $n \times m$ matrix L, L' is the transpose of L; if $n = m$, then $L > 0$ $[L \geq 0]$ implies that L is a real symmetric $(L = L')$ positive-definite [positive-semidefinite] matrix. $\mathcal{E}\{L\}$ is the expected value of L. The set of natural numbers (positive integers) will be denoted by $\mathcal{N}$. $I^{n \times n}$ (or simply I, when dimensions are clear) is the $n \times n$ identity matrix. The samples of an analog signal $f(t)$ at the points $t = kT, k = 0, 1, \ldots$, are denoted by $f(k)$, i.e., $f(k) = f(kT)$.

Consider the following linear time-invariant plant

$$\left. \begin{aligned} \dot{x}(t) &= Ax(t) + Bu(t) + Gw_c(t) \\ y(t) &= Cx(t) \end{aligned} \right\} \tag{1}$$

and the measurements

$$z(k) = y(k) + v(k) = Cx(k) + v(k) \tag{2}$$

where $x(t) \in \mathcal{R}^n$, $u(t) \in \mathcal{R}^p$, $w_c(t) \in \mathcal{R}^{\bar{p}}$ and $y(t) \in \mathcal{R}^q$. A, B, C, and G are real matrices with appropriate dimensions. The disturbances $w_c(t)$ and $v(k)$ are stationary zero-mean Gaussian noise and are independent of the initial conditions. Here, $w_c(t)$ is continuous with a spectral density $R_{w_c} \geq 0$ while $v(k)$ is discrete and its covariance is

$$\mathcal{E}[v(k)v(j)] = \begin{cases} R_v & \text{if } k = j \\ 0 & \text{otherwise} \end{cases}$$

Assume that the input vector $u(t)$ is partitioned into m subvectors, $u_i(t) \in \mathcal{R}^{p_i}$, $i = 1, 2, \ldots, m$ with $p_1 + p_2 + \cdots + p_m = p$. A block diagram of the open-loop multirate system is shown in Figure 1. The gains Ψ and $\theta_i(l_i)$, $l_i = 0, 1, \ldots, r_i - 1$, and $i = 1, 2, \ldots, m$, are defined in Eqs. (12) and (6) below, respectively. Here the sequence $\{\theta_i(l_i)\}$ is periodic with period r_i, i.e., $\theta_i(l_i + r_i) = \theta_i(l_i)$. Notice that the plant output is sampled every T

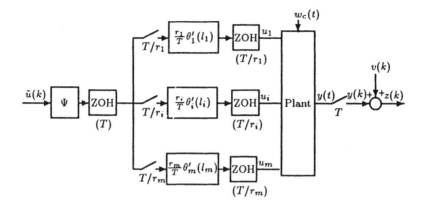

Figure 1: A block diagram of the open-loop multirate system.

seconds, while, the input to the ith channel is updated every T/r_i seconds, where r_i are positive integers and $l_i = 0, 1, \ldots, r_i - 1$. Also, notice that the input $u_i(t)$ is constrained to the following piecewise constant signal

$$u_i(t) = \frac{r_i}{T}\theta_i(l_i)\Psi\tilde{u}(k) \qquad (3)$$

for $t \in [kT + l_iT/r_i, kT + (l_i + 1)T/r_i)$, $k = 0, 1, \ldots$, and $l_i = 0, 1, \ldots, r_i - 1$. Here, $\tilde{u}(k)$ is an arbitrary input updated every T seconds.

The matrix B can be partitioned according to the partition of the input vector $u(t)$, that is

$$B = \left[B_1 \vdots B_2 \vdots \cdots \vdots B_m \right] \quad , \quad B_i \in \mathcal{R}^{n \times p_i} \qquad (4)$$

For any point $\bar{t} \in [0, T]$, define the matrix

$$
\begin{aligned}
\mathcal{H}_i(\bar{t}) &= \frac{r_i}{T} \sum_{l_i=0}^{j_i-1} \int_{l_iT/r_i}^{(l_i+1)T/r_i} e^{A(\bar{t}-\tau)} B_i \, d\tau \, \theta_i'(l_i) \\
&\quad + \frac{r_i}{T} \int_{j_iT/r_i}^{\bar{t}} e^{A(\bar{t}-\tau)} B_i \, d\tau \, \theta_i'(j_i)
\end{aligned}
\qquad (5)
$$

where j_i is such that $j_iT/r_i \le \bar{t} < (j_i + 1)T/r_i$, and where

$$\theta_i(l_i) = \int_{l_iT/r_i}^{(l_i+1)T/r_i} e^{A(T-\tau)} B_i \, d\tau \qquad (6)$$

Since A and B are constant matrices, one can write

$$\theta_i(l_i) = e^{A(r_i - l_i - 1)T/r_i}\Gamma_i \qquad (7)$$

where

$$\Gamma_i = \int_0^{T/r_i} e^{A\tau} B_i d\tau \tag{8}$$

Now, at $\bar{t} = T$, Eq. (5) defines the generalized reachability Gramian of the pair $\{A, B_i\}$ of order r_i on $[0, T]$ [17]

$$\mathcal{H}_i = \frac{r_i}{T} \sum_{l_i=0}^{r_i-1} \theta_i(l_i)\theta_i'(l_i) \tag{9}$$

Define the input multirate Gramian $\mathcal{H}$ by

$$\mathcal{H} = \sum_{i=1}^{m} \mathcal{H}_i \tag{10}$$

and let Λ be an $n \times \nu$ full rank matrix such that

$$\Lambda\Lambda' = \mathcal{H} \tag{11}$$

where ν is the rank of $\mathcal{H}$. Let Ψ be the right pseudo inverse of Λ', that is,

$$\Psi = \Lambda(\Lambda'\Lambda)^{-1} \tag{12}$$

Moreover, define the matrix

$$\Lambda(\bar{t}) = \left(\sum_{i=1}^{m} \mathcal{H}_i(\bar{t}) \right) \Psi \tag{13}$$

One can show that the state at any time $t \in [kT, kT+T]$ can be determined by the following equations:

$$x(t) = e^{A(t-kT)}x(k) + \Lambda(t - kT)\tilde{u}(k) + \int_0^{t-kT} e^{A\tau} Gw_c(t - \tau)d\tau \tag{14}$$

In particular, when $t = kT + T$, we have

$$\left.\begin{aligned} x(k+1) &= \Phi x(k) + \Lambda\tilde{u}(k) + w(k) \\ z(k) &= y(k) + v(k) = Cx(k) + v(k) \end{aligned}\right\} \tag{15}$$

where $\Phi = e^{AT}$ and

$$w(k) = \int_0^T e^{A\tau} Gw_c(kT + T - \tau)d\tau \tag{16}$$

is a stationary white Gaussian noise with zero mean and covariance

$$\mathcal{E}[w(k)w(j)] = \begin{cases} R_w & \text{if } k = j \\ 0 & \text{otherwise} \end{cases}$$

where

$$R_w = \int_0^T e^{A\tau} G R_{w_c} G' e^{A'\tau} d\tau \tag{17}$$

Equations (15) describes a time-invariant discrete system with a transfer function from the input $\tilde{u}(k)$ to the output $y(k)$ given by

$$G_d(z) = C(zI - \Phi)^{-1}\Lambda \tag{18}$$

This system is in the form of a single-rate multi-input discrete system with $\nu \leq n$ inputs. In fact, the discrete system described by Eqs. (15) can be viewed as a generalization of the single-rate sampled-data system. Assume that the matrix B has a full column rank. Then, when $r_1 = r_2 = \ldots = r_m = 1$, the multirate Gramian of Eq. (10) is reduced to

$$\mathcal{H} = \frac{1}{T}\Gamma\Gamma'$$

where

$$\Gamma = \int_0^T e^{A\tau} B d\tau$$

is the standard input matrix of a single-rate sampled-data system. Since $\mathcal{H} = \Lambda\Lambda'$, one can directly choose

$$\Lambda = \frac{1}{\sqrt{T}}\Gamma$$

and the system of Eqs. (15) is essentially the standard single-rate system.

II. CONTROLLABILITY AND STABILIZABILITY

In considering any control scheme, it is of a great importance that we study the controllability aspects of that scheme. In this section we study the controllability and stabilizability properties of the multirate system and discuss the conditions under which controllability is preserved with sampling. For the sake of completion, some of the results in this section are adapted from [4].

A. CONTROLLABILITY

We use the standard Kalman definition of controllability [18]. The following definition is useful in the sequel.

Definition 1 *An eigenvalue λ of A is said to be uncontrollable in the pair $\{A, B\}$ if there exists a left eigenvector $z' \neq 0$ of A associated with λ such that $z'A = \lambda A$ and $z'B = 0$. Otherwise, λ is said to be controllable in $\{A, B\}$. Furthermore, the pair $\{A, B\}$ is said to be controllable if all eigenvalues of A are controllable in $\{A, B\}$.*

The controllability of the multirate system is considered in Theorem 1 below. First we have the following definition.

Definition 2 *The numbers $T > 0$ and $r_i \in \mathcal{N}$, $i = 1, 2, \ldots, m$, are said to be proper if, for any two eigenvalues λ and $\bar{\lambda}$ of A, either of the following is satisfied:*

$$Im(\lambda - \bar{\lambda}) \neq \frac{2\pi\kappa r_i}{T} \quad whenever \ Re(\lambda - \bar{\lambda}) = 0 \tag{19}$$

for $i = 1, \ldots, m$ and $\kappa = \pm 1, \pm 2, \ldots$, or

$$Im(\lambda - \bar{\lambda}) \neq \frac{2\pi\kappa}{T} \quad whenever \ Re(\lambda - \bar{\lambda}) = 0 \tag{20}$$

and $\kappa = \pm 1, \pm 2, \ldots$.

We will use the two statements "$T > 0$ and $r_i \in \mathcal{N}$, $i = 1, 2, \ldots, m$, are proper" and "the sampling rates are proper" interchangeably.

Theorem 1 *Assume that $T > 0$ and $r_i \in \mathcal{N}$, $i = 1, \ldots, m$, are proper. Then the pair $\{\Phi, \Lambda\}$ is controllable if and only if the pair $\{A, B\}$ is controllable.*

Proof (i) Assume that for some $T > 0$ and $r_i \in \mathcal{N}$, $i = 1, 2, \ldots, m$, the pair $\{\Phi, \Lambda\}$ is controllable but $\{A, B\}$ is not. Then there is an eigenvalue λ of A with an associated left eigenvector $z' \neq 0$ such that $z'A = \lambda z'$ and $z'B = 0$. Thus, $z'B_i = 0$, $i = 1, 2, \ldots, m$. It follows then from Eqs. (6), (9), (10), and (11) that $z'\Lambda = 0$ (note that $z'e^{At} = e^{\lambda t}z'$ for any $t \in \mathcal{R}$). Since $e^{\lambda T}$ is an eigenvalue of Φ with an associated left eigenvector z', $e^{\lambda T}$ is uncontrollable in $\{\Phi, \Lambda\}$, which is a contradiction.

(ii) Next, assume that T and r_i, $i = 1, 2, \ldots, m$, are proper, and that $\{A, B\}$ is controllable. Since $\mathcal{H}$ and Λ have the same left null space, we only need to show that $\{\Phi, \mathcal{H}\}$ is controllable. So, assume that $\{\Phi, \mathcal{H}\}$ is not controllable. Then there exists an eigenvalue λ of Φ with a left eigenvector

$z' \neq 0$ such that $z'\Phi = \lambda z'$ and $z'\mathcal{H}z = 0$. It follows from Eqs. (9) and (10) that, for all i, $z'\theta_i(l_i) = 0$, $l_i = 0, 1, \ldots, r_i - 1$, or equivalently, from Eq. (7), $z'e^{Al_iT/r_i}\Gamma_i = 0$, for $l_i = 0, 1, \ldots, r_i - 1$. Furthermore, for any $k_i \geq r_i$, we have $e^{Ak_iT/r_i} = \Phi^{j_i}e^{Al_iT/r_i}$ for some $j_i \in \mathcal{N}$ and some $l_i \in \{0, 1, \ldots, r_i - 1\}$. Hence,

$$z'e^{Ak_iT/r}\Gamma_i = z'\Phi^{j_i}e^{Al_iT/r_i}\Gamma_i = \lambda^{j_i}z'e^{Al_iT/r_i}\Gamma_i = 0$$

Therefore, for all $k_i \geq 0$, $z'e^{Ak_iT/r_i}\Gamma_i = 0$, $i = 1, 2, \ldots, m$. Thus,

$$z'\left[\Gamma_i \vdots e^{AT/r_i}\Gamma_i \vdots \cdots \vdots e^{A(n-1)T/r_i}\Gamma_i\right] = 0$$

Since T/r_i satisfies Eq. (19) (if Eq. (20) is satisfied then so is Eq. (19)), one can use an argument similar to that of [19, proof of Theorem 12] to show that

$$z'\left[B_i \vdots AB_i \vdots \cdots \vdots A^{n-1}B_i\right] = 0$$

for $i = 1, 2, \ldots, m$. That is,

$$z'\left[B \vdots AB \vdots \cdots \vdots A^{n-1}B\right] = 0$$

contradicting the assumption that $\{A, B\}$ is controllable. This completes the proof. $\square$

Theorem 1 gives necessary and sufficient conditions for the controllability of the multirate system. It also gives sufficient conditions for choosing T and r_i, $i = 1, \ldots, m$, such that controllability is preserved with sampling. We notice from Definition 2 that if Eq. (20) is satisfied, then Eq. (19) is satisfied as well. Therefore, one can choose $T > 0$ according to Eq. (20) and controllability is preserved *for any* set of $r_i \in \mathcal{N}$, $i = 1, \ldots, m$. On the other hand, *for any* $T > 0$, one can choose the r_i's such that Eq. (19) is met. In other words, if the plant is controllable, the multirate system can be made controllable whatever the measurement rate is. In contrast, single-rate sampled-data systems may lose controllability if the sampling period T does not satisfy Eq. (20) [19].

According to Theorem 1, if the continuous plant is controllable and sampling rates are proper, we can find a real matrix $K \in \mathcal{R}^{\nu \times n}$ such that the closed-loop matrix of the overall system, $\Phi + \Lambda K$, has any desired set of self-conjugate eigenvalues. In the case when Λ is nonsingular, we can assign the whole closed-loop matrix by choosing

$$K = \Lambda^{-1}(\Phi_{cl} - \Phi)$$

where Φ_{cl} is the desired closed-loop matrix.

B. STABILIZABILITY

In the previous subsection, we studied the case of completely control-lable systems. Next, we will consider the case of stabilizable plants. The notion of stabilizability was first introduced by Wonham [20]. We say that a pair $\{F, G\}$ (continuous or discrete) is stabilizable if and only if there exist a real matrix K such that the closed-loop matrix $F+GK$ has stable eigenval-ues (i.e., those with negative real parts, if the plant is continuous, or with magnitudes less than unity, if the plant is discrete). Using Definition 1, one can also say that the pair $\{F, G\}$ is stabilizable if all uncontrollable eigenvalues of F are stable.

The next theorem gives necessary and sufficient conditions for the sta-bilizability of the multirate system.

Theorem 2 *Assume that* $T > 0$ *and* $r_i \in \mathcal{N}$, $i = 1, \ldots, m$, *are proper. Then the pair* $\{\Phi, \Lambda\}$ *is stabilizable if and only if the pair* $\{A, B\}$ *is stabi-lizable.*

Proof: (i) The necessity part ("only if") can be proved in a similar manner as in part (i) of the proof of Theorem 1 with each of the words "controllable" and "eigenvalue" to be replaced by "stabilizable" and "un-stable eigenvalue", respectively.

(ii) Next, assume that T and r_i, $i = 1, 2, \ldots, m$, are proper, and that $\{A, B\}$ is stabilizable. Again, since $\mathcal{H}$ and Λ have the same left null space, we only need to show that $\{\Phi, \mathcal{H}\}$ is stabilizable. We prove this by con-tradiction. So, assume that $\{\Phi, \mathcal{H}\}$ is not stabilizable. Without loss of generality, we assume that A and B are in the controllable-uncontrollable form:

$$A = \begin{bmatrix} A_c & A_{12} \\ 0 & A_{\bar{c}} \end{bmatrix} \quad , \quad B = \begin{bmatrix} B_c \\ 0 \end{bmatrix} \tag{21}$$

where $A_c \in \mathcal{R}^{n_c \times n_c}$, $B_c \in \mathcal{R}^{n_c \times p}$, $\{A_c, B_c\}$ is controllable, and $A_{\bar{c}}$ is sta-ble. Here, $n_c \leq n$ with $n_c = n$ if and only if $\{A, B\}$ is controllable. Let

B_{c_i}, $i = 1, 2, \ldots, m$, be the ith partition of B_c, i.e., $B_c = [B_{c_1} \vdots \cdots \vdots B_{c_m}]$. Accordingly, the matrices Φ and $\mathcal{H}$ are decomposed into

$$\Phi = \begin{bmatrix} \Phi_c & \Phi_{12} \\ 0 & \Phi_{\bar{c}} \end{bmatrix} \quad , \quad \mathcal{H} = \begin{bmatrix} \mathcal{H}_c \\ 0 \end{bmatrix} \tag{22}$$

where $\Phi_c = e^{A_c T} \in \mathcal{R}^{n_c \times n_c}$ and

$$\mathcal{H}_c = \sum_{i=1}^{m} \mathcal{H}_{c_i} \tag{23}$$

with

$$\mathcal{H}_{c_i} = \frac{r_i}{T} \sum_{l_i=0}^{r_i-1} \theta_{c_i}(l_i) \theta'_{c_i}(l_i) \tag{24}$$

$$\theta_{c_i}(l_i) = e^{A_c(r_i - l_i - 1)T/r_i} \Gamma_{c_i} \quad , \quad l_i = 0, 1, \ldots, r_i - 1 \tag{25}$$

where

$$\Gamma_{c_i} = \int_0^{T/r_i} e^{A_c \tau} B_{c_i} d\tau$$

Note that $\Phi_{\bar{z}}$ is stable since $\Phi_{\bar{z}} = e^{A_c T}$. Now, since $\{\Phi, \mathcal{H}\}$ is unstabilizable, the pair $\{\Phi_c, \mathcal{H}_c\}$ is uncontrollable. Therefore, there exists an eigenvalue λ of Φ_c with a left eigenvector $z' \neq 0$ such that $z'\Phi_c = \lambda z'$ and $z'\mathcal{H}_c z = 0$. It follows then from Eqs. (23) and (24) that, for all i, $z'\theta_{c_i}(l_i) = 0$, $l_i = 0, 1, \ldots, r_i - 1$, or equivalently, $z' e^{A_c l_i T/r_i} \Gamma_{c_i} = 0$, for $l_i = 0, 1, \ldots, r_i - 1$. Furthermore, for any $k_i \geq r_i$, we have $e^{A_c k_i T/r_i} = \Phi_c^{j_i} e^{A_c l_i T/r_i}$ for some $j_i \in \mathcal{N}$ and some $l_i \in \{0, 1, \ldots, r_i - 1\}$. Hence,

$$z' e^{A_c k_i T/r} \Gamma_{c_i} = z' \Phi_c^{j_i} e^{A_c l_i T/r_i} \Gamma_{c_i} = \lambda^{j_i} z' e^{A_c l_i T/r_i} \Gamma_{c_i} = 0$$

Therefore, for all $k_i \geq 0$, $z' e^{A_c k_i T/r_i} \Gamma_{c_i} = 0$, $i = 1, 2, \ldots, m$. Thus,

$$z' \left[\Gamma_{c_i} \vdots e^{A_c T/r_i} \Gamma_{c_i} \vdots \cdots \vdots e^{A_c(n_c - 1)T/r_i} \Gamma_{c_i} \right] = 0$$

Since T/r_i satisfies Eq. (19) (if Eq. (20) is satisfied then so is Eq. (19)), once again we can use an argument similar to that of [19, proof of Theorem 12] to show that

$$z' \left[B_{c_i} \vdots A_c B_{c_i} \vdots \cdots \vdots A_c^{n_c-1} B_{c_i} \right] = 0$$

for $i = 1, 2, \ldots, m$. That is,

$$z' \left[B_c \vdots A_c B_c \vdots \cdots \vdots A_c^{n_c-1} B_c \right] = 0$$

and the pair $\{A_c, B_c\}$ is uncontrollable. A contradiction. $\square$.

Thus, if the sampling rates are proper and the continuous plant is stabilizable, the multirate system can be made stable by the proper choice of the feedback gain matrix K. For example, we can write Φ and $\mathcal{H}$ as in Eq. (23) and define $\Lambda_c \in \mathcal{R}^{n_c \times \nu_c}$ such that $\Lambda_c \Lambda'_c = \mathcal{H}_c$, where ν_c is the rank of $\mathcal{H}_c$ (note that rank $\mathcal{H}_c$ equals rank $\mathcal{H}$, i.e., $\nu_c = \nu$). Then

$$\Lambda = \begin{bmatrix} \Lambda_c \\ 0 \end{bmatrix}$$

Accordingly, one can choose $K \in \mathcal{R}^{\nu_c \times n}$ as

$$K = [K_c \quad 0]$$

where $K_c \in \mathcal{R}^{\nu_c \times n_c}$ is such that the matrix $\Phi_c + \Lambda_c K_c$ has any desired set of self-conjugate eigenvalues.

III. MULTIRATE LQG DESIGN

We discussed, in the previous section, the conditions under which the poles of the multirate system can be arbitrarily assigned. However, since, in general, the discrete pair $\{\Phi, \Lambda\}$ represents a multi-input system, there can be many feedback gain solutions for any given set of desired poles. Such freedom can be utilized in an optimal design based on solving the continuous-time LQG problem.

Consider the open-loop multirate system of Figure 1. The optimization problem is defined by the cost function

$$J = \frac{1}{2}\mathcal{E}\left\{ \int_0^\infty [x'(t)Qx(t) + u'(t)Ru(t)]\,dt \right\} \tag{26}$$

and the dynamic constraints given by Eqs. (15) with the initial state $x(0)$ having a Gaussian distribution with a known mean $\mathcal{E}[x(0)] = \hat{x}(0)$. We assume that $Q \geq 0$ and $R > 0$ are real symmetric $n \times n$ and $p \times p$ matrices, respectively. Furthermore, for simplicity, we assume that $R =$ block-diag$\{R_1, \ldots, R_m\}$, $R_i \in \mathcal{R}^{p_i \times p_i}$. One can show that

$$J = \frac{1}{2}\mathcal{E}\left\{ \sum_{k=0}^\infty [\; x'(k) \quad \tilde{u}'(k) \;] \begin{bmatrix} \tilde{Q} & \tilde{M} \\ \tilde{M}' & \tilde{R} \end{bmatrix} \begin{bmatrix} x(k) \\ \tilde{u}(k) \end{bmatrix} \right\} \tag{27}$$

where

$$\tilde{Q} \;=\; \int_0^T e^{A'\bar{t}}Qe^{A\bar{t}}d\bar{t} \tag{28}$$

$$\tilde{M} \;=\; \int_0^T e^{A'\bar{t}}Q\Lambda(\bar{t})d\bar{t} \tag{29}$$

$$\tilde{R} \;=\; \int_0^T \Lambda'(\bar{t})Q\Lambda(\bar{t})d\bar{t} + \Psi' \sum_{i=1}^m \frac{r_i}{T} \sum_{l_i=0}^{r_i-1} \theta_i(l_i)R_i\theta_i'(l_i)\Psi \tag{30}$$

The optimization problem can then be solved using the Separation Theorem [21]. Thus, the optimal control is obtained by solving the deterministic

LQR problem (neglecting the disturbance effects and dealing with the state $x(k)$ as if it were available for measurements). Then, an estimate of the state, $\hat{x}(k)$, is obtained, using any of the classical estimation techniques, to substitute the actual state in updating the control.

Now, the optimal solution of the LQR is well-known and is given by

$$\tilde{u}_{opt}(k) = \mathbf{K} x(k) \tag{31}$$

where

$$\mathbf{K} = -(\tilde{R} + \Lambda' \tilde{P} \Lambda)^{-1}(\tilde{M}' + \Lambda' \tilde{P} \Phi) \tag{32}$$

The matrix $\tilde{P}$ is determined by the algebraic discrete Riccati equation

$$\tilde{P} = \Phi' \tilde{P} \Phi + \tilde{Q} - (\tilde{M} + \Phi' \tilde{P} \Lambda)(\tilde{R} + \Lambda' \tilde{P} \Lambda)^{-1}(\tilde{M}' + \Lambda' \tilde{P} \Phi) \tag{33}$$

The existance of a positive-definite solution of Eq. (33) and the asymptotic stability of the system are guaranteed by the next theorem.

Theorem 3 *Assume that controllability is preserved with sampling. Then, there exists a unique positive-definite solution $\tilde{P}$ of Eq. (33) if $\{A, B\}$ is controllable and $\{A, Q\}$ is observable. The corresponding closed-loop system is asymptotically stable.*

To simplify the proof of Theorem 3, we transform J of Eq. (27) into an uncoupled cost function. Define

$$\bar{Q} = \tilde{Q} - \tilde{M} \tilde{R}^{-1} \tilde{M}' \tag{34}$$

and

$$\bar{\Phi} = \Phi - \Lambda \tilde{R}^{-1} \tilde{M}' \tag{35}$$

to obtain the equivalent discrete LQR problem defined by the uncoupled cost function

$$J = \frac{1}{2} \mathcal{E} \left\{ \sum_{k=0}^{\infty} \left[x'(k) \bar{Q} x(k) + \bar{u}'(k) \tilde{R} \bar{u}(k) \right] \right\} \tag{36}$$

and the dynamic constraints

$$x(k+1) = \bar{\Phi} x(k) + \Lambda \bar{u}(k) \tag{37}$$

The optimal $\bar{u}$ that minimizes J is given by

$$\bar{u}_{opt}(k) = -(\tilde{R} + \Lambda' \tilde{P} \Lambda)^{-1} \Lambda' \tilde{P} \bar{\Phi} x(k) \tag{38}$$

where $\tilde{P}$ is the solution of the discrete algebraic Riccati equation

$$\tilde{P} = \bar{\Phi}' \tilde{P} \bar{\Phi} + \bar{Q} - \bar{\Phi}' \tilde{P} \Lambda (\tilde{R} + \Lambda' \tilde{P} \Lambda)^{-1} \Lambda' \tilde{P} \bar{\Phi} \tag{39}$$

which has the same solution $\tilde{P}$ as that of Eq. (33). The corresponding optimal $\tilde{u}$ is given by

$$\begin{aligned}
\tilde{u}_{opt}(k) &= \bar{u}_{opt}(k) - \tilde{R}^{-1}\tilde{M}'x(k) \\
&= \mathbf{K}x(k)
\end{aligned}$$

where $\mathbf{K}$ is given by Eq. (32) with the matrix $\tilde{P}$ obtained from either Eq. (33) or Eq. (39). The following lemma assures the existence of $\tilde{R}^{-1}$.

Lemma 1 $\tilde{R} > 0$ if $R > 0$ or $Q > 0$.

Proof: Referring to Eq. (30), we can write

$$\tilde{R} = \tilde{R}_Q + \tilde{R}_R \tag{40}$$

where

$$\tilde{R}_Q = \int_0^T \Lambda'(t)Q\Lambda(t)dt \tag{41}$$

$$\tilde{R}_R = \Psi' \sum_{i=1}^m \frac{r_i}{T} \sum_{l_i=0}^{r_i-1} \theta_i(l_i)R_i\theta_i'(l_i)\Psi \tag{42}$$

(i) First assume that $Q > 0$. Also, let y be an arbitrary vector in $\mathcal{R}^\nu$ such that $y'\tilde{R}_Q y = 0$. It follows from Eq. (41) that $\Lambda(t)y = 0$ for all $t \in [0, T]$ ($\Lambda(t)$ is continuous on $[0, T]$). In particular, at $t = T$, $\Lambda(T)y = \Lambda y = 0$. Hence, $y = 0$ since Λ has full column rank. Therefore, $\tilde{R}_Q > 0$, and so, $\tilde{R} > 0$.

(ii) Next, assume that $R > 0$. We claim that $\tilde{R}_R > 0$, and hence, $\tilde{R} > 0$. In fact, if $\tilde{R}_R$ is singular, then $\exists y \neq 0$ such that $y'\tilde{R}_R y = 0$. Now, since $R_i > 0$ $\forall i$, it follows then from Eq. (42) that

$$\theta_i'(l_i)\Psi y = 0 \quad \forall l_i \text{ and } i \Leftrightarrow \frac{r_i}{T} \sum_{l_i=0}^{r_i-1} \theta_i(l_i)\theta_i'(l_i)\Psi y = \mathcal{H}_i \Psi y = 0 \quad \forall i$$

Therefore, $\sum_{i=1}^m \mathcal{H}_i \Psi y = \Lambda y = 0$, contradicting the fact that Λ has full (column) rank. □

Thus, since we have assumed that $R > 0$, Lemma 1 guarantees the non singularity of $\tilde{R} + \Lambda' \tilde{P}\Lambda$. Moreover, although there is a minus sign in the expression of $\bar{Q}$ of Eq. (34), $\bar{Q} \geq 0$ always. This is shown in the next lemma.

Lemma 2 $\bar{Q} \geq 0$

Proof: Let z be an arbitrary vector in $\mathcal{R}^n$ and choose $x(0) = z$ and

$$u_i(t) = \frac{r_i}{T}\theta_i'(l_i)\Psi\tilde{u}(0)$$

for $l_i T/r_i \leq t < (l_i+1)T/r_i$, $l_i = 0, 1, \ldots, r_i - 1$, and $i = 1, 2, \ldots, m$, where $\tilde{u}(0) = -\tilde{R}^{-1}\tilde{M}'z$. The state $x(t)$ on $[0, T]$ is determined by Eq. (14). Then, with some manipulations one can show that

$$\int_0^T [x'(t)Qx(t) + u'(t)Ru(t)]\, dt = z'\bar{Q}z$$

Therefore, $z'\bar{Q}z \geq 0$ since the L.H.S. is nonnegative. $\square$

Proof of Theorem 3: It is well known that Eq. (39), and hence, Eq. (33) has a unique positive-definite solution $\tilde{P}$, and the corresponding closed-loop system is asymptotically stable, if $\{\bar{\Phi}, \Lambda\}$ is controllable and $\{\bar{\Phi}, \bar{Q}\}$ is observable. Thus, it suffices to show that:

(a) $\{\bar{\Phi}, \Lambda\}$ is controllable if and only if $\{A, B\}$ is controllable, and

(b) $\{\bar{\Phi}, \bar{Q}\}$ is observable if and only if $\{A, Q\}$ is observable.

(a) Let λ be an uncontrollable eigenvalue in the pair $\{\bar{\Phi}, \Lambda\}$ and let $z' \neq 0$ be an associated left eigenvector. Then, $z'\bar{\Phi} = \lambda z'$ and $z'\Lambda = 0$. Now, $\lambda z' = z'\bar{\Phi} = z'(\Phi - \Lambda\tilde{R}^{-1}\tilde{M}') = z'\Phi$. Therefore, the pair $\{\Phi, \Lambda\}$ is uncontrollable. Similarly, one can show that if $\{\Phi, \Lambda\}$ is uncontrollable, then $\{\bar{\Phi}, \Lambda\}$ is uncontrollable. Statement (a) then follows from Theorem 1.

(b) First assume that $\{A, Q\}$ is unobservable. Then there exists an eigenvalue λ of A and a vector $z \neq 0$ such that $Az = \lambda z$ and $Qz = 0$. Thus, $Qe^{At}z = e^{\lambda t}Qz = 0$ for all $t \geq 0$. It follows then from Eqs. (28) and (29) that $\tilde{Q}z = 0$ and $\tilde{M}'z = 0$. Referring to Eqs. (34) and (35), we have $\bar{Q}z = 0$ and $\bar{\Phi}z = e^{\lambda T}z$. Therefore, $e^{\lambda T}$ is unobservable in $\{\bar{\Phi}, \bar{Q}\}$.

Next, assume that $\{A, Q\}$ is observable but $\{\bar{\Phi}, \bar{Q}\}$ is not. Let λ be an unobservable eigenvalue of $\bar{\Phi}$ in $\{\bar{\Phi}, \bar{Q}\}$ and let $z \neq 0$ be an associated eigenvector. Then, $\bar{\Phi}z = \lambda z$ and $\bar{Q}z = 0$. Consider the initial state $x(0) = z$ and the inputs $u_i(t) = \frac{r_i}{T}\theta'_{l_i}\Psi\tilde{u}(0)$ for $l_i T/r_i \leq t < (l_i + 1)T/r_i$, $l_i = 0, 1, \ldots, r_i - 1$, $i = 1, 2, \ldots, m$, where $\tilde{u}(0) = -\tilde{R}^{-1}\tilde{M}'z$. One can show that

$$\int_0^T [x'(t)Qx(t) + u'(t)Ru(t)]\, dt = z'\bar{Q}z = 0$$

Since $R > 0$ and $Q \geq 0$,

$$\int_0^T x'(t)Qx(t)dt = \int_0^T u'(t)Ru(t)dt = 0 \qquad (43)$$

It can be shown that

$$\int_0^T u'(t)Ru(t)dt = z'\tilde{M}\tilde{R}^{-1}\tilde{R}_R\tilde{R}^{-1}\tilde{M}'z \qquad (44)$$

where $\tilde{R}_R > 0$ as shown in the proof of Lemma 1. But since the L.H.S. of Eq. (44) is zero and $\tilde{R}_R > 0$, $\tilde{M}'z = 0$. Therefore,

$$0 = z'\bar{Q}z = z'(\tilde{Q} - \tilde{M}\tilde{R}^{-1}\tilde{M}')z = z'\tilde{Q}z$$

It follows from Eq. (28) that $Qe^{At}z \equiv 0$, which is equivalent to say that the pair $\{A, Q\}$ is unobservable. This completes the proof. $\square$

The next theorem gives necessary and sufficient conditions for the existence of a unique nonnegative-definite solution of the Riccati equation and the asymptotic stability of the overall system.

Theorem 4 *Assume that controllability is preserved with sampling. Then, there exists a unique positive-semidefinite solution $\tilde{P}$ of Eq. (33) and the corresponding closed-loop system is asymptotically stable if and only if $\{A, B\}$ is stabilizable and $\{A, Q\}$ is detectable.*

The proof is left as an exercise.

As we mentioned before, the state $x(k)$ is actually not available for measurements. Therefore, an estimate $\hat{x}(k)$ should be used in Eq. (31), instead. We will consider the current estimation-type Kalman filter for the estimation of $x(k)$. Assume that the plant ($\{A, C\}$) is observable [detectable] and that T satisfies Eq. (20). Then, the pair $\{\Phi, C\}$ is also observable [detectable]. Furthermore, assume that the pair $\{A, GR_{w_c}G'\}$ is controllable [stabilizable]. It is easy to show that the pair $\{\Phi, R_w\}$ is, also, controllable [stabilizable]. It is well known (e.g., see [22]) that, under these conditions, the optimal estimate of the state, $\hat{x}(k)$, that minimizes the steady-state error covariance $\hat{P} = \mathcal{E}[e(\infty)e'(\infty)]$, where $e(k) = x(k) - \hat{x}(k)$, is given by

$$\hat{x}(k+1) = \check{x}(k+1) + \mathbf{L}[z(k+1) - C\check{x}(k+1)] \tag{45}$$

where

$$\check{x}(k+1) = \Phi\hat{x}(k) + \Lambda\tilde{u}(k) \tag{46}$$

is the prediction at time $t = kT$ of $x(k+1)$ based on $\hat{x}(k)$ and $\tilde{u}(k)$. The filter gain $\mathbf{L}$ is given by

$$\mathbf{L} = \check{P}C'(R_v + C\check{P}C')^{-1} \tag{47}$$

where $\check{P}$ is the solution of the discrete algebraic Riccati equation

$$\check{P} = R_w + \Phi\check{P}\Phi' - \Phi\check{P}C'(R_v + C\check{P}C')^{-1}C\check{P}\Phi' \tag{48}$$

The error covariance $\hat{P}$ is given by

$$\hat{P} = (I - \mathbf{L}C)\check{P} \tag{49}$$

The following is a summary of the overall LQG problem and solution: The cost function to be minimized is

$$J = \frac{1}{2}\mathcal{E}\left\{\sum_{k=0}^{\infty} [\ x'(k)\quad \tilde{u}'(k)\ ]\begin{bmatrix} \tilde{Q} & \tilde{M} \\ \tilde{M}' & \tilde{R} \end{bmatrix}\begin{bmatrix} x(k) \\ \tilde{u}(k) \end{bmatrix}\right\} \tag{50}$$

subject to the dynamics

$$\left.\begin{aligned} x(k+1) &= \Phi x(k) + \Lambda\tilde{u}(k) + w(k) \\[2mm] z(k) &= Cx(k) + v(k) \end{aligned}\right\} \tag{51}$$

where $\tilde{Q}$, $\tilde{M}$, and $\tilde{R}$ are obtained from Eqs. (28)–(30). The initial state $x(0)$ is assumed to have a Gaussian distribution with a known mean $\mathcal{E}[x(0)] = \check{x}(0)$. The corresponding optimal $\tilde{u}$ is given by

$$\tilde{u}_{opt}(k) = \mathbf{K}\hat{x}(k) \tag{52}$$

where $\mathbf{K}$ is determined from Eq. (32). The optimal state estimate is given by

$$\hat{x}(k+1) = \check{x}(k+1) + \mathbf{L}[z(k+1) - C\check{x}(k+1)] \tag{53}$$

where

$$\check{x}(k+1) = (\Phi + \Lambda\mathbf{K})\hat{x}(k) \tag{54}$$

and $\mathbf{L}$ is obtained from Eq. (47). The optimal average cost is given by

$$J_{av_{opt}} = \frac{1}{2T}\text{trace}\left\{\tilde{P}R_w + \hat{P}\mathbf{K}'(\tilde{R} + \Lambda'\tilde{P}\Lambda)\mathbf{K}\right\} \tag{55}$$

where $\tilde{P}$ can be computed from Eq. (33).

We should note that, although the covariance matrix $\hat{P}$ is minimum, it can still be large if either R_v or R_w is large. In particular, if T is relatively large and A is unstable, R_w can also be relatively large as the integrand in Eq. (16) is nonnegative. Thus, unlike the noise-free case [4], reducing T might be necessary to obtain a better performance.

IV. CONTROL OF RESONANCE: A DESIGN PROBLEM

In this section, we utilize the proposed multirate scheme to design a controller for a plant with measurements taken at the output only. Here, an LQR solution together with a reduced-order estimator [23] are combined to obtain a compensator to control the resonance in disk drive systems.

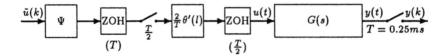

Figure 2: Open-loop multirate system.

The control of resonance is a generic problem associated with the control of any electromechanical system that exhibits a relatively high degree of structural flexibility between the sensor and the actuator [23]. Of particular importance is the control of the magnetic read and write head in the disk servomechanism of computer magnetic memory systems. In this section, we utilize the multirate scheme for such systems to achieve the required design closed-loop requirements.

Consider a typical transfer function

$$G(s) = \frac{\omega_n^2}{s^2(s^2 + 2\zeta\omega_n s + \omega_n^2)} \quad , \quad \omega_n = 2\pi f_n \tag{56}$$

with the following nominal values: natural frequency $f_n = 2.5$ KHz and a resonance peak in the magnitude of the frequency response (at $f = f_n$) of 25 db corresponding to $\zeta = 0.0281$ (resonance peak $= 20 \log(1/2\zeta)$ db). The actual value of f_n and the resonance peak are assumed to range between 2.25 to 2.75 KHz, and 20 to 30 db, respectively. The latter variation corresponds to a variation in ζ from 0.05 to 0.0158.

The problem is to find a digital compensator that satisfies the following closed-loop requirements:

- stability margin ≥ 6 db,

- phase margin $\geq 40°$, and

- bandwidth (at the zero db cross-over frequency) $= 300$ Hz.

The measurements are taken at a 4-KHz rate, which corresponds to $T = 0.25$ ms.

As shown below, a controller operating at $r = 2$ (control updated every 0.125 ms) can be found that meets these requirements. First, we find $G_d(z)$, the open-loop transfer function of the discrete system of Figure 2. A direct way to do this is to find a state-space realization, $\{A, B, C\}$, of the plant $G(s)$ and then compute the pair $\{\Phi, \Lambda\}$. The discrete transfer function is then given by

$$G_d(z) = C(zI - \Phi)^{-1}\Lambda \tag{57}$$

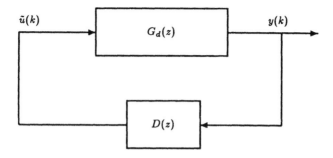

Figure 3: Closed-loop multirate system.

Next, we find a controller $D(z)$ such that the closed-loop system of Figure 3 has the desired specifications.

Since the plant parameters ζ and f_n can take any values in the given ranges, we need to design $D(z)$ for the worst case. This translates to $\zeta = 0.0158$ (lowest damping) and $f_n = 2.25$ KHz (closest to half the sampling frequency). In this case, the open-loop system parameters are

$$\Phi = \begin{bmatrix} -0.8681 & -0.0256 & 0.0094 & 0.0014 \\ 5.1114 & -0.8796 & -0.0214 & 0.0100 \\ 0 & 0 & 1.0000 & 0.2500 \\ 0 & 0 & 0 & 1.0000 \end{bmatrix},$$

$$\Lambda = \begin{bmatrix} 0.0428 & 0.0349 \\ 0.5643 & 0.3345 \\ 12.5392 & 6.1487 \\ 99.9268 & -0.7735 \end{bmatrix},$$

and

$$C = \begin{bmatrix} 1 & 0 & 0 & 0 \end{bmatrix}.$$

The corresponding transfer function is

$$G_d(z) = \frac{\begin{bmatrix} n_1(z) & n_2(z) \end{bmatrix}}{z^4 - 0.2523z^3 - 1.6011z^2 - 0.0409z + 0.8943} \tag{58}$$

where

$$n_1(z) = 0.0428z^3 + 0.1913z^2 + 0.1821z + 0.0391$$

and

$$n_2(z) = 0.0349z^3 + 0.0091z^2 - 0.0145z - 0.0331$$

The controller $D(z)$ is obtained by combining a multirate LQR solution (recall that the system is deterministic) and a reduced-order estimator. Thus, the design parameters that determines $D(z)$ are the cost matrices Q and R of the LQR cost function, and the three poles of the estimator. After few experimentation with these parameters, the following compensator is obtained

$$D(z) = \frac{-1.31 \begin{bmatrix} n_3(z) \\ n_4(z) \end{bmatrix}}{z^3 + 0.0798z^2 + 0.1923z + 0.0057} \tag{59}$$

where

$$n_3(z) = 1.0294z^3 + 0.8757z^2 - 0.6931z - 0.8257$$

and

$$n_4(z) = 1.0966z^3 + 0.9954z^2 - 0.6290z - 0.8236$$

which corresponds to estimator poles at $z = 0.4$, $z = 0.3$, and $z = 0.3$, and cost matrices $R = 100$ and

$$Q = \begin{bmatrix} 0 & 0 & 0 & 0 \\ 0 & 0 & 0 & 0 \\ 0 & 0 & 0.5 & 0 \\ 0 & 0 & 0 & 1 \end{bmatrix}$$

Controller $D(z)$ of Eq. (59) provides a 6.5-db gain margin (GM), a 40° phase margin (PM), a 303-Hz bandwidth (BW), and a 32.7-db stability margin at the resonance frequency f_n. For this controller to be acceptable, however, it should also satisfy the design specifications when f_n and ζ are different from the nominal values. In fact, simulation has shown that $D(z)$ of Eq. (59) satisfies the design requirements for all combinations of $2.25 \leq f_n \leq 2.75$ and $0.0158 \leq \zeta \leq 0.05$. Table I shows some of the simulation results.

Table I: Desired design specifications.

ζ	f_n (KHz)	GM (db)	PM	BW (Hz)	Stability Margin at f_n (db)
0.0158	2.25	6.5	40°	303	32.7
0.0158	2.50	6.6	40°	302	14.3
0.0158	2.75	6.6	40°	301	9.0
0.0281	2.25	6.5	40°	303	38.8
0.0281	2.50	6.5	40°	302	20.6
0.0281	2.75	6.6	40°	301	16.2
0.05	2.25	6.4	40°	303	46.2
0.05	2.50	6.5	40°	302	28.5
0.05	2.75	6.6	40°	301	27.5

Figures 4 through 6 show the frequency response of the loop gain $G_d(z)D(z)$, for $G_d(z)$ corresponding to $f_n = 2.25$ KHz and $\zeta = 0.0158$ (design case),

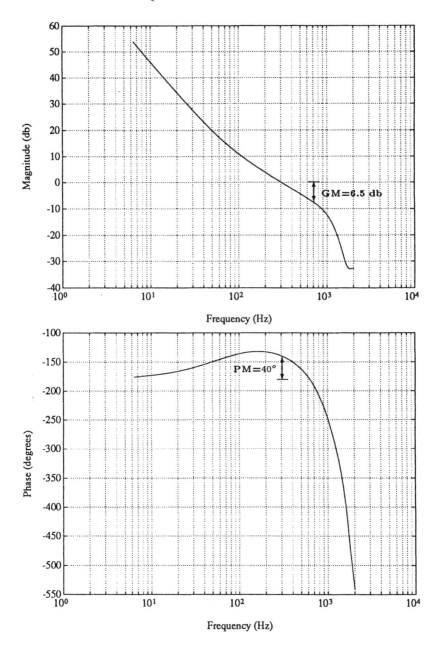

Figure 4: Gain and phase of the frequency response of $G_d D$, for $f_n = 2.25$ KHz and $\zeta = 0.0158$.

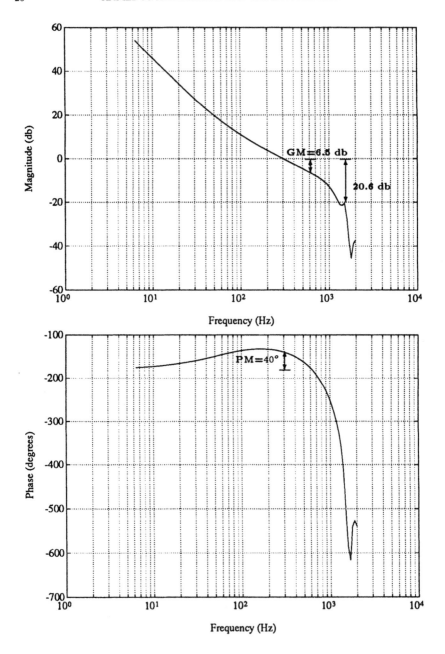

Figure 5: Gain and phase of the frequency response of $G_d D$, for $f_n = 2.50$ KHz and $\zeta = 0.0281$.

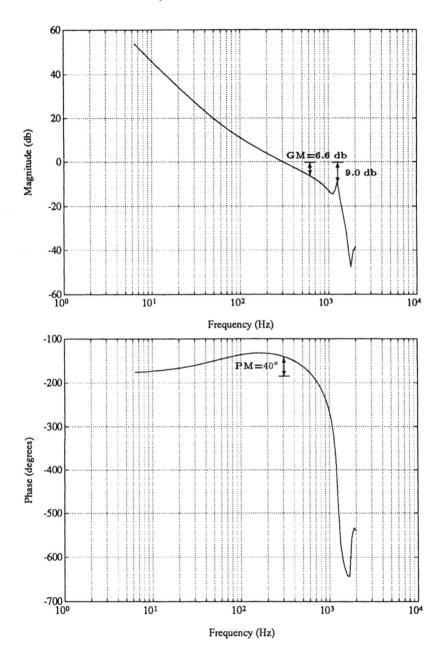

Figure 6: Gain and phase of the frequency response of $G_d D$, for $f_n = 2.75$ KHz and $\zeta = 0.0158$.

$f_n = 2.50$ KHz and $\zeta = 0.0281$ (nominal case), and $f_n = 2.75$ KHz and $\zeta = 0.0158$ (worst case of stability margin). Notice that the stability margin (at the resonance frequency) in the last case is the lowest (9 db) among all combinations of f_n and ζ – see Table I. Also, note that the resonance in the first case is eliminated. This elimination is due to the fact that the controller $D(z)$, which is designed for this case, has complex zeros that almost cancel the complex poles of $G_d(z)$.

V. REFERENCES

1.   M. J. Er, B. D. O. Anderson, and W. Yan, "Gain Margin Improvement Using Generalized Sampled-Data Hold Function Based Multirate Output Compensator", *Automatica* **30**, pp. 461-470 (1994).

2.   S. Longhi, "Structural Properties of Multirate Sampled-Data Systems", *IEEE Transactions on Automatic Control* **AC-39**, pp. 692-696 (1994).

3.   N. Yen and Y. Wu, "A Multirate Controller Design of Linear Periodic Time Delay Systems", *Automatica* **28**, pp. 1261-1266 (1992).

4.   H. M. Al-Rahmani and G. F. Franklin, "Multirate Control: A New Approach", *Automatica* **28**, pp. 35-44 (1992).

5.   H. M. Al-Rahmani and G. F. Franklin, "A New Optimal Multirate Control of Linear Periodic and Time-Invariant Systems", *IEEE Transactions on Automatic Control* **AC-35**, pp. 406-415 (1990).

6.   M. Berg, N. Ami, and J. Powell, "Multirate Digital Control System Design", *IEEE Transactions on Automatic Control* **AC-33**, pp. 1139-1150 (1988).

7.   M. de la Sen and C. Lopez-Abadia, "Multirate Adaptive Control with Applications to the Lateral Dynamics of Aircrafts", *Int. J. Control* **45**, pp. 379-382 (1987).

8.   M. Araki and K. Yamamoto, "Multivariable Multirate Sampled-Data Systems: State-Space Description, Transfer Characteristics, and Nyquist Criterion", *IEEE Transactions on Automatic Control* **AC-31**, pp. 145-154 (1986).

9.   D. P. Glasson, "Development and Application of Multi-Rate Digital Control", *IEEE Control System Magazine* **3**, pp. 2-8 (1983).

10. D. P. Glasson, "A New Technique for Multirate Digital Control Design and Sample Rate Selection ", *AIAA Journal Guid. Contr.* **5**, pp. 379-382 (1982).

11. A. B. Chammas and C. T. Leondes, "On the design of linear time-invariant systems by periodic output feedback– Parts I and II", *Int. J. Control* **27**, pp. 885-903 (1978).

12. W. H. Boykin and B. D. Frazier, "Multirate Sampled-Data Systems Analysis via Vector Operators", *IEEE Transactions on Automatic Control* **AC-20**, pp. 548-551 (1975).

13. R. A. Meyer and C. S. Burrus, "A Unified Analysis of Multirate and Periodically Time-Varying Digital Filters", *IEEE Transactions on Circuits and Systems* **CAS-22**, pp. 162-168 (1975).

14. D. C. Flowers and J. L. Hammond, "Simplification of the Characteristic Equation of Multirate Sampled-Data Systems", *IEEE Transactions on Automatic Control* **AC-17**, pp. 249-251 (1972).

15. E. I. Jury, "A Note on Multirate Sampled-Data Systems", *IEEE Transactions on Automatic Control* **AC-12**, pp. 319-320 (1967).

16. L. A. Gimpelson, "Multirate Sampled-Data Systems", *IRE Transactions on Automatic Control* **AC-5**, pp. 30-37 (1960).

17. H. M. Al-Rahmani and G. F. Franklin, "Linear Periodic Systems: Eigenvalue Assignment Using Discrete Periodic Feedback", *IEEE Transactions on Automatic Control* **AC-34**, pp. 99-103 (1989).

18. R. E. Kalman, P. L. Falb, and M. A. Arbib, *Topics in Mathematical System Theory*, McGraw-Hill, New York, N.Y. (1969).

19. R. E. Kalman, Y. C. Ho, and K. S. Narendra, "Controllability of Linear Dynamical Systems", *Contributions to Differential Equations* **1**, pp. 189-213 (1963).

20. W. M. Wonham, "On a Matrix Riccati Equation of Stochastic Control", *SIAM J. Control* **6**, pp. 681-697 (1968).

21. T. F. Gunckel and G. F. Franklin, "A General Solution for Linear Sampled-Data Control Systems", *Transactions of the ASME, Journal of Basic Engineering* **85-D**, pp. 197-201 (1963).

22. R. F. Stengel, *Stochastic Optimal Control*, John Wiley & Sons, New York, N.Y. (1986).

23. G. F. Franklin, J. D. Powell, and M. L. Workman, *Digital Control of Dynamic Systems*, Addison-Wesley, Reading, MA (1990).

The Design of Digital Pole Placement Controllers

Michel Kinnaert

Youbin Peng

Laboratoire d'Automatique, C.P.165, Université libre de Bruxelles
50, Avenue F.D. Roosevelt, 1050-Brussels, Belgium

I. INTRODUCTION

The basic idea of pole placement control is to assign the closed-loop poles at desired locations. Although this idea is very simple, there exist many variants of pole placement in well-known text books such as [1-5]. The differences lie in the way to use and to choose the design specifications for the purposes of reference tracking and disturbance rejection. These design specifications are: the desired closed-loop polynomial, the observer polynomial, the internal model and the prefilter. In order to compare these variants, we present a unified pole placement control algorithm which allows us to investigate all the design aspects of the method.

However, the pole placement technique has a drawback in that many design parameters are not directly related to engineering specifications. Although some design guidelines are provided in [6-9], they are not complete, and are sometimes implicit. Here, we will describe a more comprehensive set of design rules to develop an assisted pole placement control. Using this assisted method to solve a tracking problem, users only need to choose some engineering specifications such as the rise time, the settling time and the percentage of overshoot of a desired reference step response. The performances for load disturbance, output disturbance and measurement noise rejection are automatically adjusted in the proposed approach. However, as their tuning relies on a single parameter, further adjustment can be performed easily by the designer. The resulting controller is in the form of a digital two-degree-of-freedom compensator and a digital prefilter. The design method assures that the closed-loop system has a suitable intersample behavior.

CONTROL AND DYNAMIC SYSTEMS, VOL. 70

The windup problem caused by actuator nonlinearities, such as saturation, is considered next. The design of the anti-windup compensator is presented using two interpretations. First, the discrepancy between the actual system input and the controller output is considered as a special load disturbance[10]. Hence, the anti-windup compensator plays the role of a feedforward controller for rejecting such a disturbance. Second, the so-called notion of the realizable reference signal[11] is clarified. This allows us to analyze the properties of the anti-windup compensator. The above discussion sheds light on the reason why the conditioning technique derived in [11] is a natural choice to prevent windup in our assisted pole placement design.

The issue of controller validation is also discussed. In particular the influence of the location of the controller poles and zeros on the time and frequency domain performances with respect to the disturbance and measurement noise rejection is analyzed. The tools to check a posteriori the stability robustness of the control system are reviewed to give a complete overview of the validation procedure.

The remainder of this chapter is organized as follows. In section II, the unified pole placement control is described. In section III, a set of tools to be used in the sequel are reviewed. In section IV, some design guidelines are provided to develop an assisted pole placement control. In section V, the anti-windup compensator is discussed. In section VI, the controller validation is considered. In section VII, numerical examples are given to illustrate the proposed method. Finally, the conclusions are given in section VIII.

II. UNIFIED POLE PLACEMENT CONTROL

The following linear strictly proper discrete-time single-input single-output(SISO) model is used in the development of the algorithm:

$$y(t) = \frac{B(q)}{A(q)}\big(u(t)+v(t)\big)+\xi(t) \ , \qquad (1)$$

where $A(q)$ and $B(q)$ are polynomials of degrees $degA$ and $degB$ in the forward shift operator q, $y(t)$, $u(t)$, $v(t)$, and $\xi(t)$ are the system output, input, load disturbance, and output disturbance respectively. $degA > degB$ since the system is strictly proper. This model is supposed to describe a sampled-data system here. We also assume that $A(q)$ and $B(q)$ are coprime. In the sequel, z will denote the z-transform variable associated to the operator q.

Consider a linear discrete-time proper controller of the form

$$u(t) = \frac{T(q)}{R(q)}\frac{F_n(q)}{F_d(q)}w(t) - \frac{S(q)}{R(q)}\big(y(t)+\varepsilon(t)\big),\qquad (2a)$$

where $R(q)$, $S(q)$, $T(q)$, $F_n(q)$ and $F_d(q)$ are polynomials of suitable degrees in q. $w(t)$ and $\varepsilon(t)$ denote the reference signal and the measurement noise respectively. The properness of the controller implies that

$$\deg R \geq \deg T,\qquad (2b)$$

$$\deg R \geq \deg S,\qquad (2c)$$

$$\deg F_d \geq \deg F_n.\qquad (2d)$$

This control structure is the so-called two-degree-of-freedom RST controller with a prefilter, represented in Fig.1. The introduction of the prefilter provides the possibility of making the tracking performance independent of the disturbance and measurement noise rejection performances.

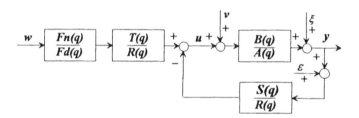

Fig.1. The closed-loop system

The closed-loop transfer functions between $w(t)$ and $y(t)$, $v(t)$ and $y(t)$, $\xi(t)$ and $y(t)$, and $\varepsilon(t)$ and $y(t)$ can be derived from Eqs.(1) and (2a). This yields respectively:

$$H_w(q) = \frac{B(q)T(q)}{A(q)R(q)+B(q)S(q)}\frac{F_n(q)}{F_d(q)},\qquad (3a)$$

$$H_v(q) = \frac{B(q)R(q)}{A(q)R(q)+B(q)S(q)},\qquad (3b)$$

$$H_\xi(q) = \frac{A(q)R(q)}{A(q)R(q) + B(q)S(q)} \ , \tag{3c}$$

$$H_\varepsilon(q) = \frac{-B(q)S(q)}{A(q)R(q) + B(q)S(q)} \ . \tag{3d}$$

The task of the controller design is twofold. First, we try to shape $H_w(q)$ in order to have a reference tracking performance as close as possible to the desired one. Second, we also attempt to shape $H_v(q)$, $H_\xi(q)$ and $H_\varepsilon(q)$ so that the load disturbance, output disturbance and measurement noise rejection performances correspond to the desired ones.

To simplify the above task, we may like to cancel some system poles and/or zeros by controller zeros and /or poles. To illustrate this idea, the polynomials $A(q)$ and $B(q)$ are factorized as follows:

$$A(q) = A^+(q)A^-(q) \ , \tag{4a}$$

$$B(q) = B^+(q)B^-(q) \ , \tag{4b}$$

where $A^+(q)$ and $B^+(q)$ will be cancelled by $S(q)$ and $R(q)$ respectively. That is to say

$$R(q) = B^+(q)\overline{R}(q) \ , \tag{5a}$$

$$S(q) = A^+(q)S'(q) \ . \tag{5b}$$

To ensure the internal stability, $A^+(z)$ and $B^+(z)$ must be stable, i.e. their roots are inside the open unit circle in the z-plane. Moreover, for reasons which will be clarified in section IV, $A^+(z)$ and $B^+(z)$ should be well-damped, i.e. their roots lie inside a specific region in the z-plane.

The internal model principle[12] can be introduced to improve the tracking performance and the disturbance rejection. The idea is to include a prescribed polynomial $M(q)$ in $R(q)$. Since $M(q)$ appears in the numerators of $H_v(q)$ and of $H_\xi(q)$, it can be used to cancel the known disturbance dynamics in order to improve the disturbance rejection. In the case where $T(q)=S(q)$ and no prefilter is added, $M(q)$ also appears in the tracking error transfer function $1-H_w(q)$; thus it can be used to cancel the known reference dynamics so as to improve the tracking performance. However, in a more general case, this operation is not sufficient, since for $R(q)$ to appear in $1-H_w(q)$, we need to impose more conditions on $T(q)$ and $F_n(q)/F_d(q)$[13]. At this stage, we take the internal model principle into account by introducing

$$R(q) = B^+(q)M(q)R'(q) .\qquad(6)$$

To perform the pole placement, the closed-loop characterstic polynomial is made equal to a desired polynomial $A_m(q)A_o(q)$, i.e.

$$A(q)R(q) + B(q)S(q) = A_m(q)A_o(q) ,\qquad(7a)$$

with $$A_m(q) = B^+(q)A'_m(q) ,\qquad(7b)$$

$$A_o(q) = A^+(q)A'_o(q) .\qquad(7c)$$

Here, using $A_m(q)A_o(q)$ does not change anything although it perhaps seems redundant. The reason for using two polynomials and the meaning of these two polynomials will be given shortly.

Using Eqs.(5) and (7b,c), Eq.(7a) can be reduced to the following Diophantine equation.

$$A^-(q)M(q)R'(q) + B^-(q)S'(q) = A'_m(q)A'_o(q) .\qquad(8)$$

From Eq.(2c) and $degA>degB$, we have $degA+degR>degB+degS$ and $degA^- +degM+degR'> degB^-+degS'$. This implies that, in Eq.(7a),

$$\deg A + \deg R = \deg A_m + \deg A_o ,\qquad(9a)$$

and in Eq.(8),

$$\deg A^- + \deg M + \deg R' = \deg A'_m + \deg A'_o .\qquad(9b)$$

Hence, the Diophantine equation (8) amounts to $degA^-+degM+degR'+1$ independent linear equations to be solved for the $degR'+degS'+2$ unknown coefficients of $R'(q)$ and $S'(q)$. The independence of the equations is assured by the coprimeness of $A(z)$ and $B(z)$, and by the fact that the roots of $M(z)$ are typically on the unit circle and do not coincide with any roots of $B(z)$. The system of equations has a unique solution if and only if

$$\deg S' = \deg A^- + \deg M - 1 ,\qquad(10a)$$

i.e. $$\deg S = \deg A - 1 .\qquad(10b)$$

The solution of Eq.(8) can be obtained by Sylvester resultant method[4,5] or Euclid's method[1,3]. However, if $\deg A^- + \deg M = 0$, then Eq.(8) has an infinite number of solutions. In this case, we assign

$$\deg S' = 0 ,\qquad(11a)$$

i.e.
$$\deg S = \deg A \,, \tag{11b}$$
and we choose
$$S'(q) = A'_m(1) A'_o(1) / B^-(1) \,, \tag{12a}$$

so that
$$R'(q) = \frac{A'_m(q) A'_o(q) - B^-(q) A'_m(1) A'_o(1) / B^-(1)}{A^-(1) M(1)} \,. \tag{12b}$$

Equation (12b) imposes $R'(1)=0$, and thus introduces an integrating action in $R(q)$.

From Eqs.(5)-(8), we can rewrite the transfer functions (3) as follows: :

$$H_w(q) = \frac{B^-(q) T(q)}{A'_m(q) A'_0(q) A^+(q)} \frac{F_n(q)}{F_d(q)} \,, \tag{13a}$$

$$H_v(q) = \frac{B(q) M(q) R'(q)}{A'_m(q) A'_0(q) A^+(q)} \,, \tag{13b}$$

$$H_\xi(q) = \frac{A^-(q) M(q) R'(q)}{A'_m(q) A'_0(q)} \,, \tag{13c}$$

$$H_\varepsilon(q) = \frac{-B^-(q) S'(q)}{A'_m(q) A'_0(q)} \,. \tag{13d}$$

Now, let us clarify the meaning of $A_m(q)$ and $A_o(q)$ by investigating Eqs.(7b), (7c) and (13). We denote $A_o(q)$ the part of the desired closed-loop characteristic polynomial which should not influence the reference tracking, and thus should disappear in $H_w(q)$. We denote $A_m(q)$ the remaining part of the desired closed-loop characteristic polynomial. In [3], $A_m(q)$ is called the desired closed-loop polynomial, and $A_o(q)$ is interpreted as the observer polynomial. Here, for the sake of notation coherence, we keep the same symbols, but these two polynomials have a more general meaning in our context.

For $A_o(q)$ not to appear in $H_w(q)$, we assign

$$T(q) = A'_o(q) A^+(q) B_m(q) \,, \tag{14}$$

where $B_m(q)$ is a free polynomial to be chosen by the designer.

The controller must be causal for implementation. The two controller causality conditions (2b,c) impose two implicit conditions for $degA'_m$ and $degA'_o$. Using Eqs.(9) and (14), we translate Eq.(2b) into

$$\deg A'_m \geq \deg A + \deg B_m - \deg B^+ . \tag{15}$$

Using Eqs.(9) and (10), or Eqs.(9) and (11), we translate Eq.(2c) into

$$\deg A'_o \geq \deg A + \deg A^- + \deg M - \deg B^+ - \deg A'_m - 1 , \tag{16a}$$

if

$$\deg A^- + \deg M > 0 ,$$

or

$$\deg A'_o \geq \deg A - \deg B^+ - \deg A'_m , \tag{16b}$$

if

$$\deg A^- + \deg M = 0 .$$

Finally, we give the closed-loop relationship between $u(t)$ and $w(t)$, $v(t)$, $\xi(t)$, and $\varepsilon(t)$. It can be easily derived from Eqs.(1) and (2a), and it will be used in the sequel.

$$u(t) = \frac{AT}{AR + BS} \frac{F_n}{F_d} w(t) - \frac{BS}{AR + BS} v(t) - \frac{AS}{AR + BS} \left(\xi(t) + \varepsilon(t) \right) . \tag{17a}$$

In Eq.(17a), the explicit dependence on q is suppressed for the sake of simplicity. From Eqs.(5)-(8) and Eqs.(14) and (17a), we can obtain the transfer functions between $w(t)$ and $u(t)$, $v(t)$ and $u(t)$, $\xi(t)$ and $u(t)$, and $\varepsilon(t)$ and $u(t)$ respectively:

$$G_w(q) = \frac{A(q)B'_m(q)}{A'_m(q)B^+(q)} \frac{F_n(q)}{F_d(q)} , \tag{17b}$$

$$G_v(q) = \frac{-B^-(q)S'(q)}{A'_m(q)A'_0(q)} , \tag{17c}$$

$$G_\xi(q) = G_\varepsilon(q) = \frac{-A(q)S'(q)}{A'_m(q)A'_0(q)B^+(q)} . \tag{17d}$$

Now, let us summarize the unified pole placement design.

Step 1: Choose the strategy for pole-zero cancellation, i.e. the strategy to factorize $A(q)$ and $B(q)$ into $A^+(q)A^-(q)$ and $B^+(q)B^-(q)$ respectively;

Step 2: Choose the prescribed polynomial $M(q)$;

Step 3: Choose $degA'_m$, $degB_m$ and $degA'_o$ to satisfy the controller causality conditions.

Step 4. Choose $A'_m(q)$, $B_m(q)$, $A'_o(q)$, $F_n(q)$ and $F_d(q)$ to shape $H_w(q)$, $H_v(q)$, $H_\xi(q)$, and $H_g(q)$.

Step 5. Solve the Diophantine equation (8) (use the solution (12) in the degenerated case where $degA^- + degM = 0$).

The above procedure contains a lot of design parameters. Our next objective is to provide some guidelines for their choice. To this end, we first review a set of tools that will be used to justify the design rules and to perform the controller validation.

III. PREREQUISITES

III.A. THE NATURAL FREQUENCY OF A DISCRETE POLE OR ZERO

The notion of natural frequency of a discrete pole (or zero) will be used extensively in the sequel, both in the design guidelines and in the controller validation sections. Its definition is based on the use of the zero-pole mapping technique[14], in a non-calssical way. Indeed, instead of deducing the discrete poles and zeros from their continuous-time equivalent, the converse is done here[16].

Consider a strictly proper discrete-time transfer function $H_d(z)$. Suppose that $H_d(z)$ corresponds to a sampled-data system, and let $H_c(s)$ denote its continuous-time equivalent. The plant input sequence is transformed into a continuous-time signal via a zero-order hold, and a suitable anti-aliasing filter is used before sampling the plant output signal. Hence, the non-zero poles of $H_d(z)$ are related to the poles of $H_c(s)$ according to $z = e^{sT}$, where T is the sampling period. The idea of the zero-pole mapping technique is that the map $z = e^{sT}$ can also be applied to the non-zero zeros, although this yields an approximate solution.

If we let $a_z + jb_z$ denote a non-zero discrete pole (or zero), and $a_s + jb_s$ denote its continuous equivalent, we can write

$$a_z + jb_z = e^{(a_s + jb_s)T} . \tag{18}$$

Thus, a_s and b_s can be computed as follows:

$$a_s = \frac{1}{T} \ln\left(\sqrt{a_z^2 + b_z^2} \right) , \qquad (19a)$$

$$b_s = \frac{1}{T} \text{atan2}(b_z, a_z) , \qquad (19b)$$

where $\text{atan2}(b_z, a_z)$ denotes arc tangent of b_z/a_z in the range $-\pi$ to π.

The natural frequency corresponding to a given pole (or zero) is defined as

$$\omega_n = \sqrt{a_s^2 + b_s^2} . \qquad (20)$$

In the sequel, when we consider the natural frequency of a non-zero discrete pole (or zero), we assume that the mapping (19) has been performed to deduce the corresponding ω_n from Eq.(20).

III.B. PARAMETRIZATION OF SOME PROTOTYPE CONTINUOUS-TIME TRANSFER FUNCTIONS IN TERMS OF THEIR STEP RESPONSE OVERSHOOT, RISE TIME AND SETTLING TIME

In this section, we first define the percentage of overshoot, rise time and settling time of a step response. Next, we show how to obtain the parameters of some prototype continuous-time transfer functions from their step response overshoot, rise time and settling time, by exact or approximate expressions.

Consider the step response $s(t)$ of a continuous-time system described by the transfer function $H(s)$. The percentage overshoot σ is defined by:

$$\sigma = \frac{\max_{t \geq 0} |s(t)| - |s(\infty)|}{|s(\infty)|} 100\% . \qquad (21)$$

For oscillatory systems, the rise time t_r is the time for $s(t)$ to rise from $s(0)$ to $s(\infty)$, and the settling time t_s is the time for the transient to die away within some specified tolerance band (here 1%). For non-oscillatory systems, the rise time t_r is the time for $s(t)$ to rise from $0.1s(\infty)$ to $0.9s(\infty)$, and the settling time t_s is considered to be the same as the rise time t_r.

For some prototype models, we can easily compute the exact or approximate values of t_r, t_s and σ in terms of the model parameters, and vice versa. Here, we investigate three cases.

Model A. The first-order prototype system:

$$H(s) = \frac{\omega_n}{s + \omega_n} \, . \tag{22a}$$

From $s(t)$ and some standard computations, we can deduce the following values

$$t_r = t_s = \frac{2.2}{\omega_n} \, , \tag{22b}$$

and
$$\sigma = 0 \, . \tag{22c}$$

Hence, the model parameter can be obtained by

$$\omega_n = \frac{2.2}{t_r} \, . \tag{22d}$$

Model B. The second-order prototype system:

$$H(s) = \frac{\omega_n^2}{s^2 + 2\zeta\omega_n s + \omega_n^2} \, . \tag{23a}$$

It is known that small values of ζ would yield short rise time[15]. Yet, too small a ζ yields a large overshoot and a large settling time. A general accepted range of ζ for satisfactory all-around performance is between 0.5 and 1, which corresponds to so-called underdamped systems. In this case, standard computations and some approximations yield the following values[4]:

$$t_r \approx \frac{2.5}{\omega_n} \, , \tag{23b}$$

$$t_s \approx \frac{4.6}{\zeta\omega_n} \, , \tag{23c}$$

$$\sigma = e^{\frac{-\pi\zeta}{\sqrt{1-\zeta^2}}} \, . \tag{23d}$$

A second-order model whose step response exhibits approximately given t_r, t_s and σ can thus be obtained by choosing:

$$\zeta = \frac{-\ln\sigma}{\sqrt{\pi^2 + \ln^2\sigma}} \,, \tag{23e}$$

$$\omega_n = \max\left(\frac{4.6}{\zeta t_s}, \frac{2.5}{t_r}\right). \tag{23f}$$

Model C. The k-order prototype system (k (>2) is an integer):

$$H(s) = \frac{\left(5\zeta\omega_n\right)^{k-2}\omega_n^{\,2}}{\left(s + 5\zeta\omega_n\right)^{k-2}\left(s^2 + 2\zeta\omega_n s + \omega_n^{\,2}\right)}. \tag{24}$$

The rise time, settling time and overshoot of such a model are approximately given by the expressions obtained for model B, since the influence of the non-dominant poles is very small. Eqs.(23e,f) can thus be used to compute ζ and ω_n in Eq.(24) so as to obtain a system with predefined step response characteristics. The resulting t_r and t_s will be a little larger than the desired ones while the real σ will be a little smaller.

III.C. INFLUENCE OF THE RELATIVE POLE/ZERO LOCATION ON THE STEP RESPONSE OVERSHOOT AND/OR UNDERSHOOT

Consider a strictly proper stable discrete-time transfer function $H(z)$ with non-zero steady state gain and well-damped poles. If $H(z)$ has positive real zeros outside the closed unit disc, the step response of the system will exhibit undershoot. This undershoot might be very large if the natural frequency of one of those zeros happens to be much smaller than the natural frequency ω_d of the dominant poles of $H(z)$. If $H(z)$ has positive real zeros inside the closed unit disc, but with natural frequencies smaller than ω_d, then the step response of $H(z)$ exhibits an overshoot. This overshoot might be very large if the natural frequency of one of those zeros happens to be much smaller than ω_d. If $H(z)$ has complex zeros with natural frequencies smaller than ω_d, the step response of the system, $s(t)$, exhibits a large total variation $T_v(s)$, as defined below, and illustrated in Fig.2[16].

$$T_v(f)\underline{\Delta}\sup_{0\le t_1\le\ldots\le t_N}\sum_{i=1}^{N-1}\left|f(t_i) - f(t_{i-1})\right|. \tag{25}$$

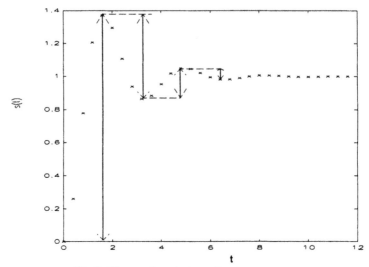

Fig.2. The total variation of a step response

From Fig.2, the total variation of a function $f(t)$, $T_V(f)$ is seen to be the sum of all consecutive peak-to-valley differences in $f(t)$.

The above statements are deduced in [17] from a set of theoretical and experimental results.

Note that the total variation of the step response $s(t)$ of a proper stable discrete-time linear system described by $H(z)$ can also be expressed in terms of the impulse response of this system, $h(t)$ via the following relationship:

$$T_v(s(t)) = \|h(t)\|_1 , \tag{26}$$

where the l_1-norm of $h(t)$ is defined as:

$$\|h(t)\|_1 \triangleq \sum_{i=0}^{\infty} |h_i| . \tag{27}$$

This result is given in [16] for continuous-time systems, but its validity for discrete-time systems can easily be proven.

III.D. LINK BETWEEN TIME AND FREQUENCY DOMAIN PROPERTIES

Consider a proper and stable rational discrete-time system with transfer function $H(q)$, and with impulse response $h(t)$. A classical relation between the peak gain and the root-mean-square (RMS) gain of this system is[18]:

$$\|H(z)\|_{rms.gn} \le \|H(z)\|_{pk.gn} , \tag{28}$$

where the RMS gain can be computed via

$$\|H(z)\|_{rms.gn} = \|H(z)\|_{\infty} = \sup_{\Phi} \left|H(e^{j\Phi})\right| , \tag{29}$$

and the peak gain via

$$\|H(z)\|_{pk.gn} = \|h(t)\|_1 . \tag{30}$$

The names of these gains are justified in [18]. From Eq.(29), the value of the RMS gain is seen to be equal to the maximum value of the magnitude in the Bode plot of $H(z)$. On the other hand, the peak gain is the maximal achievable absolute value of the ratio between the largest peaks in the output and input signals. The maximum is taken over the class of bounded inputs. From Eqs.(26), (28) and (30), we deduce:

$$\|H(z)\|_{rms.gn} \le T_v(s(t)) . \tag{31}$$

Hence, if $H(z)$ has a large RMS gain, the total variation of its step response is large too.

III.E. PROPERTIES OF THE SENSITIVITY AND COMPLEMENTARY SENSITIVITY FUNCTIONS

$H_\zeta(z)$ (see Eq.(3c)) is classically called the sensitivity function, and $-H_\varepsilon(z)$ (see Eq.(3d)), the complementary sensitivity function. Their time and frequency domain characteristics are fundamental in the analysis of the closed-loop performance. They are commonly denoted by S and T respectively. However, since the letters R, S and T are also somewhat sanctioned notations for the controller structure we are using, we shall stick to $H_\zeta(z)$ and $-H_\varepsilon(z)$.

III.E.1. Link between the sensitivity and the complementary sensitivity functions

From Eqs.(3c,d), the following link is obtained directly:

$$H_\xi(z) + \left(-H_\varepsilon(z)\right) = 1 . \tag{32}$$

Multiplying the expression by $\dfrac{z}{z-1}$ and taking the inverse z-transform yields

$$s_\xi(t) + \left(-s_\varepsilon(t)\right) = 1 , \qquad \forall t \geq 0 , \tag{33}$$

where $s_\xi(t)$ and $s_\varepsilon(t)$ are the step responses of $H_\xi(z)$ and $H_\varepsilon(z)$ respectively.

III.E.2. Bode's integral theorem for discrete-time systems

Therom 1[9,19]: Provided the open-loop discrete-time system is strictly proper, and provided the closed-loop feedback system is asymptotically stable, the sensitivity function $H_\xi(z)$ satisfies the following integral constraint :

$$\sum_{i=1}^{m} \ln|\beta_i| = \frac{T}{\pi} \int_0^{\pi/T} \ln \left| H_\xi(e^{j\omega T}) \right| d\omega , \tag{34}$$

where β_i, $i=1, ...,m$, are the unstable poles of the open-loop system, and T is the sampling period.

□□□

Equation (34) expresses the fact that, if $\left| H_\xi(e^{j\omega T}) \right|$ is made smaller than one in a given frequency band, it will necessarily be larger than one in some other frequency band. The larger the number of unstable open-loop poles and the larger their magnitude, the larger the peak in the sensitivity function and/or the larger the frequency range where the sensitivity magnitude exceeds one. This peak induces a significant amplification of the output disturbance $\xi(t)$ in the corresponding frequency band. Besides, it also decreases the so-called modulus margin, ΔM, which is the distance from the open-loop Nyquist curve to the critical point $(-1, j\, 0)$ in the complex plane. Indeed,

$$\Delta M \triangleq \frac{1}{\sup\limits_{\omega} \left| H_\xi(e^{j\omega T}) \right|} . \tag{35}$$

This margin is a more reliable indicator of stability robustness than the classical gain and phase margins.

IV. ASSISTED POLE PLACEMENT CONTROL

We now present a set of guidelines which can be followed to design a pole placement controller on the basis of only three specfications. Besides a linear time-invariant model of the plant described by Eq.(1), the required data are the desired rise time, settling time and overshoot of the step response of the closed-loop system with respect to the reference signal.

Guideline 1: Place the closed-loop poles in part at a point on the real axis, located in the open interval $(0,1)$, and in part at the origin in the z-plane.

For a sampled-data system with a time-delay dT (where T is the sampling period), there are $d-1$ poles at the origin. It is reasonable to keep the same number of poles at the origin for the closed-loop system. We place the remaining closed-loop poles at a single location for the sake of simplicity. In this way, we reduce the number of tuning knobs as much as possible. $A'_m(q) A'_o(q)$ will thus be made of a product of monomials q and $(q - e^{-T\beta})$.

The later corresponds to the continuous-time factor $(s + \beta)$. The larger β, the faster the disturbance rejection (see Eqs.(13b-d)). However, by looking at Eqs.(17c,d), we realize that β should not be much larger than the smallest natural frequency corresponding to the roots of $B^-(z)$ and $A^-(z)$. Indeed, from prerequisite III.C, too large a β would result in a large total variation of the control signal with respect to load and/or output disturbance steps. This will typically induce actuator saturation. Besides, if β is too large with respect to the smallest natural frequency of the roots of $B^-(z)$, the step respose of $H_\varepsilon(z)$, $s_\varepsilon(t)$, will exhibit a large total variation by Eq.(13d) and prerequiste III.C. $T_v(s_\varepsilon(t))$ will be large as well, due to Eq.(33).

Hence, as a rule of thumb, we propose to choose

$$\beta = \min\left(\frac{\omega_z}{0.8}, \frac{\omega_p}{0.8} \right), \qquad (36)$$

where ω_z and ω_p are the smallest natural frequencies corresponding respectively to the roots of $B(z)$ and $A(z)$.

Guideline 2: Factorize $A(q)$ into $A^+(q)A^-(q)$ so that $A^+(q)$ is well-damped, and $A^-(q)$ is not well-damped.

Here, by $A^+(q)$ is well-damped, we mean that the continuous equivalents of all its roots lie in the dotted region shown in Fig.3. Thus, by an abuse of language, in our context, the term well-damped characterizes the roots which are not only well-damped in the usual sense (i.e $x=0$ and $T=0$ in Fig.3), but also have a sufficiently large natural frequency.

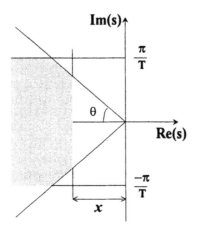

Fig.3. The region of well-damped roots

The reason why only well-damped poles can be cancelled is as follows. From Eq.(13b), we note that $A^+(q)$ appears in the denominator of $H_v(q)$. Thus, if the load disturbance rejection is the concern, $A^+(q)$ should be well-damped to avoid too oscillatory and/or too slow step response with respect to $v(t)$.

A rule of thumb to choose the dotted region of Fig.3 for pole cancellation is to set $\theta = \pi/4$ and to choose x equal to the natural frequency corresponding to the dominant closed-loop poles, β.

Guideline 3: Factorize $B(q)$ into $B^+(q)B^-(q)$ so that $B^+(q)$ is well-damped, and $B^-(q)$ is not well-damped.

The reason why badly-damped zeros should not be cancelled is as follows. From Eqs.(17b,d), we note that $B^+(q)$ appears in the denominators of $G_w(q)$, $G_\xi(q)$ and $G_\delta(q)$. Thus, if the badly-damped zeros were cancelled, the control action would be too oscillatory or too slow. Moreover, the oscillation in the control signal induces intersample ripples in the output signal. However, it is not observed in the output at the sampling times, as the corresponding modes are unobservable. Besides, from Eq.(13b) we note that $B(q)$ always appears in the numerator of $H_v(q)$. This means that the influence of $B^+(q)$ on the load disturbance rejection will not be removed even when $B^+(q)$ is cancelled by the controller.

Hence, as a rule of thumb, the region where zero cancellations are performed can be chosen identically to the region for pole cancellation defined in guideline 2.

Remark 1: Note that we have not handled the sampling zeros separately. Applying the zero-pole mapping procedure to the sampling zeros does not make sence, as they correspond to the zeros at infinity of the continuous-time system. However, it turns out that, for a typical choice of sampling period (see for instance the rule of thumb mentioned in section VII), the continuous-time equivalents of those zeros are located outside the dotted region of Fig.3. Hence, the sampling zeros are not cancelled. This is a commonly accepted rule, as such a cancellation would create intersample ripples in the output signal. The reason for such a phenomenon is that the sampling zeros are located on the negative real axis for sufficiently fast sampling period[3].

□□□

Guideline 4: Choose $M(q)$ according to the Internal Model Principle.

If the dynamics of $w(t)$, $v(t)$ and $\xi(t)$ are known, $M(q)$ should be their least common multiple. More conditions on $T(q)$ and $F_n(q)/F_d(q)$ should be imposed in order to assure the asymptotic tracking performance[13]. However, as for most cases of process control, we can simply introduce an integrating action by setting $M(q)=1-q$ here. In this situation, the static gains of $T(q)$ and $F_n(q)/F_d(q)$ should be adjusted so that $H_w(1)=1$, in order to guarantee a zero steady-state error.

Guideline 5: Choose $degA'_m$, $degB_m$ and $degA'_o$ by taking the equalities in Eqs.(15) and (16a) (or (16b)).

The above choice assures $degR=degT$ and $degR=degS$. This implies that the resulting controller has no time delay.

Guideline 6: Choose $A'_m(q) = q^{d-1}\overline{A}_m(q)$ with $\overline{A}_m(q) = \left(q - e^{-T\beta}\right)^{\deg A_m' - d + 1}$,

and $A'_o(q) = \left(q - e^{-T\beta}\right)^{\deg A_o'}$.

Note that by Eq.(15), $degA_m'$ is necessarily larger than $d-1$. Thus the above structure of $A'_m(q)$ is always valid.

As already mentioned in guideline 1, setting all the roots of $\overline{A}_m(q)A_o(q)$ at the same location is an option that we take for simplicity. Another possibility could be to choose two well-damped poles as the dominant poles, and to fix the remaining non-dominant poles so as to optimize some cost functions. The choice of a criterion and the study of the possible advantage of such an approach require further research.

Guideline 7: Choose $B_m(q) = \dfrac{A'_m(1)}{B^-(1)}$.

The role of $B_m(q)$ is to introduce complementary zeros in $H_w(q)$. As this operation will increase the controller degree, we propose to use $F_n(q)$ instead of $B'_m(q)$ for this purpose. Hence, we only use $B'_m(q)$ to introduce a normalization factor such that the closed-loop static gain without prefilter is equal to one.

Guideline 8: Choose $F_n(q) = \overline{A}_m(q)$.

Since $\overline{A}_m(q)$ has been chosen by only considering the disturbance and measurement noise rejection, we cancel it and use $F_d(q)$ to improve the tracking performance.

Guideline 9: Design $F_d(q)$ so that the step response of $H_w(q)$ has the the desired rise time, settling time, and overshoot.

To this end, note that

$$H_w(q) = \frac{B^-(q)\overline{A}_m(1)}{q^{d-1}F_d(q)B^-(1)}, \tag{37}$$

with $deg\, F_d = deg\, A'_m - d + 1$.

We associate to $H_w(q)$ a $deg\, F_d$-order continuous-time prototype model $H(s)$ of the form (22a) if $degF_d$ is equal to 1, and of the form (23a) or (24) otherwise. The parameters ω_n and ζ of this model are computed from t_r, t_s, and σ according to Eq.(22c) in the first case, and Eqs.(23e,f) in the second case. We next apply the zero-pole mapping technique[14] with a slight modification to $H(s)$. Let $H_{pt}(q)$ denote the resulting discrete-time transfer function. By a slight modification, we mean that all the zeros of $H(s)$ at $s=\infty$ are mapped into $z=\infty$ instead of $z=-1$. This assures that $\dfrac{1}{H_{pt}(q)}$ is a polynomial of degree $deg\ F_d$. Note that the zero-pole mapping technique assures $H_{pt}(1)=1$ due to $H(0)=1$. We finally set

$$F_d(q) = \frac{\overline{A}_m(1)}{H_{pt}(q)} \ . \tag{38}$$

The first step of this procedure amounts to designing a kind of approximate continuous-time equivalent of $H_w(q)$, whose step response has the required t_r, t_s, and σ. In this approximation, the role of $B^-(q)$ is not considered. This is due to the fact that there is no simple way to link the parameters of a transfer function with zeros to its step response properties, t_r, t_s, and σ. When $degB^-\neq0$, the above choice of $F_d(q)$ will often have to be adjusted to obtain the required t_r, t_s, and σ. Should the overshoot (or undershoot) and settling time be too large, one could, for instance, slow down the step response by decreasing the natural frequencies of both dominant and non-dominant roots of $F_d(z)$. This would also induce an increase in t_r, of course.

Remark 2: It is also possible to determine the structure of a prototype model for $H_w(q)$ by minimizing a cost function. The Integral of the Absolute Error (IAE) criterion, the Integral of the Square Error (ISE) criterion, and the Integral of Time multiplied Absolute Error (ITAE) criterion are often used to this end [20]. However, the relationship between the engineering specifications and the prototype model is not explicit, and the inclusion of fixed zeros (due to $B^-(q)$) is not considered in the existing procedures either.

□□□

We now turn to the problem of actuator saturation, which may occur for large reference changes, when a relatively small rise-time has been imposed by the designer. To guarantee that the controller keeps satisfactory performances in such circumstances, an anti-windup compensator has to be introduced.

IV. ANTI-WINDUP COMPENSATOR

In practice, it often happens that the actual system input is temporarily different from the controller output. This might be, for instance, due to an

actuator nonlinearity (such as a saturation). In such circumstances, the
phenomenon of controller windup can be observed. It typically occurs when the
controller contains an integrating action. If the control signal gets saturated for
some time, the integrating action continues to accumulate the error, and it can
reach quite large values if the error keeps the same sign. Once the error
changes sign, a long time period is then needed for the controller output to
leave the saturation, due to the large absolute value of the integral in the
control law. Note that during the saturation of the control signal, the controlled
system and the controller actually work in open-loop.

In order to diminish the effects of the windup phenomenon, an anti-windup
compensator is added to the controller. Many anti-windup schemes have been
given in the literature[3,11,21,22]. Here, we will give our interpretation of this
problem, which allows us to select one of these possibilities.

In this section, we will suppress the disturbance and the measurement noise
for the sake of simplicity. Without loss of generality, let us consider the
saturation nonlinearity. It is defined by the function "sat" as follows:

$$sat(u) = \begin{cases} u_{max} & u > u_{max} \\ u & \text{if} \quad u_{min} \leq u \leq u_{max} \\ u_{min} & u < u_{min} \end{cases} . \tag{39}$$

The real input of the controlled system is $u^r = sat(u)$ as shown in Fig.4. The
difference between u and u^r, denoted by $\delta(t) = u^r(t) - u(t)$, can be interpreted
as a special load disturbance as shown in Fig.5. However, we should note that
$\delta(t)$ is not an independent load disturbance, due to the fact it depends on $u(t)$.

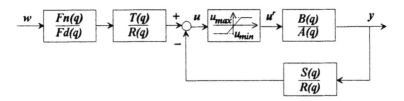

Fig.4. The constrained closed-loop system

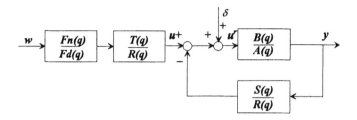

Fig.5. The constrained closed-loop system using the δ-interpretation

Our interpretation of the windup problem is as follows. Suppose one has chosen a $H_w(q)$ which has a fast step response, and a large positive step reference signal occurs, then a large positive input $u(t)$ is computed by the controller. Here the positive character of $u(t)$ is due to an implicit assumption, namely the direct feedthrough term between $w(t)$ and $u(t)$ is positive. If $u(t)$ exceeds u_{max}, the actuator will saturate and a negative load disturbance $\delta(t)$ will be produced. The amplitude of $\delta(t)$ depends on $u(t)$ and u_{max}. During saturation, the constrained system temporarily works in open-loop, i.e.

$$y(t) = \frac{B(q)}{A(q)} u_{max} .$$ (40a)

Referring to Fig.5, Eq.(40a) can be rewritten as

$$y(t) = \frac{BT}{AR + BS} \frac{F_n}{F_d} w(t) + \frac{BR}{AR + BS} \delta(t) ,$$ (40b)

or

$$y(t) = \frac{BT}{AR + BS} \frac{F_n}{F_d} w^r(t) ,$$ (40c)

where

$$w^r(t) = w(t) + \frac{R(q)F_d(q)}{T(q)F_n(q)} \delta(t) .$$ (40d)

In Eqs.(40b,c), the explicit dependence on q is suppressed for the sake of simplicity. $w^r(t)$ is called the realizable reference signal in [11]. The realizable reference is the signal that the output $y(t)$ tracks all the time (during and out of saturation) through $H_w(q)$, in the same way as $y(t)$ tracks $w(t)$ in the absense of actuator saturation. The realizable reference thus means the reference which can be realized (tracked by the controlled system output) within the linear zone of the actuator. This notion was initially defined in the restricted context of the conditioning technique[11]. The present approach shows that it can be generalized in a global constrained control loop framework.

With the above interpretation, the windup phenomenon can be considered as the problem caused by the fact that $w^r(t)$ is too far from $w(t)$ and varies too much for a long time. Indeed, by investigating Eq.(40d), we find that $w^r(t)$ is

linked to $\delta(t)$ through $\dfrac{R(q)F_d(q)}{T(q)F_n(q)}$. As $R(q)F_d(q)$ contains some roots, especially one root at point $(1, j\ 0\ )$ in the z-plane, with natural frequencies smaller than the natural frequency of the dominant roots of $T(q)F_n(q)$,

$\dfrac{R(q)F_d(q)}{T(q)F_n(q)}\delta(t)$ will produce a large variation in $w^r(t)$, and make it far from $w(t)$ even when the control signal leaves the saturation.

To compensate for the effect of $\delta(t)$, a natural way is to use a feedforward compensator, as shown in Fig.6.

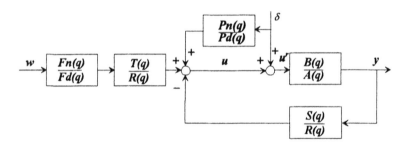

Fig.6 The constrained closed-loop system with an anti-windup compensator

Here, we denote the compensator by $P_n(q)/P_d(q)$. Since $\delta(t)$ is first produced by the controller and can only be compensated for later, the feedforward compensator should not contain a direct feedthrough. Hence, we must impose that $degP_d > degP_n$. With the feedforward compensator, $w^r(t)$ becomes

$$w^r(t) = w(t) + \frac{R(q)F_d(q)}{T(q)F_n(q)}\left(1 + \frac{P_n(q)}{P_d(q)}\right)\delta(t) . \qquad (41)$$

In fact, it can be shown that most anti-windup compensators use $1 + P_n(q)/P_d(q)$ to cancel $R(q)$ in Eq.(41) in order to have a $w^r(t)$ closer to $w(t)$. Among these schemes, the conditioning technique[11,21] suggests a special choice for $P_n(q)/P_d(q)$, namely

$$\frac{P_n(q)}{P_d(q)} = \frac{T(q)F_n(q)}{R(q)F_d(q)} \frac{R(\infty)F_d(\infty)}{T(\infty)F_n(\infty)} - 1 .$$ (42)

Equation (42) assures that $\deg P_d > \deg P_n$ and leads to

$$w^r(t) = w(t) + \frac{R(\infty)F_d(\infty)}{T(\infty)F_n(\infty)} \delta(t)$$ (43)

Remember that, without loss of generality, we assumed $\dfrac{R(\infty)F_d(\infty)}{T(\infty)F_n(\infty)} > 0$ at the beginning of the discussion. From Eq.(43), we see that when $u(t)$ is larger than u_{max}, $w^r(t)$ is smaller than $w(t)$. This is logical since $y(t)$ can not track $w(t)$ as in the absense of saturation. Once the control signal leaves saturation, $w^r(t)$ is equal to $w(t)$, and $y(t)$ can thus track $w(t)$ right away.

Remark 3: To make $y(t)$ track $w(t)$ just after desaturation is sometimes too optimistic, it may produce another saturation in the opposite direction if $u_{max} - u_{min}$ is too small. This can be interpreted as a "short-sightedness" problem[23]. The so-called filtered or generalized conditioning technique[23,24] provides additional design parameters for $P_n(q)/P_d(q)$, when needed. The resulting improvement is due to the fact that $w^r(t)$ is still a little smaller than $w(t)$ after desaturation, and the opposite saturation is thus avoided.

□□□

To use the conditioning technique, the following conditions should be satisfied

$$\deg R + \deg F_d = \deg T + \deg F_n ,$$ (44a)
$$T(q) \text{ and } F_n(q) \text{ are stable.}$$ (44b)

Equation (44a) assures that $\dfrac{R(\infty)F_d(\infty)}{T(\infty)F_n(\infty)}$ exists. Property (44b) is a sufficient condition for stability, provided the controlled system is asymptotically stable. The reason is as follows.

During saturation, as the constrained system temporarily works in open-loop, the control law is

$$u(t) = \frac{T(q)F_n(q)}{R(q)F_d(q)} w(t) - \frac{S(q)}{R(q)} y(t) + \frac{P_n(q)}{P_d(q)} \left(u^r(t) - u(t) \right), \quad (45a)$$

or $\quad \left(1 + \frac{P_n(q)}{P_d(q)} \right) u(t) = \frac{T(q)F_n(q)}{R(q)F_d(q)} w(t) - \frac{S(q)}{R(q)} y(t) + \frac{P_n(q)}{P_d(q)} u^r(t). \quad (45b)$

To assure the boundedness of $y(t)$, the controlled system should be asymptotically stable. For $u(t)$ to be bounded, $P_d(q)$ should contain the unstable roots of $R(q)$, and $P_n(q)+P_d(q)$ should be stable. Applying this result to Eq.(42) yields property (44b).

Our assisted pole placement control assures that Eq.(44a) and property (44b) are fulfilled. Hence, the conditioning technique should be used except in the case where u_{max}-u_{min} is too small. In that situation, $H_w(q)$ should be "slowed down", or a more general anti-windup compensator should be used. If we know the maximum amplitude of the step changes in the reference signal, we prefer the first solution.

VI. CONTROLLER VALIDATION

Controller validation can, and must be performed by simulation, in order to observe the transient behavior of the controlled and the control signals subject to realistic reference changes, disturbances and measurement noise. However, a priori information can be gained by inspecting the location of the roots of $R(q)$ and $S(q)$ with respect to the chosen closed-loop poles. This will be one of the points developed in this section.

Another one is the robustness issue. This problem is not dealt within the formulation of the pole placement design. However, the stability robustness of the closed-loop system with respect to non-parametric plant perturbations can be checked a posteriori, using tools which are briefly reviewed here.

VI.A. INFLUENCE OF THE CONTROLLER POLES AND ZEROS ON THE DISTURBANCE AND MEASUREMENT NOISE REJECTION PERFORMANCES

Here, we do not consider the feedforward part of the controller as it does not play any role in the disturbance and measurement noise rejection performances. Thus when we refer to the controller zeros, we mean the roots of $S(q)$. In pole placement, the designer does not master directly the position of the controller poles and zeros, as the polynomials $R(q)$ and $S(q)$ are obtained by computing a particular solution of a Diophantine equation. It turns out that, when the guidelines of section IV are followed to design the controller, adequate disturbance and measurement noise rejection performances are most often obtained, for the typical transfer functions encountered in the process industry [2]. However, exceptions might occur, and it is useful to have simple rules allowing to detect possible problems. Such rules can be derived by analyzing the relative location of the roots of $R(q)$ and $S(q)$ with respect to the desired closed-loop poles.

VI.A.1. Influence of the roots of $R(q)$

a. Effects of the roots outside the closed unit disc

From Eq.(34), we conclude that such roots will generally increase the peak in the sensitivity function, and hence decrease the modulus margin (see III.E.2). By Eq.(31), an increase in $\left\| H_\xi(z) \right\|_{rms.gn}$ typically induces an increase in $T_v(s_\xi(t))$. Hence the roots of $R(q)$ outside the closed-unit disc tend to deteriorate the response to step disturbances in $\xi(t)$. Indeed, a large $T_v(s_\xi(t))$ is most often undesirable. Such roots should thus be avoided. When is that possible? We now turn to this question.

b. Avoiding the roots outside the closed unit disc

We will only give a necessary and sufficient condition on the plant poles and zeros for being able to stabilize this process with a stable controller. Such a condition is obviously necessary for achieving pole placement with a stable controller.

The results to be stated next are expressed in an easier way if we describe the plant transfer function in terms of the backward shift operator q^{-1} , or the associated transform variable λ. Thus, let

$$P(q) = \frac{B(q)}{A(q)} = \frac{B^*(q^{-1})}{A^*(q^{-1})} = \frac{B^*(\lambda)}{A^*(\lambda)} = P^*(\lambda) , \qquad (46)$$

where X^* is the reciprocal polynomial of X (and P^* is defined in an obvious way for rational fractions). The stability region in the λ–plane is the complement of the closed unit disc.

Theorem 2 [25]: A plant $P^*(\lambda)$ can be stabilized with a stable controller if and only if the number of real roots of $A^*(\lambda)$ between every pair of real roots of $B^*(\lambda)$ in the interval $[-1, 1]$ is even.

$$\square\square\square$$

Such a plant is said to be strongly stabilizable, and the condition satisfied by its poles and zeros is known as the parity interlacing property.

Of course, most often, in order to fulfil the internal model principle, the controller has some poles on the unit circle (e.g. an integrating action). Yet, the above theorem can be used to check whether additional unstable poles in the controller are needed to stabilize the plant. Indeed, it suffices to replace the polynomial $A^*(\lambda)$ by $A^*(\lambda)M^*(\lambda)$ in $P^*(\lambda)$. Here M^* is the reciprocal polynomial of M defined in section II.

VI.A.2. Influence of the roots of $S(q)$

a. Effect of the roots outside the closed unit disc

From section III.C, and from the definition of the complementary sensitivity function, we conclude that positive real roots of $S(q)$ outside the closed unit disc will cause a large undershoot in $-s_g(t)$ if their natural frequency is much smaller than the natural frequency of the dominant closed-loop poles (roots of $A_m(q)A_0(q)$). Hence the step response of $H_g(q)$, $s_g(t)$, will exhibit a large total variation, and so will $s_\xi(t)$, due to Eq.(33). This results again in a poor disturbance rejection performance. Frequency domain arguments based on an integral relation which is not reviewed here[18], also indicate that

$\left\| H_\xi(z) \right\|_{rms.gn}$ will be large if one requires that the sensitivity be small in the frequency range corresponding to the roots of $S(q)$ outside the unit disc. This amounts to having unstable roots of $S(q)$ with natural frequencies much smaller than the natural frequency of the dominant closed-loop poles. The large RMS gain of $H_\xi(z)$ yields a poor modulus margin as already mentioned in VI.A.1.a.

b. Effect of stable roots outside the disc determined by the dominant roots of $A_m(q)A_0(q)$

By the expression "the disc determined by the dominant roots of $A_m(q)A_0(q)$", we mean the disc of radius β, in terms of our design guidelines.

From section III.C, and from the definition of the complementary sensitivity function, $s_\varepsilon(t)$ will exhibit a large total variation, and so will $s_\zeta(t)$ if such stable roots of $S(q)$ are present. Thus the controller will have a poor performance for rejecting step disturbances in $\xi(t)$.

Remark 4: Experience has also shown that the large total variation observed in $s_\varepsilon(t)$ in the situations mentioned in VI.A.2.a and b corresponds to a large value of the magnitude of $H_\varepsilon(e^{j\omega T})$ above the natural frequencies of the "bad" zeros. By "bad" zeros, we mean the zeros whose effect is discussed in VI.A.2.a and b[17]. This will induce poor measurement noise rejection, and poor robustness with respect to multiplicative non-parametric uncertainties on the plant model. The last point is clarified in the next section.

$\qquad\qquad\qquad\qquad\qquad\qquad\qquad\qquad\qquad\qquad\qquad\qquad\qquad\qquad$ ☐☐☐

VI.B. ROBUST STABILITY

We have only analyzed the performance of the controller for the nominal model up to now. However, there exist modelling uncertainties. One way to take them into account is to assume that the true plant Nyquist curve lies in a given envelope around the nominal Nyquist curve. This defines a family of plant models. The controller is said to assure the robust stability of the closed-loop system if it stabilizes the entire class of plants. The uncertainties defined in this way are called non-parametric uncertainties. They can be of additive or multiplicative type.

To formalize the above considerations, we let $B^\circ(q)/A^\circ(q)$ denote the true plant transfer function, and we assume that it has the same number of unstable poles as the model $B(q)/A(q)$. For additive non-parametric uncertainties, the true plant is supposed to fulfil the following inequality :

$$\left| \frac{B^\circ(e^{j\omega T})}{A^\circ(e^{j\omega T})} - \frac{B(e^{j\omega T})}{A(e^{j\omega T})} \right| < \left| \gamma_a(e^{j\omega T}) \right| , \quad \forall \omega , \qquad (47)$$

where γ_a is a prespecfied stable rational function.

In the case of multiplicative uncertainties, we assume that

52 MICHEL KINNAERT AND YOUBIN PENG

$$\frac{B^{o}(q)}{A^{o}(q)} = (1+L(q))\frac{B(q)}{A(q)} \ , \tag{48a}$$

$$|L(j\omega T)| < |\gamma_m(j\omega T)| \ , \ \forall\omega \ , \tag{48b}$$

where γ_m is a prespecfied stable rational function.

Necessary and sufficient conditions for a controller to robustly stabilize the class of plants (47) and (48) were derived in [26,27], in a continuous-time framework. For single-input single-output discrete-time systems, in our pole placement context, they can be stated in the following form[3]:

Theorem 4 : The closed-loop system is robustly stable if and only if the controller stabilizes the nominal plant $B(q)/A(q)$ and

$$\left|\frac{A(e^{j\omega T})S(e^{j\omega T})}{A(e^{j\omega T})R(e^{j\omega T})+B(e^{j\omega T})S(e^{j\omega T})}\right| < |\gamma_a(e^{j\omega T})| \ , \ \forall\omega \ , \tag{49}$$

where γ_a satisfies Eq.(47).

□□□

The left side of Eq.(49) is sometimes called the input sensitivity function.

Theorem 5 : The closed-loop system is robustly stable if and only if the controller stabilizes the nominal plant $B(q)/A(q)$ and

$$\left|\frac{B(e^{j\omega T})S(e^{j\omega T})}{A(e^{j\omega T})R(e^{j\omega T})+B(e^{j\omega T})S(e^{j\omega T})}\right| < |\gamma_m(e^{j\omega T})| \ , \ \forall\omega \ , \tag{50}$$

where γ_m satisfies Eq.(48).

□□□

The left side of Eq.(50) is called the complementary sensitivity function, as mentioned in section III.D.

Equations (49) and (50) can be used for two purposes in our development. First, when a design has been done, the left sides of Eqs.(49) and (50) can be computed and plotted. The frequency domain conditions on the model uncertainty tolerance can thus be derived. Second, when the frequency domain uncertainties are known a priori, we can look for the range of the closed-loop poles which assures the robust stability. This last type of approach has been further developed in [28], where an iterative method is proposed to simultaneously place the dominant closed-loop poles, and shape the sensitivity function in order to achieve predefined robustness properties.

VII. NUMERICAL EXAMPLES

Example 1: Consider a fourth-order continous-time system with unit d.c. gain, characterized by

$$\frac{B_c(s)}{A_c(s)} = \frac{(1-4s)(1+2s)}{(1+s)^2(1+10s)^2} \cdot \qquad (51)$$

Its step response is plotted in Fig.7. The rise time is approximately equal to *50* secs. Our aim is to achieve a closed-loop system whose tracking performance with respect to a reference step is characterized by a rise time of *20* secs, a settling time of *40* secs and a *10%* overshoot.

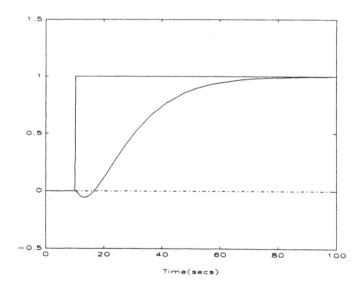

Fig.7. The open-loop step response

To choose the sampling period, we use the following heuristic rule [8]:

$$\omega_d T \approx 0.2 - 0.6 , \tag{52}$$

where ω_d is the natural frequency of the dominant closed-loop poles. This last parameter can be obtained directly by applying guideline 1 on the continuous-time model. This yields ω_d=0.125. Here, we choose a sampling period T=2 secs. The corresponding sampled-data system is described by :

$$\frac{B(q)}{A(q)} = \frac{-0.0463(q-0.3681)(q-1.6587)(q+0.2762)}{(q-0.1353)^2(q-0.8187)^2} . \tag{53}$$

The different design guidelines yield successively the following results :

1. $\beta = 0.125$;

2. $A^+(q) = (q-0.1353)^2$ and $A^-(q) = (q-0.8187)^2$;

3. $B^+(q) = q-0.3681$ and $B^-(q) = -0.0463(q-1.6587)(q+0.2762)$;

4. $M(q) = 1-q$;

5. $\deg A_m' = 3$, $\deg B_m' = 0$ and $\deg A_o' = 2$;

6. $A_m'(q) = (q-0.7788)^3$ and $A_o'(q) = (q-0.7788)^2$;

7. $B_m(q) = 0.2781$;

8. $F_n(q) = (q-0.7788)^3$;

9. $F_d(q) = 0.3852(q^3 - 2.1679q^2 + 1.5514q - 0.3554)$.

After solving the Diophantine equation (8), we end up with the following controller polynomials :

$$R(q) = q^4 - 2.6056q^3 + 2.4973q^2 - 1.0523q + 0.1606 ,$$

$$S(q) = 0.4122q^4 - 0.7862q^3 + 0.4662q^2 - 0.0871q + 0.0051 ,$$

$$T(q) = 0.2871q^4 - 0.5084q^3 + 0.2910q^2 - 0.0536q + 0.0031 .$$

Figure 8 shows the resulting closed-loop response with respect to a unit reference step. The rise time and settling time are slightly larger than the desired ones. This is due to the presence of a non-minimum phase zero. It cannot be cancelled of course, and it introduces a discrepancy between the closed-loop prototype model and the actual closed-loop system.

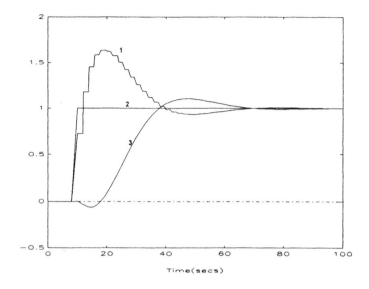

Fig.8. The closed-loop step response
Control signal (1), reference signal (2) and output signal (3)

To show the disturbance rejection performance, we introduce a step load disturbance, and a step output disturbance from the time *100* secs, and *200* secs respectively. Both steps have a magnitude of *0.1*. The results are plotted in Fig.9. As expected, the integrating action assures zero steady state error.

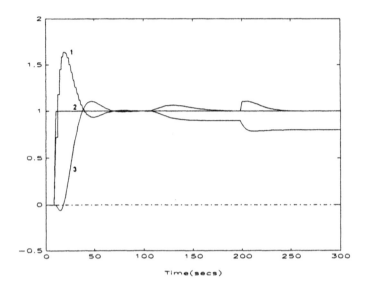

Fig.9. The closed-loop step response with step load and output disturbances
Control signal (1), reference signal (2) and output signal (3)

Next, the interest of an anti-windup compensator is illustrated by introducing a saturation element ($u_{max}=1.2$, and $u_{min}=-1.2$) between the controller output and the plant input. Figure 10 illustrates the tracking performance in the absence of anti-windup compensator, while Fig.11 is obtained after introducing such a compensator. The conditioning technique is used for the design of the anti-windup element. It brings a strong improvement of the tracking performance as demonstrated by the decrease in the settling time and overshoot of the response in Fig.11, compared to Fig.10.

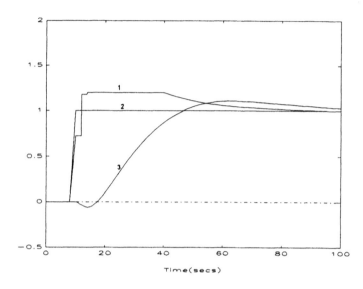

Fig. 10. The constrained closed-loop step response without anti-windup
compensator
Constrained control signal (1), reference signal (2) and output signal (3)

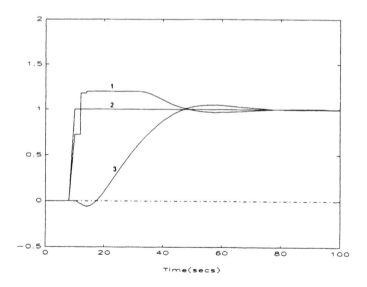

Fig. 11. The constrained closed-loop step response using the conditioning
technique
Constrained control signal (1), reference signal (2) and output signal (3)

Finally, Fig.12 and 13 can be used to analyze the robustness properties of the closed-loop system. From the magnitude Bode plot of the (output) sensitivity function, we deduce that the modulus margin is equal to *0.6751*.

From the magnitude Bode plots of the complementary sensitivity and the input sensitivity functions, we can also deduce the allowable multiplicative and additive non-parametric modelling uncertainties respectively.

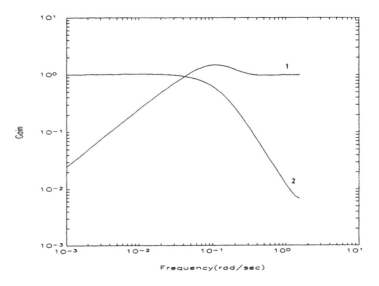

Fig. 12. The frequency responses of the sensitivity (1) and complementary sensitivity (2) functions

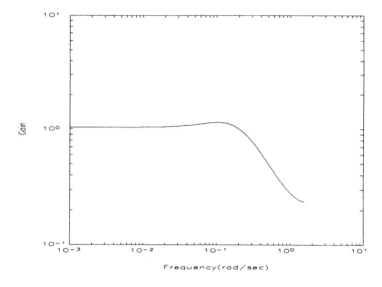

Fig. 13. The frequency responses of the input sensitivity

We now illustrate some of the validation tools based on the location of the poles and the zeros of the controller. In particular, we demonstrate the effect of an unstable pole in the controller on the time and frequency domain performances of the closed-loop system. To this end, we consider a simple second order system.

Example 2: Consider a system described by the following continuous-time transfer function

$$\frac{B_c(s)}{A_c(s)} = \frac{1.5}{(s+0.5)(s+3)} \ . \tag{54}$$

A digital controller with a sampling period of *0.4* sec is first designed using the guidelines of section IV. Next, a second design is performed, for which we keep the same desired closed-loop poles as previously, but we introduce a fixed unstable pole at $z=1.1$ in the controller. Let C_1 and C_2 denote the compensator obtained respectively in the first and the second cases Figure 14(Fig. 15) shows the magnitude Bode plots of the sensitivity and the complementary sensitivity functions for the closed-loop system corresponding to $C_1(C_2)$. The increase in the peak of the sensitivity function due to the unstable pole at $z=1.1$ appears clearly in Fig. 15. The large peak in the complementary sensitivity function corresponding to C_2 is due to a root in the controller numerator $S_2(z)$, whose natural frequency, *0.1517*, is much smaller than the natural frequency of the dominant closed-loop poles, namely $\omega_d = 0.625$ rad/sec (see table 1).

Table 1: Natural frequencies corresponding to the roots of $S_1(z)$ and $S_2(z)$

$S_1(z)$	$S_2(z)$
0.5037 3.0000	0.1517 0.4997 3.0000

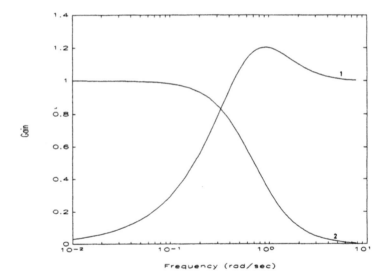

Fig.14: The frequency responses of the sensitivity (1) and complementary sensitivity (2) functions obtained with the controller C_1

Finally, Fig.16 shows the step responses of the sensitivity functions obtained with C_1 and C_2. The link between the total variation of a step response, and the corresponding peak sensitivity function (Eq.(31)) is clearly demonstrated.

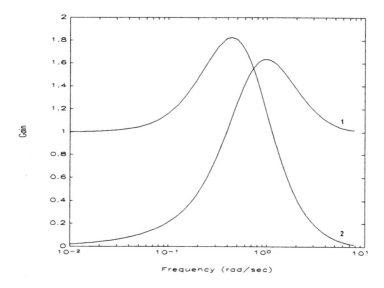

Fig. 15: The frequency responses of the sensitivity (1) and complementary sensitivity (2) functions obtained with the controller C_2

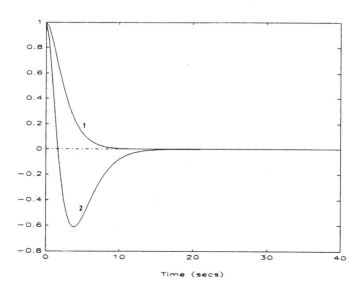

Fig 16: Step response of the sensitivity function corresponding to C_1 (1) and to C_2 (2)

VIII. CONCLUSIONS

An assisted pole placement method has been proposed to help the designers in their choice of the design parameters. Once the desired rise time, settling time and percentage of overshoot of the step response of the closed-loop system with respect to a reference change have been specified, the computation of the controller is performed automatically. The design guidelines are obtained from a mixture of theoretical and heuristic results. They should yield adequate performance for most classical systems encountered in the process industry. Should the control problem be a difficult one (system with several poorly damped modes, and stringent robustness requirements), the proposed approach could provide a first design which could then be refined via the method recently proposed in [28], for instance.

A thorough discussion of the influence of the controller poles and zeros on the performance of the closed-loop system has been presented. Roughly speaking, the conclusion is that a satisfactory time and frequency domain behavior can only be obtained when the natural frequencies of the controller zeros are close to or larger than the natural frequency of the dominant closed-loop poles. Moreover, unstable controller poles should be avoided. As the controller is computed by solving a Diophantine equation, one does not master the location of the controller poles and zeros. However, by adequate adaptation of the design parameters one should be able to obtain a satisfactory location of these poles and zeros for realistic problems.

Given the importance of the location of the roots of $R(q)$ and $S(q)$, one could look for polynomials $R(q)$ and $S(q)$ with roots in a specific region, such that the dominant closed-loop poles correspond to some desired ones. This route has been investigated in [29], where either the roots of $R^*(\lambda)$ or the roots of $S^*(\lambda)$ are guaranteed to belong to a disk of radius $\alpha<1$ in the λ-plane. This approach has several shortcomings. Besides the fact that only the roots of $R^*(\lambda)$ or $S^*(\lambda)$ are mastered, it relies on a pole-zero cancellation strategy that is not adequate for all sampled-data systems. In particular, no sampling zero should be located outside the disk of radius α, as they would be cancelled in the proposed design. Moreover, the solution involves the design of a unit which solves a given interpolation problem, and this might lead to a high order controller. Although a controller reduction procedure could be used, it is not guaranteed that the poles and zeros of the reduced order compensator would be adequately located.

The problem of guaranteeing that both the roots of $R(q)$ and $S(q)$ be located in a specific region while achieving desired dominant closed-loop poles is extremely complex. Indeed, a related issue is the stabilization of a plant with a stable and inverse stable controller. This is known to be equivalent, roughly speaking, to the problem of simultaneous stabilization of three plants[30], and it is conjectured, in this last reference, that there do not exist tractable necessary and sufficient conditions for this problem. The term 'tractable' means that the conditions can actually be implemented.

ACKNOWLEDGEMENT

This work is partly supported by a grant from l'Institut pour l'Encouragement de la Recherche scientifique dans l'Industrie et l'Agriculture (I.R.S.I.A.).

IX. REFERENCES

1. V. Kucera, *Discrete Linear Control: the Polynomial Equation Approach*, John Wiley & Sons, New York(1979).
2. R. Isermann, *Digital Control Systems*, Springer-Verlag, Berlin(1981).
3. K.J. Aström and B. Wittenmark, *Computer Controlled Systems, Theory and Design*, Prentice Hall, Englewood Cliffs, New Jersey(1984).
4   G.C. Goodwin and K.S. Sin, *Adaptive Filtering, Prediction and Control*, Prentice-Hall, Englewood Cliffs, New Jersey(1984).
5. C.-T. Chen, *Linear System Theory and Design*, Holt, Rinehart and Winston, New York(1984).
6. C.C. Hang, "The Choice of Controller Zeros", *IEEE Control Systems Magazine*, 9(1), pp.72-75(1989).
7. C.-T. Chen and B.Seo, "Application of Linear Algebraic Method for Control System Design", *IEEE Control Systems Magazine*, 10(1), pp.43-47(1990).
8. K.J. Åström, "Some Aspects on Pole-placement Design", *Preprints of the 11th IFAC World Congress*, Tallinn, vol.2, pp.167-172(1990).
9. R.H. Middleton and G.C. Goodwin, *Digital Control and Estimation, A Unified Approach*, Prentice Hall, Englewood Cliffs, New Jersey(1990).
10. S. Ronnback, K.S. Walgama and J.Sternby, "An Extensiton to the Generalized Anti-windup Compensator", *Preprints of the 13th IMACS World Congress on Computation and Applied Mathematics*, Dublin, vol.3, pp.1192-1196(1991).
11. R. Hanus, M. Kinnaert and J.L. Henrotte, "Conditioning Technique, a General Anti-windup and Bumpless Transfer Method", *Automatica*, 23(6), pp.729-739(1987).
12. B.A. Francis and W.M. Wonham, "The Internal Model Principle of Control Theory", *Automatica*, 12(5), pp.457-465(1976).

13. Y. Peng, *On Adaptive Control: Pole-Zero Placement and Generalized Predictive Control*, Ph.D. Thesis, Department of Control Engineering, The Free University of Brussels, Brussels(1991).

14. G.F. Franklin, J.D. Powell and M.L. Workman, *Digital Control of Dynamic Systems*, Addison-Wesley, New York(1990).

15. B.C. Kuo, *Automatic Control Systems*, Prentice-Hall, Englewood Cliffs, New Jersey(1991).

16. S.P. Boyd and C.H. Barratt, *Linear Controller Design: Limits of Performance*, Prentice-Hall, Englewood Cliffs, New Jersey(1991).

17. B.A. León de la Barra, *Zeros and their Influence on Time and Frequency Domain Properties of Scalar Feedback Systems*, Ph.D. Thesis, Department of Elctrical and Computer Engineering, The University of Newcastle, Newcastle(1992).

18. S.P. Boyd and J. Doyle, "Comparison of Peak and RMS Gains for Discrete-time Systems", *Systems and Control Letters*, 9, pp.1-6(1987).

19. C. Mohtadi, "Bode's Integral Theorem for Discrete-time Systems", *IEE Proceedings, Part D*, 137, pp.57-66(1990).

20. B. Shahian and M. Hassul, *Control System Design Using Matrix$_x$*, Prentice-Hall, Englewood Cliffs, New Jersey(1992).

21. R. Hanus, "A New Technique for Preventing Control Windup", *Journal A* 21(1), pp15-20(1980).

22. R. Hanus, "Anti-windup and Bumpless Transfer: a Survey", *Computing and Computers for Control Systems*(P. Borne et al., editors), J.C.Baltzer AG, Scientific Publishing Co. and IMACS, pp.3-9(1989).

23. K.S. Walgama, S.Ronnback and J.Sternby, "Generalisation of Conditioning Technique for Anti-windup Compensators", *IEE Proceedings, Part D*, 139, pp.109-118(1992).

24. R. Hanus and Y. Peng, "Modified Conditioning Technique for Controller with Nonminimum Phase Zeros and/or Time-delays", *Preprints of the 13th IMACS World Congress on Computation and Applied Mathematics*, Dublin, vol.3, pp.1188-1189(1991).

25. M. Vidyasagar, *Control System Synthesis, a Factorization Approach*, MIT Press, MA(1987).

26. J.C. Doyle and G. Stein, "Multivariable Feedback Design: Concepts for a Classical / Modern Synthesis", *IEEE Transactions on Automatic Control*, AC-23(1), pp.4-16(1981).

27. M.J. Chen and C.A. Desoer, "Necessary and Sufficient Conditions for Robust Stability of Linear Distributed Feedback Systems", *International Journal of Control*, 35, pp.255-267(1982).

28. I.D. Landau, C. Cyrot and0 D. Rey, "Robust Control Design Using the Combined Pole Placement/Sensitivity Function Shaping Method",

Proceedings of the 2nd European Control Conference, Groningen, pp.1693-1698(1993).

29. M. Kinnaert and V. Blondel, "Discrete-time Pole Placement with Stable Controller", *Automatica*, 28(5), pp.935-943(1992).

30. V. Blondel, *Simultaneous Stabilization of Linear Systems: Mathematical Solutions, Related Problems and Equivalent Formulations*, Ph.D.Thesis, Centre for Systems Engineering and Applied Mechanics, Université Catholique de Louvain, L.L.N.(1992).

Linearization Techniques for Pulse Width Control of Linear Systems

F. Bernelli-Zazzera

P. Mantegazza

Dipartimento di Ingegneria Aerospaziale
Politecnico di Milano, Milano, ITALY

I. INTRODUCTION

Discontinuously operating actuators are simple, rugged, and effective control devices that allow an easy exploitation of relatively large powers. They operate by switching among discrete output power levels; a fairly general model is represented by an actuator that is either off or operate at any of its two discrete working points. The control action is then a simple switching among its three outputs and such a discontinuous operation impede an input-output proportionality, thus making any affected system response non linear. Because of their chattering operation there can generally be no true steady state but a limit cycling periodic response. However by adopting a sufficiently fast switching rate and appropriate techniques in the implementation of the controller, the system has often enough low pass filtering capability to ensure a sufficiently smooth response, thus allowing the satisfaction of even the strictest specifications. There are nonetheless limitations to their applications that are difficult to overcome, e.g.: when a significant actuating power must be transferred between moving parts and it its not possible to limit their wearing rate to an acceptable level; or if stringent specifications are set on state derivatives.

Other considerations can be related to: ease of implementation, intrinsic limitations of the available actuators, constraints imposed by the mathematical modeling of the system. A typical aerospace application (due to the background of the authors, most examples will deal with mechanical/aerospace examples) is given by thrusters used for shape and attitude control of space structures [1], which can produce time varying forces only if their input is modulated in time and if they are coupled in two opposite directions, to allow both positive and negative control forces.

Except for the simplest control tasks, discontinuous actuators make it necessary to design a switching law that can insure an appropriate degree of proportionality in relation to the control policy being implemented. This can be achieved through different implementation and design/modeling techniques, such as: use of a dithering signal and/or auto-oscillation [2,3], variable structure systems (VSS) [4-8], pulse width modulation (PWM) [9], pulse frequency modulation (PFM) [9] and pulse width-pulse frequency modulation (PWPFM) [9].

It can be seen that the above techniques use the duration of a constant input, rather than its amplitude, as control variable. Clearly these type of controls are intrinsically non-linear and discrete in time and fall into the broader class of pulse modulated control systems.

The related literature covers different aspects and implementation strategies, including the analysis of stability conditions [9], modeling issues [10], and presents many significant applications to spacecraft attitude [11,12,13] and robot control [14] systems. Among the above cited techniques, in many aerospace applications [15-27], a greater attention has been devoted to the wider class of VSS, in the sequel they will be called variable structure or sliding mode controls [4-8], of which a pulse width modulated control can be interpreted as a sub domain. In fact it has been demonstrated that, under the assumption of a very high, theoretically infinite, sampling rate, a sliding mode control system can be equivalent to a pulse width modulated controller [28,29]. However in the sequel, for clarity and conciseness, only discrete time pulse modulated controls, with finite, relatively low, sampling rates will be taken into account except for a very concise mention to sliding mode controls.

A: GENERALITIES ON PULSE MODULATORS

Generally speaking, a pulse modulator is a device capable of transforming a continuous input signal into pulses whose amplitude, frequency and duration follow a prescribed law, defining the type of modulator. Following the definitions of Ref. [9], pulse modulators can be divided into two main groups.

A first group controls independently the pulse amplitude, frequency and duration on the basis of the sampled input function, according to

$$m(t) = \begin{cases} M\big[u(t_k)\big] & \text{for} & t_k < t \le t_k + \delta\big[u(t_k)\big] \\ 0 & \text{for} & t_k + \delta\big[u(t_k)\big] < t \le t_k + T\big[u(t_k)\big] \end{cases} \tag{1}$$

$$t_{k+1} = t_k + T\big[u(t_k)\big] \tag{2}$$

$$\delta\big[u(t_k)\big] \le T\big[u(t_k)\big] \tag{3}$$

where m is the pulse amplitude, δ the pulse width, T the duty cycle, i.e. the inverse of the pulse frequency, and u the pulse control input. In general the three functions $M(.)$, $\delta(.)$ and $T(.)$ are non-linear, and the pulse width δ is bounded by the duty cycle time T.

The second group of modulators fixes the pulse shape, i.e. its amplitude and width, and controls the frequency and sign of the pulses, on the basis of the input function

$$m(t) = \begin{cases} Mp_{k-1} & \text{for} & t_k \le t \le t_k + \delta \\ 0 & \text{for} & t_k + \delta < t \le t_k + \max(T, T_k) \end{cases} \tag{4}$$

where T is an arbitrary constant such that $T \le \delta \le 0$, and T_k is the smallest positive number such that

$$I(T_k) = \int_{t_k}^{t_k + T_k} g(T_k + t_k - t)u(t)dt = \pm s \tag{5}$$

with s being arbitrary and non negative, $g(.)$ a general non-linear function and $p_k = sgn[l(T_k)]$.

In some particular cases the definition of the previously encountered functions, i.e. $M(.)$, $\delta(.)$, $T(.)$ and $g(.)$, simplifies the modulator and leads to more familiar controllers. Despite its apparent major complication, most applications of pulse modulated control deal with modulators of the first group.

1: Pulse Amplitude Modulators

An extreme case of the first group is represented by the linear sample and hold element, also known as Pulse Amplitude Modulator (PAM), which is the basis of most of the discrete time control algorithms; it is defined by letting $T(.)=\Delta$, positive and constant, $\delta(.)=\Delta$ and $M[u(k\Delta)]=Mu(k\Delta)$, with M constant.

2: Pulse Width Pulse Frequency Modulators

Still in the first group we find the Pulse Width Pulse Frequency Modulator (PWPFM), by selecting two appropriate functions $F(.)$ and $f(.)$ and two appropriate constants c_0, c_1 such that

$$M\big[u(t_k)\big] = M\cdot sgn\big[u(t_k)\big] \tag{6}$$

$$T\big[u(t_k)\big] = \begin{cases} F\big[u(t_k)\big] \ge T_{min} & \text{for} \quad \big|u(t_k)\big| < c_1 \\ T_{min} & \text{for} \quad \big|u(t_k)\big| \ge c_1 \end{cases} \tag{7}$$

$$\delta\big[u(t_k)\big] = \begin{cases} \delta_{max} = T_{min} & \text{for} \quad \big|u(t_k)\big| \ge c_1 \\ f\big[u(t_k)\big] \le \delta_{max} & \text{for} \quad 0 \le c_0 < \big|u(t_k)\big| < c_1 \\ 0 & \text{for} \quad \big|u(t_k)\big| \le c_0 \end{cases} \tag{8}$$

3: Pulse Width Modulators

A further simplification is the Pulse Width Modulator (PWM), obtained by imposing $T(.)=\Delta$ positive constant, $M[u(k\Delta)]=M\cdot sgn[u(k\Delta)]$ and

$$\delta\big[u(k\Delta)\big] = \begin{cases} \beta\big|u(k\Delta)\big| & . \quad \text{if} \quad \big|u(k\Delta)\big| \le \Delta/\beta \\ \Delta & \text{if} \quad \big|u(k\Delta)\big| > \Delta/\beta \end{cases} \tag{9}$$

4: Variable Structure/Sliding Mode Control

A detailed exposition of the theory of Variable Structure Control (VSC), also called Sliding Modes Control (SMC), is beyond the scope of this presentation; a brief introduction will be given in any case for a single input system, since this type of control presumes some kind of input commutation closely related to PWM controls; the interested reader is referred to Refs. [4-8] for further details.

Let consider a single input, non-linear, multivariable system, whose state vector {x} evolves according to

$$\{\dot{x}\} = f(\{x\}) + g(\{x\}) \cdot u \tag{10}$$

with the input u being a two valued function $u \in U = \{0,1\}$, and let $s(\{x\})$ be a scalar function of the state vector. It is possible to commute the input function u from 0 to 1 according to

$$u = \begin{cases} 1 & \text{if} & s(\{x\}) > 0 \\ 0 & \text{if} & s(\{x\}) \le 0 \end{cases} \tag{11}$$

so that the function $s(\{x\})$ defines a hyper surface determining the activity of the actuator. If the motion is stable around the surface $s(\{x\})$, then it is said to be "sliding" on $s(\{x\})$. The conditions for the existence of this motion can be determined by noting that, for stability, the time derivative of the function $s(\{x\})$ must be

$$\begin{cases} \dot{s}(\{x\}) < 0 & \text{for} & s(\{x\}) > 0 \\ \dot{s}(\{x\}) > 0 & \text{for} & s(\{x\}) \le 0 \end{cases} \tag{12}$$

with, calling ∇ the gradient operator,

$$\dot{s}(\{x\}) = \nabla s \cdot \{\dot{x}\} = \nabla s \cdot \left[f(\{x\}) + g(\{x\}) u \right] \tag{13}$$

Recalling the system dynamics, i.e. Eq. (10), and the control input function, i.e. Eq. (11), the condition for the existence of sliding modes becomes

$$\lim_{s\to 0^+} \nabla s \cdot \left[f(\{x\}) + g(\{x\})u \right] < 0$$
$$\lim_{s\to 0^-} \nabla s \cdot \left[f(\{x\}) \right] < 0 \tag{14}$$

A further condition for their existence is that is possible to determine a bounded control function $0 < u_{eq} < 1$, called equivalent control, so that the ideal system represented by

$$\{\dot{x}\} = f(\{x\}) + g(\{x\}) \cdot u_{eq}(\{x\}) \tag{15}$$

evolves on the same sliding surface $s(\{x\})$.

A remarkable characteristic of any sliding motion is its insensitivity a certain class of disturbances and/or system parameter uncertainties.

An interesting equivalence between VSC and PWM control systems [28,29], can be found by modulating a PWM control with sampling time Δ, according to

$$u = \begin{cases} 1 & \text{for} \quad t_k < t \le t_k + D[x(t_k)]\Delta \\ 0 & \text{for} \quad t_k + D[x(t_k)]\Delta < t \le t_k + \Delta \end{cases} \tag{16}$$

where $D(.)$ defines the duty ratio function. Taking the limits for $\Delta \to 0$, the discrete time system will evolve, on the average, according to

$$\{\dot{x}\} = f(\{x\}) + D(x) \cdot g(\{x\}) \tag{17}$$

which, compared to Eq. (15), tells us that the equivalent control function u_{eq} of a VSC can be used as a duty ratio function for a PWM control, and in this case the resulting motion will be a sliding mode.

It is important to remark that in the case of a linear system, controlled by an on-off controller, a VSC can be regarded as a suitable way to determine an equivalent linear closed loop behavior, provided that the full state of the system is available [2,4-6]. However in practical applications, especially when a relatively large order system is involved, the full state is seldom available and a state observer is required. Thus, even if a VSC can be easily designed by using many of the methodologies available for the design of linear control systems, a large order compensator is generally needed. This can cause a loss of robustness and does not allow the adoption of controllers with a constrained structure. It can then be impossible to adopt effective controllers that can be more easily implemented because of a decentralized input-output and compensation topology, determined a priori by the designer on the base of its understanding of the system. On the contrary the techniques to be presented in the second part of the work lead to a linear equivalent of a PWM controlled system that allows a larger freedom in the design of the controller structure.

5: Implementation issues

In practice, as already mentioned in the introductory presentation of discrete time controls, the evaluation of the input function $u(.)$ at a finite rate, combined with measurement noise, can lead to a continuous triggering of the actuators even at a null steady state operating point with negligible disturbances, causing a useless waste of power. This is especially true for fixed frequency controllers, such as PWM modulators. In all the cases in which this has to be avoided, the simplest solution comes from the adoption of a dead zone, i.e., a range of values of the input function $u(.)$ for which the output is zero. A more sophisticated solution, called Schmitt trigger [12], has a hysteretical logic for switching on and off the actuators. A representation of the input-output behavior of these two devices is given in Fig. 1.

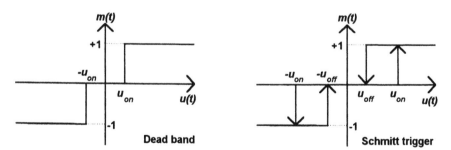

Figure 1: input-output characteristics of a dead band and Schmitt trigger

According to the switching values selected, limit cycles could appear in any case, especially if disturbances are present.

It must be noted that even if the adoption of a dead zone, with or without hystheresis, can pose considerable limitations on the performances achievable, being the system uncontrolled for small errors and/or disturbances, the results obtainable without it are often worse.

II: LINEAR SYSTEMS AND PWM CONTROL

The previous section has pointed out the main motivations for the use of pulse modulated control, and it appears that the strongest reasons for this are dictated by the actuators rather than by control theories. In fact the design of systems controlled by discontinuous actuators can require the use of optimization methods that are extremely difficult and cumbersome when applied to real systems. On the other hand many effective design techniques are available for linear systems, either continuous or discrete in time. This contrast is the basis for the development of suitable models to "linearize" PWM controls.

A remarkable approach is developed in Ref. [10], and since it represents the basis of the technique exposed further on, the main assumptions and results will be reported here.

The linear system to be controlled is assumed governed by a system of time invariant linear differential equations of the form

$$\{\dot{x}\} = [A]\{x\} + [B]\{u\} \tag{18}$$

where [A] and [B] are constant matrices. It is further assumed that, over a fixed sampling period Δ, each component of the control input vector $\{u\}$ can take only two values, U_{max} and U_{min}. The control is implemented by varying, starting from the middle of the control period, the fraction δ_n of the sampling period during which the control is at U_{max}, as shown in Fig. 2.

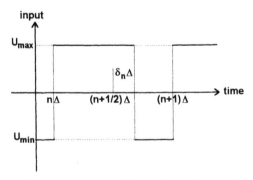

Figure 2: pulse width modulated control

Assuming, without loss of generality, the presence of a single input to the system, that $U_{max}=-U_{min}=1$, and that δ_n is small enough to allow the approximation

$$\left[e^{-A\delta_n\Delta}\right] = [I] - [A]\delta_n\Delta \tag{19}$$

the system dynamics can be written in terms of a new state vector $\{z\}$, as

$$\{z\}_{n+1} = \left[e^{A\Delta}\right]\{z\}_n + \left[\overline{\Gamma}\right]\delta_n \tag{20}$$

with, defining $\{x\}_n$ as the average state over the sampling period,

$$\{z\}_n = \{\overline{x}\}_n - 2[A]^{-1}\left(\left[e^{A\Delta/2}\right]-[I]\right)[B]\delta_n \tag{21}$$

$$\left[\overline{\Gamma}\right] = 2\left[e^{A\Delta/2}\right][A]^{-1}\left(\left[e^{A\Delta}\right]-[I]\right)[B] \tag{22}$$

which, despite its appearance, is valid even for a singular [A] matrix.

The analysis presented in Ref. [10] shows also a more general formulation in which the desired state is not the origin, the maximum and minimum input values are not equal in magnitude and a disturbance acts on the system. This case needs not to be reported here since its main concepts remain unchanged with respect to those summarized above. Furthermore the connection between the measurements taken at an arbitrary time and the average state vector is also established in Ref. [10], but once more this is not reported since the formulation to be presented in this work assumes that all the measurements are taken at the same frequency and sampling instants. The interested reader is referred to the original paper for further details.

One important characteristic of the above PWM system is that the input can never be identically zero, even when there is no excitation, but only its average value can be nulled. As already mentioned, this can lead to an unacceptable waist of power and can be a severe limitation in those systems with limited power available, e.g. aerospace applications. The adoption of a dead zone can solve the problem, but such a type of control is not suitable for unstable systems, since a non zero input can cause an unbounded output.

However the previous considerations and modeling method set up the basis of a different technique for the linearization of PWM controls applied to linear systems.

III: LINEAR EQUIVALENT OF A PWM CONTROL

The alternative method proposed is based on the evaluation of the difference between the response of a discrete linear time invariant system, subject to an arbitrary PAM control, and that of a PWM control of equal impulse over each sampling period, applied with a constant delay referred to the sampling instants. The response error, between the parent PAM and its equivalent PWM system, is seen to depend on the input delay and it is proven that there exists a fixed delay that minimizes the error itself. So, with the application of an equivalent PWM control, the system response matches that of its parent driven by a linear PAM control, with obvious advantages in the design of the control laws. Moreover the input can be made identically zero and therefore suitable for unstable systems [30].

A: FORMULATION

The basic idea is then to determine a suitable PAM control law for a linear, time invariant system represented in state space form, and to apply it as a PWM control. The system dynamics is again written as a system of first order linear time invariant differential equations

$$\{\dot{x}\} = [A]\{x\} + [B]\{u\} \tag{23}$$

where $\{x\}$ and $\{u\}$ are the n states and m inputs to the system, $[A]$ and $[B]$ matrices of appropriate dimensions. In the linear discrete control form, i.e. PAM, the input $\{u\}$ is considered constant over each sampling period Δ and applied with no delay with respect to the sampling instant t, whereas in the current PWM formulation the amplitude $\{u_m\}$ is fixed and the unknown variables become the periods δ_i during which each u_{m_i} is applied, called firing times, and the delays τ_i between the sampling instant and the input application, called firing delays. This is depicted in Fig. 3, where the delay τ, as it will be demonstrated in the sequel, is supposed constant.

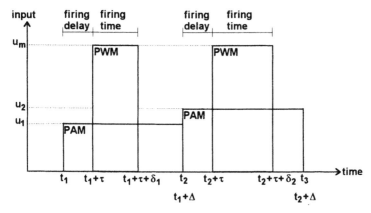

Figure 3: PAM and its equivalent PWM control mode

The system response to the PAM control is then determined by the superposition of the effects of each input

$$\{x(t+\Delta)\} = [\Phi(\Delta)]\{x(t)\} + \sum_{i=1}^{m}[\Psi(\Delta)]\{B_i\}\,u_i\Delta \qquad (24)$$

having defined the matrices

$$[\Psi(d)] = \sum_{i=0}^{\infty}\frac{[A]^i}{(i+1)!}d^i \qquad (25)$$

$$[\Phi(d)] = \left[e^{Ad}\right] = [I] + [A][\Psi(d)]d = \sum_{i=0}^{\infty}\frac{[A]^i}{i!}d^i \qquad (26)$$

For this same system, an equivalent PWM control response will now be searched.

B: RESPONSE TO A PWM CONTROL INPUT

Considering the PWM input depicted in Fig. 3, the system response becomes

$$\{x(t+\Delta)\} = [\Phi(\Delta)]\{x(t)\} + \sum_{i=1}^{m}[\Phi(\Delta-\tau)][\Psi(-\delta)]\{B_i\}\,u_{m_i}\delta_i \qquad (27)$$

For conciseness only a single input will be hereafter considered as, if the response to a single PWM input matches the corresponding PAM response, the global PWM response will match any discrete linear response because of the system linearity. The suffix i will then be dropped in the sequel.

The first assumption made concerns the firing time δ, evaluated in such a way to keep the total impulse of the PAM and PWM control equal over each sampling period Δ

$$\delta = u\Delta/u_m \qquad (28)$$

The firing time δ is also assumed brief enough to allow the following approximation

$$\left[e^{-A\delta}\right] = [I] - [A]\delta \qquad (29)$$

so that the PWM response of Eq. (27) becomes

$$\{x(t+\Delta)\} = [\Phi(\Delta)]\{x(t)\} + [\Phi(\Delta-\tau)]\{B\}\,u_m\delta \qquad (30)$$

The comparison of Eqs. (30) and (24) shows no differences in the state transition matrix, since the unforced dynamics is obviously not affected by the way the input is applied, while the control term appears different. Recalling the definitions of the $[\Phi]$ matrix and Eq. (28), the difference between the PAM response, defined as the desired response, and the PWM response, can be expressed through the error vector $\{E\}$, whose expression is

$$\{E\} = -\sum_{k=1}^{\infty}\frac{[A]^k\left((k+1)(\Delta-\tau)^k - \Delta^k\right)\{B\}u\Delta}{(k+1)!} =$$
$$= -\sum_{k=1}^{\infty}\frac{[A]^k\left((k+1)(\Delta-\tau)^k - \Delta^k\right)\{B\}u_m\delta}{(k+1)!} \qquad (31)$$

The above error vector is directly proportional to the integral of the input over a sampling period, with a proportionality coefficient dependent, in a non-linear way, on the delay τ and the system matrices [A] and [B]. It is therefore useful to evaluate that particular delay $\bar{\tau}$ which minimizes, in some sense, the error, while maintaining the approximation of a small δ. Prior to this evaluation, it is useful to note that such a $\bar{\tau}$ actually exists, since the evaluation of the error for three particular delays, precisely $\tau=0$ (impulse applied at the beginning of the sampling period), $\tau=\Delta-\delta$ (impulse applied at the end of the sampling period) and $\tau=\Delta/2$ leads to

$$
\begin{aligned}
\tau = 0 \quad &\rightarrow \quad \{E\} = -\sum_{k=1}^{\infty}\frac{k[A]^{k}\Delta^{k}\{B\}u_{m}\delta}{(k+1)!} \\
\tau = \Delta - \delta \quad &\rightarrow \quad \{E\} = -\sum_{k=1}^{\infty}\frac{[A]^{k}\big((k+1)(\delta\Delta)^{k}-\Delta^{k}\big)\{B\}u_{m}\delta}{(k+1)!} = \\
&\qquad\qquad\quad \cong +\sum_{k=1}^{\infty}\frac{[A]^{k}\Delta^{k}\{B\}u_{m}\delta}{(k+1)!} \\
\tau = \Delta/2 \quad &\rightarrow \quad \{E\} = -\sum_{k=2}^{\infty}\frac{[A]^{k}\big((k+1)(\Delta/2)^{k}-\Delta^{k}\big)\{B\}u_{m}\delta}{(k+1)!}
\end{aligned}
\tag{32}
$$

So it is obvious that the error vector reverses its sign as the delay τ increases from 0 to $\Delta-\delta$, and in particular for $\tau=\Delta/2$ the first term of the summation becomes zero. This consideration allows to evaluate the optimal delay $\bar{\tau}$ by equating to zero the norm of the error vector, i.e., by solving $\{E\}^{T}\{E\}=0$. Writing $\bar{\tau}=\Delta/2+\tilde{\tau}$, indicating with $/\tau$ the derivative with respect to τ, evaluating the derivatives at $\tau=\Delta/2$ and introducing the notation

$$
[\Theta] = [\Phi(\Delta/2)] = \left[e^{A\Delta/2}\right] \quad ; \quad [\Xi] = [\Psi(\Delta)] = [A]^{-1}\left[e^{A\Delta}-I\right]/\Delta \tag{33}
$$

$$
\{E\}_{\tau=\Delta/2} = \{\varepsilon\} = [\Xi-\Theta]\{B\}u\Delta = [\Xi-\Theta]\{B\}u_{m}\delta \tag{34}
$$

the application of the first step of the Newton-Raphson method allows to write

$$
\tilde{\tau}\cdot\left(\{E\}^{T}\{E\}\right)\Big|_{/\tau}\Big|_{\tau=\Delta/2} = -\{\varepsilon\}^{T}\{\varepsilon\} \tag{35}
$$

$$\{\varepsilon\}^{T}\{\varepsilon\} \; = \; \{B\}^{T}[\Xi\text{-}\Theta]^{T}[\Xi\text{-}\Theta]\{B\}u_{m}^{2}\delta^{2} \qquad (36)$$

$$\left(\{E\}^{T}\{E\}\right)\Big/_{\tau}\Big|_{\tau=\Delta/2} = \{B\}^{T}\left([\Theta]^{T}[A]^{T}[\Xi\text{-}\Theta] + [\Xi\text{-}\Theta]^{T}[A][\Theta]\right)\{B\}u_{m}^{2}\delta^{2} \quad (37)$$

$$\tilde{\tau} \; = \; -\frac{\{B\}^{T}[\Xi\text{-}\Theta]^{T}[\Xi\text{-}\Theta]\{B\}}{\{B\}^{T}\left([\Theta]^{T}[A]^{T}[\Xi\text{-}\Theta] + [\Xi\text{-}\Theta]^{T}[A][\Theta]\right)\{B\}} \qquad (38)$$

$$\bar{\tau} \; = \; \frac{\Delta}{2} \; -\frac{\{B\}^{T}[\Xi\text{-}\Theta]^{T}[\Xi\text{-}\Theta]\{B\}}{\{B\}^{T}\left([\Theta]^{T}[A]^{T}[\Xi\text{-}\Theta] + [\Xi\text{-}\Theta]^{T}[A][\Theta]\right)\{B\}} \qquad (39)$$

In this approximation the optimal delay for each of the inputs to the system does not depend on the actual input value and can then be evaluated independently from the control law. Moreover, compared to the discrete linear response, this conversion technique does not affect the controlled system dynamics, including stability, so that the controller synthesis can be performed on the PAM discrete system, which is linear.

C: RESPONSE TO A PWM REFERENCE COMMAND

The next step is to determine the equivalent δ_{c} for the same system subject to a constant reference input u_{c}. Once more a single input will be considered, with the obvious consideration that, in presence of many reference commands, each one should be treated separately.

The response of the discrete linear system to the reference input u_{c} is

$$\{x(t+\Delta)\} \; = \; [\Phi(\Delta)]\{x(t)\} \; +[\Psi(\Delta)]\{B\}\, u_{c}\Delta \qquad (40)$$

and, at steady state

$$\{x(t+\Delta)\} \; = \; \{x(t)\} \; = \; \{x_{s}\} \; = \; -[A]^{-1}\{B\}u_{c} \qquad (41)$$

If the input includes both a reference command and a control command, then the response becomes

$$\{x(t+\Delta)\} = [\Phi(\Delta)]\{x(t)\} + [\Psi(\Delta)]\{B\} (u+u_c)\Delta \tag{42}$$

Transforming u into δ by using Eq. (28), and u_c into a yet unknown δ_c, and applying the command u_m for a time $\delta+\delta_c$, with delay τ, still supposing δ to be small, we get

$$\begin{aligned}\{x(t+\Delta)\} &= [\Phi(\Delta)]\{x(t)\} + [\Phi(\Delta-\tau)][\Phi(-\delta_c)]\{B\} u_m\delta \\ &+ [\Phi(\Delta-\tau)][\Psi(-\delta_c)]\{B\} u_m\delta_c\end{aligned} \tag{43}$$

At steady state $\delta=0$ and the goal is to obtain $\{x(t+\Delta)\}=\{x(t)\}$, according to

$$[\Phi(\Delta) - I]\{x(t)\} = -[\Phi(\Delta-\tau)][\Psi(-\delta_c)]\{B\} u_m\delta_c \tag{44}$$

With a PWM reference command the state vector can never be constant over a sampling period even at steady state; for the equivalence with the PAM response the optimal PWM command will then cause an average state $\{\overline{x}\}$ as close as possible to the value $\{x_s\}$ over each sampling period. The average state is

$$\{\overline{x}\} = \frac{1}{\Delta}\int_0^\Delta \{x(t+\lambda)\}d\lambda \tag{45}$$

and the integral must be evaluated considering, according to Fig. 3, that the vector $\{x(t+\lambda)\}$ assumes different values depending on λ; more precisely

$$\begin{aligned}0 < \lambda \le \tau &\rightarrow [\Phi(\lambda)]\{x(t)\} \\ \tau < \lambda \le \tau+\delta_c &\rightarrow [\Phi(\lambda)]\{x(t)\}+[\Phi(\lambda-\tau)][\Psi(\tau-\lambda)]\{B\} u_m(\tau-\lambda) \\ \tau+\delta_c < \lambda \le \Delta &\rightarrow [\Phi(\lambda)]\{x(t)\}+[\Phi(\lambda-\tau)][\Psi(-\delta_c)]\{B\} u_m\delta_c\end{aligned} \tag{46}$$

$$\{\overline{x}\} = \frac{1}{\Delta}[A]^{-1}\left([\Phi(\Delta-\tau)][\Psi(\delta_c)]\{B\} u_m\delta_c - \{B\} u_m\delta_c + [\Phi(\Delta)-I]\{x(t)\}\right) \tag{47}$$

which, by virtue of Eq. (44) becomes

$$\{\overline{x}\} = -\frac{1}{\Delta}[A]^{-1}\{B\} u_m\delta_c \tag{48}$$

Then, comparing Eq. (48) with Eq. (41), it can be stated that to obtain $\{\bar{x}\}=\{x_s\}$, each discrete command u_c must be transformed, regardless of the delay τ, into a pulsed command of duration

$$\delta_c = u_c \Delta / u_m \qquad (49)$$

which is equivalent to that of the control commands. This should have been expected since, if a steady state condition is actually reached, the effect of different delays would simply be a time shift of the response, with no impact on the integral of the response over a sampling period.

In conclusion, having evaluated a control strategy suitable for the linear system of Eq. (23), the PWM implementation must be performed by modifying the input according to Eqs. (28) and (49), and applying the resulting command with a delay τ given by Eq. (39).

D: CLOSED LOOP STABILITY ANALYSIS

The most important point in designing active controls is to ensure the closed loop stability of the system. Supposing a direct output feedback control law of the kind

$$\{u\} = [K][C]\{x(t)\} \qquad (50)$$

$[C]\{x(t)\}$ being the measured output, the closed loop responses of the same system subject respectively to a PAM and PWM control input are

$$\{x(t+\Delta)\} = ([\Phi(\Delta)] + [\Psi(\Delta)]\{B\}[K][C]\Delta)\{x(t)\} = [N_{PAM}]\{x(t)\} \qquad (51)$$

$$\{x(t+\Delta)\} = ([\Phi(\Delta)] + [\Phi(\Delta-\tau)]\{B\}[K][C]\Delta)\{x(t)\} = [N_{PWM}]\{x(t)\} \qquad (52)$$

If the delay τ is computed according to Eq. (39), then, due to the assumptions made to compute $\bar{\tau}$, the following relations hold

$$[\Phi(\Delta-\bar{\tau})]\{B\}u_m\delta = [\Psi(\Delta)]\{B\}u\Delta \qquad (53)$$

$$[\Phi(\Delta-\bar{\tau})]\{B\}\Delta = [\Psi(\Delta)]\{B\}\Delta \qquad (54)$$

$$[N_{PAM}] = [N_{PWM}] \tag{55}$$

and therefore the closed loop stability properties of the PAM controlled system pertain also to the PWM controlled one, both for open loop stable and unstable systems. Once more this allows the computation of the feedback gain matrix [K] by referring to a PAM system, with great computational advantage since many efficient algorithms are available for this purpose.

For any other value of the firing delay τ, the closed loop stability of the discretely controlled system does not automatically guarantee the stability of the pulse width controlled system. In fact, if the actual delay is $\tau = \bar{\tau} + \xi$, the relation between the closed loop state matrix of the PWM and PAM controlled system is given by

$$[N_{PWM}]_{(\tau=\bar{\tau}+\xi)} = [N_{PAM}] + [\Phi(\xi)-I][\Psi(\Delta)]\{B\}[K][C]\Delta \tag{56}$$

and the eigenvalues of the matrix [N$_{PWM}$] can differ substantially from those of [N$_{PAM}$]. This obviously holds only if no control saturations occur and sets a hard constraint on the precision of the implementation of the delay $\bar{\tau}$. In PWM control the input gets saturated if δ becomes greater than the sampling time Δ, a condition equivalent to the amplitude saturation in the discrete case.

E: SOME INTERESTING CONSIDERATIONS

Before showing some applications, a few remarkable properties of this particular linearized PWM implementation will be discussed.

- First of all, with the same control transposition from discrete to pulse width, the average steady state and the transient response at each sampling instant are preserved.
- The approximations made, i.e. Eq. (29), concern the firing time δ, but never the sampling time Δ, although the two are somehow related due to the way in which δ is evaluated. This means that in this framework PWM actuators may be "linearized" even for relatively low control frequencies, a rather unusual, remarkable and convenient property.
- The result of Eq. (39) shows that the optimal firing delay does not depend upon the actual input. This means that, in practice, the maximum and

minimum force values are absolutely independent; if different force signs are required, two force levels of opposite signs will be necessary, $u_{max}>0$ and $u_{min}<0$, but it can be $|u_{max}| \neq |u_{min}|$. The fundamental point is to consider, in the evaluation of the firing time δ, i.e. Eq. (28), the appropriate force value depending on the sign of the discrete input u.

- Another fundamental property is that the equivalence is established for every sampling instant independently from the PAM input to be modulated into its equivalent PWM. This means that even non-linear discrete time control laws can be converted with the same logic, provided the assumptions pertaining to Eq. (29) are acceptable. The particular emphasis given to linear controllers simply comes from the simpler stability and performance analyses.

- In practical applications, the firing time δ is evaluated, for each control, at the sampling times, and then applied with its proper delay. The entity of the duration can determine three different situations at the next sampling instant, shown in Fig. 4, and precisely:

a) saturation: the duration δ is greater than the sampling time Δ, so the pulsed input is applied across two subsequent sampling periods. At the next sampling time the new duration is evaluated and applied with the proper delay, with the old input still active. This is not strictly related to the theoretical formulation adopted since, as it can be inferred from Eq. (27), if $\delta=\Delta$, the equivalence between PAM and PWM controls is obtained with $\tau=0$. Nonetheless, for simplicity, it is preferred to apply with the optimal delay even a saturated input since, if the control stays saturated for more than one sampling period, the difference will appear only during the first period. It is also possible that the subsequent duration has a different sign, which would simply mean that the input must be applied in the opposite direction but still with the optimal delay.

b) overlap: the duration δ is shorter than the sampling time Δ, but greater than $\Delta-\tau$, the time between the firing and the next sampling instant. In this case again the new duration is evaluated with the old input still active, but, being applied with the proper delay, there will be a short fraction of time with no input, since the old input will cease before the activation of the newer one.

c) no overlap: the duration δ is shorter than $\Delta-\tau$, so each pulse is contained in one single sampling period.

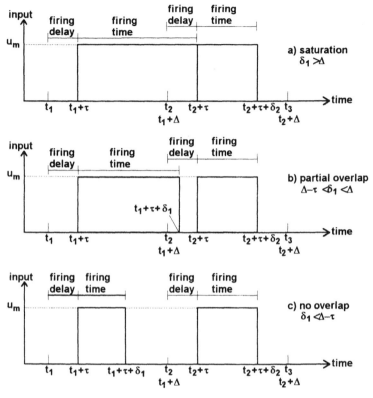

Figure 4: possible operating conditions of a PWM control

IV. EXAMPLES

A: CONTROL OF A DC ELECTRIC MOTOR

Even if most of the applications to be presented will be related to mechanical and aerospace fields, this example is set forth to demonstrate the general viability of the method just developed and is related to the speed control of a permanent magnet direct current (DC) motor.

It is well known that medium to high power DC motor are nowadays mainly controlled through PWM power drives to minimize the losses within the power amplifier. However to simplify the implementation the modulation is mostly applied to a square wave, i.e. to a driving signal having just two output states

whose theory is presented in Ref. [10] and briefly recalled in section II. Another general approach that can be adopted to design a two output states PWM power drive for DC is the sliding mode control approach [7].

Here we will clearly adopt a three state output, i.e. a power drive including an off command. This type of drive has the drawback of a slightly more complex implementation but on the other hand it reduces the current ripple and is often used to drive PWM controlled step motors. We will then assume that the speed control is to be implemented through a PID controller thus leading to the following mathematical model:

$$\frac{di}{dt} = -\frac{1}{\tau_e}i - \frac{k_e}{\tau_e R}\omega + \frac{1}{\tau_e R}V_c$$

$$\frac{d\omega}{dt} = \frac{R}{\tau_m K_t}i - \frac{R}{\tau_m K_t^2}T_d \qquad (57)$$

$$\frac{d\theta}{dt} = \omega$$

for the motor [31], and

$$V_c = G_p(\omega_c - \omega) + G_i(\theta_c - \theta) + G_d i \qquad (58)$$

for the control, where: i is the current, ω and ω_c are the rotational and commanded rotational speeds, θ and θ_c the rotation and commanded rotation, V_c is the control voltage, T_c the disturbing torque, K (0.33 Nm/A-Vs) is the motor speed/torque constant, R (.2 Ω) the overall winding + brush resistance, τ_e=L/R (50 ms) and τ_m=RJ/K^2 (10 ms) are the electrical and mechanical time constant, L and J being the total smoothing + winding inductance and the total moment of inertia of the load + motor respectively. The parenthesized numbers indicate the actual values used.

For the implementation of the PID control i is available through an appropriately placed sensing resistor, ω is measured by a tachometer and θ by a synchro-resolver or encoder angular rotation sensor; a set of measurements which is usually available in standard high performance motor drives.

The design of the above control law has been carried out by pole assignment, on the discrete time difference equations corresponding to Eqs. (57,58), in such a way to optimize the response speed and to minimize the overshoot to a step disturbance. This has been obtained through the following poles: $100 \pm j100$, -50, and led to the following gains: $G_p = -12.6$, $G_i = -4.46$, $G_d = -.216$

The sampling time adopted is 1 ms, the firing resolution is 10 μs. Figure 5 shows the results obtained in a non-linear simulation when a 7.5% variation of the reference speed is imposed against a disturbing torque of 60% the nominal motor torque, when the maximum allowed current in the winding is twice the nominal current and the voltage across the transistor power bridge is 125% the nominal voltage. The reference nominal values are $i_n = 10$ A, $V_n = 100$ V, $\omega_n = 300$ rad/s and $T_n = 3.3$ Nm/A. An anti windup control is imposed on the integral feedback term by limiting the correction of the control voltage related to G_i to 17% of the maximum value. It can be noticed that due to the limitation imposed on the instantaneous value of the current the control voltage is never saturated and the linearly designed control behaves fairly well despite the significant non-linearities introduced by current saturation and anti-windup of the integral control.

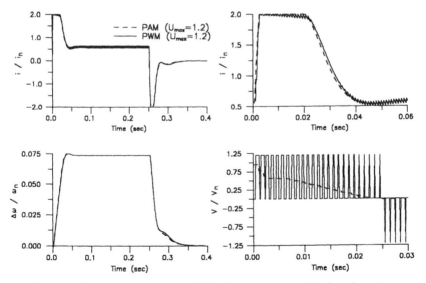

Figure 5: linear and equivalent PWM response of a DC electric motor

It is recalled that the improved capability of modern power transistors in fast and reliably switching high current has allowed the use of higher switching frequencies, i.e. a very low sampling time in the present formulation, of the order of the Khz. This, even if it introduces higher losses in the magnet, has allowed to avoid the annoying noise typical of earlier power drives. The adoption of a relatively low switching frequency is nonetheless adopted here as it strains the PAM to PWM equivalence determined through Eq. (28) much more than the highest frequencies now adopted that make the use of the above equation more and more close to the ideal case.

B: CONTROL OF A SECOND ORDER SYSTEM

This example demonstrates the effects of the firing delay τ, the sampling time Δ and the maximum input amplitude u_m on both the transient and steady state response of a mechanical system modeled as a simple spring-mass-damper system. In some sense it can be considered as preliminary to the experimental results presented further on.

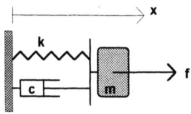

Figure 6: scheme of a second order system

The dynamics of the system having mass m, damping coefficient c, elastic spring constant k (see Fig. 6), is represented in the state form by

$$\left\{\begin{matrix} \ddot{x} \\ \dot{x} \end{matrix}\right\} = \begin{bmatrix} -2\zeta\omega & -\omega^2 \\ 1 & 0 \end{bmatrix}\left\{\begin{matrix} \dot{x} \\ x \end{matrix}\right\} + \begin{bmatrix} 1/m \\ 0 \end{bmatrix}\{f\} \qquad (59)$$

and for the particular case analyzed the system parameters are ω=6.28, ζ=0.1 and m=1. The input control force f is given by

$$\{f\} = k_v\{\dot{x}\} + k_e\{x-r\} + m\omega^2\{r\} \qquad (60)$$

where {r} is the desired set point. It can be noted that the feed forward control, i.e. $m\omega^2\{r\}$, is capable of achieving the desired set point in case of perfect model knowledge. Thus the feedback terms are added to improve the response performances and to take into account uncertainties in the model knowledge.

With these system characteristics, a controller sampling time of 0.1s appears appropriate, and the desired closed loop performances are expressed only in terms of damping factor, to be as close as possible to 0.7. The design leads to the gain values reported in Tab. I, which shows also the performances of the same system analyzed with a smaller sampling time. The controller has not been evaluated for this shorter sampling time, since its performances are already satisfactory even when the same gains of the longer sampling time are used, and since the main goal of the example is to compare the PAM and PWM responses of analogous systems, with the sampling time as a distinctive parameter of the system configuration.

Δ (s)	$\bar{\tau}$ (s)	ω (rad/s)	ζ	k_v	k_e
0.1	0.0477	4.35	0.71	-2.67	24.2
0.05	0.0241	4.08	0.60	-2.67	24.2

Table I: controlled second order system parameters

The step responses of the system to a unit reference command, both for PAM and PWM, with different firing delays and maximum input amplitudes, are reported in Figs. 7,8,9.

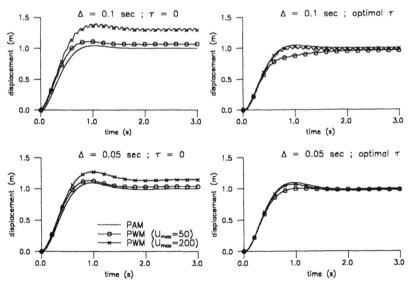

Figure 7: displacement response of a second order system (PAM and PWM)

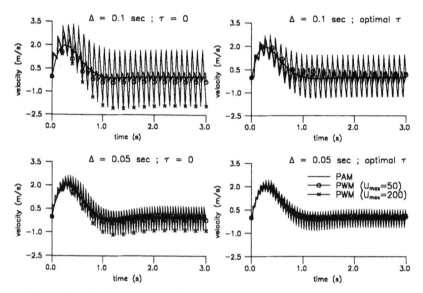

Figure 8: velocity response of a second order system (PAM and PWM)

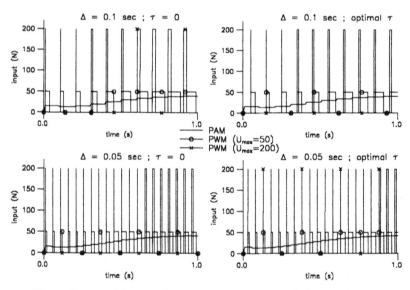

Figure 9: control inputs of a second order system (PAM and PWM)

Again the analysis of the step responses is in good agreement with the theory, but some considerations should be pointed out.

- Figures 7 and 8 show the peculiar response of PWM controlled systems, which tend to oscillate around the average condition even at "steady state", due to the on/off input type. This can be a serious limitation if strict performances are desired on state derivatives or on not sufficiently smoothed states, in the present case the velocity. It is seen however that the displacement response is fairly smooth and actually reaches a "steady state" condition if the delay is the optimal one. The experimental results presented further on demonstrate that if the desired steady state condition is a complete rest, then the PWM control works in an excellent way.

- If the firing delay is not the optimal one, the error in the PWM controlled response relative to the "desired" PAM response is proportional to the input amplitude u_m, and this clearly appears from the steady state biases in Fig. 7. In this particular example the value $u_m=50$ is quite close to the steady state discrete input, precisely 39.48, so it is obvious that the steady state value of δ will be very close to Δ, leading to a continuous overlap of the input impulses with those of the subsequent sampling period (case b of Fig. 4). This does not match rigorously the assumptions made to establish the equivalence between PAM and PWM controls, and explains why the performances of the system subjected to a PAM input and to a PWM input with no delay are quite similar. In fact, as already stated in section III.E, if $\delta=\Delta$, the equivalence between PAM and PWM control is obtained with $\tau=0$, and this closely resembles the present case.

- The optimal firing delay τ guarantees that the steady state response of the PAM and PWM controlled systems are the same. The difference in the transient response depends on the input amplitude. Higher amplitudes mean smaller pulse firing durations δ, and so the approximation of Eq. (29) becomes indeed true, and Eq. (30) represents the real dynamic behavior of the system, so that the optimal firing delay effectively minimizes the response error. The same holds also for smaller sampling periods since δ is also proportional to Δ.

C: CONTROL OF THERMALLY INDUCED VIBRATIONS ON A FLEXIBLE BOOM

This further example shows an application of a PWM controller to an unstable coupled thermoelastic structural system, in which the heating input depends upon the actual structural shape resulting from thermally induced displacements. This leads to a thermo-structural feedback that can cause unstable vibrations of a slender flexible boom in space. It represents a challenging test for a variety of reasons: the system is multi-input and unstable, and the control action is either unidirectional or zero. Furthermore this example will demonstrate that the equivalent linearized PWM system, obtained with the technique developed in this work, allows the adoption of a fully decentralized controller.

The structure under test represents a simplified model of the boom of the OGO-IV spacecraft, which was known to be affected by sustained oscillations related to the crossings from the Earth's shadow into sunlight and vice versa. A detailed presentation of the coupled thermoelastic model can be found in Refs. [32,33]; according to this method, the dynamics of the system is represented in state space form as

$$
\begin{Bmatrix} \ddot{q} \\ \dot{q} \\ \dot{h} \end{Bmatrix} = \begin{bmatrix} -2\zeta\omega & -\omega^2 & \Phi^T Z\Psi \\ I & 0 & 0 \\ 0 & \Psi^T\gamma & -\lambda \end{bmatrix} \begin{Bmatrix} \dot{q} \\ q \\ h \end{Bmatrix} + \begin{bmatrix} 0 \\ 0 \\ \Psi^T \end{bmatrix} \{u\} + \begin{Bmatrix} 0 \\ 0 \\ \Psi^T P^* \end{Bmatrix} \tag{61}
$$

where $\{q\}$ and $\{h\}$ are generalized structural and thermal modal coordinates, $[\Phi]$ and $[\Psi]$ are the structural and thermal mode shapes, $[Z]$ and $[\gamma]$ are the thermo-elastic and elasto-thermal coupling terms, ω, ζ and λ are the structural natural frequencies, damping factors and thermal eigenvalues, $\{u\}$ is the vector of control thermal loads and $\{P^*\}$ is the part of sunlight thermal input which does not depend on the structural displacements. The structural and thermal modes are evaluated on a double finite element scheme (Fig. 10), one for the structural displacements and one for the temperature distribution, which are only requested to be interfaceable, i.e., the two models must have some common grid points but not necessarily the same connections between these points.

Table II reports the slowest and fastest structural eigenvalues, and the slowest and fastest thermal eigenvalues used in the model of Eq. (61), and how these eigenvalues are modified by the coupling effects. The thermal eigenvalues are hardly affected by the coupling and, neglecting structural damping, the lowest structural modes become unstable due to the thermal coupling.

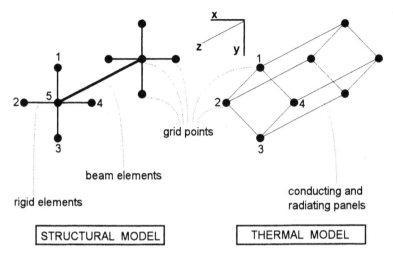

Figure 10: finite element models of the flexible boom

	Uncoupled		Coupled	
	Real	Imag	Real	Imag
1st str.	0.0	±1.28	+1.5e-4	±1.28
10th str.	0.0	±31.8	-1.4e-5	±31.8
1st therm.	-0.654	0.0	-0.654	0.0
38th therm.	-0.658	0.0	-0.658	0.0

Table II: eigenvalues of the mathematical model of the flexible boom

The control system makes use of six equally spaced sensors and actuators, positioned along the structure as shown in Fig. 11. It is assumed that the sensors measure local displacements and velocities, or allow their easy and fast reconstruction so that the observer dynamics need not be modeled, while the actuators are heating layers placed on the shaded surface of the boom. Each actuator is driven only by its nearest sensor, in a fully decentralized control scheme. The main drawback of this thermal control, in terms of actuator complexity, is represented by the necessity of cooling, so a purely heating control law would be highly desirable. This is achieved, for each heater, with a control law of the kind

$$\{u\} = [\mathrm{diag}(K_x)]\{m_x\} + [\mathrm{diag}(K_{\dot{x}})]\{m_{\dot{x}}\} + \{r\} \qquad (62)$$

where the constant vector $\{r\}$ is chosen so as to eliminate the temperature difference between the sunlit and shaded surfaces (nodes 4 and 2 in Fig. 10) in the uncontrolled case.

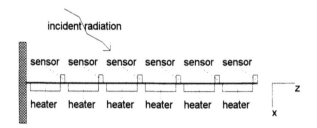

Figure 11: layout of the flexible boom control system

The controller was designed referring to the discrete equivalent of Eq. (61), sampled at 20 Hz, and the system response, shown in Fig. 12, was evaluated both with a PAM and corresponding equivalent PWM control, supposing each heater capable of transmitting 15 W, at most, to the structure, and with each PWM input applied with its proper optimal delay, as shown in Table III.

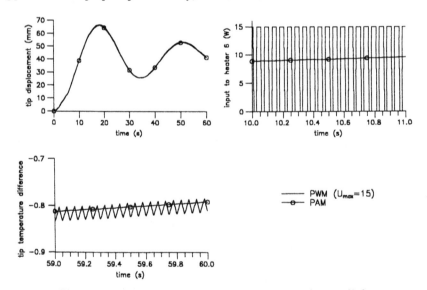

Figure 12: response of the flexible boom exposed to sunlight

sampling	optimal firing delay (s)					
time (s)	heater 1	heater 2	heater 3	heater 4	heater 5	heater 6
0.05	0.0298	0.0281	0.0269	0.0277	0.0274	0.0290

Table III: PWM firing delays for each heater

These results show that the PWM response matches the PAM response, even for an open loop unstable system. The control action, when active, is a pure heating, and shows that the present PWM implementation works perfectly well even with multi-input systems and with $|u_{max}| \neq |u_{min}|$. In the present example the real goal is to stop the structure, and this is obtained via a thermal coupling which acts as a low pass filter on the actual input, so that even in PWM mode the structural response is smooth and could meet stringent requirements even at steady state; the temperature instead shows the typical oscillations produced by an on-off control system, and is, in this case, regarded as being not influent onto the overall performances. Its average value is in fact very close to the parent PAM response.

D: SUPPRESSION OF THE VIBRATIONS OF A FLEXIBLE BEAM

This final example presents the results of a simple experiment designed to verify the applicability of the proposed PWM control to real mechanical systems, comparing the results with a different control strategy still making use of a pulsed control. Once more the structure of the controller to be used will be assigned a priori on the base of the available measures and of the physical understanding of the system.

1: Hardware description

The experiment has been conducted on a uniform steel beam, clamped at one end and free at the other. The beam is 0.7 m long, 0.06 m wide and 0.002 m thick. The free end holds an aluminum T profile to which two micro-electrovalves are fixed, back-to-back. The micro-electrovalves receive pressurized air from the same reservoir and tubing, and represent a bi-directional actuator; they are never switched on simultaneously and the force applied to the structure depends on which valve is active, so in the sequel the term "active valve" will be used to indicate the direction of the (constant) force applied. Two piezoelectric accelerometers are placed respectively at the free end and in the middle of the beam. A scheme of the layout is shown in Fig. 13, while Fig. 14 shows a picture of the beam tip with the actuator and Fig. 15 shows the first two bending modes of the instrumented beam, at frequencies of 1.92 and 15.3 Hz. The third mode, not shown for clarity, has a node at the free end, due to the mass of the actuator, so it appears immediately that this and the second mode will be poorly controllable. So it is decided to filter the acceleration signal at 6 Hz prior to its numerical integration with an appropriate band pass filter [34] thus limiting the authority of the controller to the first bending mode. This has the added beneficial effect of eliminating spurious accelerations caused by the impact of the shutter of the switching electrovalve (see Fig. 16). A block scheme of the experimental set-up is shown in Fig. 17.

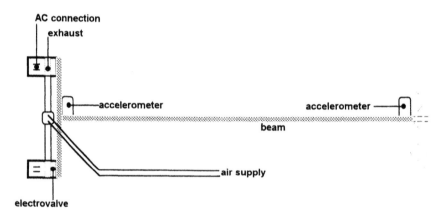

Figure 13: sensors and actuator layout

Figure 14: picture of the beam tip with electrovalves

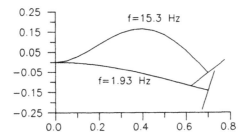

Figure 15: first bending modes of the instrumented beam

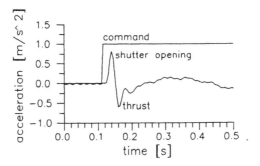

Figure 16: valve response to a step command

Before testing the two controllers, the most important characteristics of the actuator, i.e., its delay and force, were evaluated in a very simple, although not rigorous, way. The delay was determined by giving a step command to the valve and recording, starting from rest, the unfiltered acceleration of the free end. The response shows two peaks, displayed in Fig. 16, the first in the opposite direction of the second; the first peak, due to the transverse sensitivity of the accelerometer, represents the effect of the opening of the valve shutter, and the second is the actual acceleration due to the fully developed air jet. The same happens when the valve is closed, with a slightly shorter delay since the valve is mono-stable, i.e., the only stable position is the closed one. The average recorded delay, it is not constant since the electrovalve is driven in AC, is 0.055 seconds, independent from the pressure of the air. The force developed was determined by identifying, through a best fit, the experimentally recorded response resulting from a defined sequence of commands. At the reservoir selected pressure of 8 bar the identified force is 0.35 N, a rather low value for this class of valves. This is not very important for this test, and it is primarily due to the severe pressure drop caused by the narrow tubing adopted, which was however required to avoid excessive modifications of the beam dynamics, and also to the lack of a properly designed discharge nozzle.

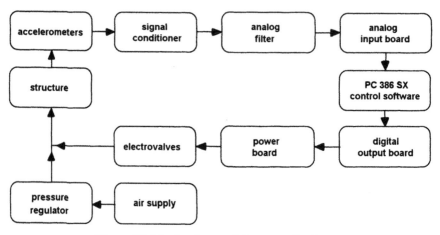

Figure 17: block scheme of the controller layout

2: Controller evaluation and software implementation

For the application of the PWM controller, two different designs were performed, both with a direct feedback of the acceleration and velocity signals. The first design uses a co-located feedback, while the second is based on measurements taken at the middle of the beam. Both controllers are designed, referring to a parent PAM system, by numerically minimizing a sampled time quadratic cost function in which the performance vector is composed by the tip displacement and velocity [30,35]. In order to face the robustness issue dictated by possible pressure drops during the experiment, the design was carried out by a simultaneous optimization of three different systems: two of them correspond to the maximum forces at pressures of 6 and 8 bars, taken as the reference condition, while the third refers to a force of 0.5 N. These designs are not meant to be the best possible, but they anyway allow a sensible interpretation of the results and an insight into the peculiarities of this particular method of PWM design. Table IV reports the gain values of the two controllers, k_a and k_v being the gains on the acceleration and velocity signals, the frequency and damping factor of the first mode, predicted on the basis of the PAM design, and the firing delay $\bar{\tau}$, invariant with the sensor position, evaluated according to Eq. (39), for the nominal condition.

accelerometer position	Δ (s)	$\bar{\tau}$ (s)	ω_1 (Hz)	ζ_1 (%)	k_a	k_v
tip	0.1	0.0523	1.93	10.7	-0.043	-0.728
middle	0.1	0.0523	1.92	10.8	-0.151	-2.268

Table IV: PWM controller parameters

It is noticed that $\bar{\tau}$ is almost equal to the actuator response delay, so in the control cycle no delay has to be imposed, since the actuator itself provides the almost optimal one. This can suggest a very practical way of selecting the sampling time Δ for a given system, if all the optimal firing delays are of similar values, as that particular sampling time for which the actuator characteristic time response matches the firing delays.

The actual implementation of this algorithm can be carried out in many different ways; in any case it is compulsory to design a device, either by hardware or software, capable of keeping a digital output activated, during the control period, for a time equal to the impulse duration evaluated on the basis of the sampled measurements. Two possible strategies for a correct implementation of the control law are presented., both making use of inexpensive commercially available hardware devices and a PC computer.

- Probably, the most logical solution requires the generation, with an appropriate circuit, of a square wave delayed by $\bar{\tau}$ with respect to the call to the control interrupt routine. This requires a perfect synchronization of the delayed wave with the interrupt routine that can be obtained only if the two events are triggered by the same clock, otherwise the firing delay will be altered. The algorithm will then be implemented in two phases (see Fig. 18a):

1) the first call to the interrupt routine is used to generate the delayed square wave: it is then necessary to program a counter, wait for the firing delay and then program it again as a square wave with the same frequency as the acquisition interrupts. After these operations two out of phase periodic events are available: the delayed square wave will be used as triggering signal for the actuation, whose duration is evaluated by the control routine;

2) the subsequent calls to the interrupt routine are used for acquisition and control; once the direction and duration of the input are evaluated, a timer has

to be programmed in the *one-shot* mode and its output is used as command for the appropriate valve. To this aim it is fundamental that the firing delay is long enough to allow the execution of all the operations, otherwise the timer will not be programmed and the input applied will be the one evaluated during the previous cycle.

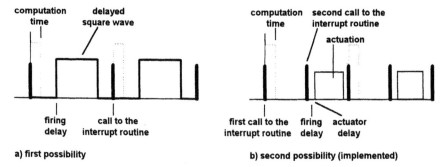

a) first possibility   b) second possibility (implemented)

Figure 18: interrupt cycles for the PWM implementation

• The solution actually adopted is based on the fact that the optimal firing delay is really close to half of the control period, which is set at .1 s. Thus by using a double interrupt at 0.05 s the steps required are the following (see Fig. 18b):

1) the first interrupt call is used to acquire the acceleration, integrate it to recover the velocity, evaluate the control and program the timer with the care required to take into account possible saturations;

2) the second interrupt call activates the *one-shot* timer to command the actuator, so the actual command is delayed by half the sampling time with respect to the acquisition time.

All this would not take into account the actuator dynamics. In practice it has been verified that the actual force is already delayed with respect to the actuator command by a fraction of time fairly close to the optimal firing delay. So in the present application the first call to the interrupt routine does all the operations required, including the actuator activation, and the actuator dynamics provides the optimal delay; the second call to the interrupt routine is kept only because the PC clock does not allow an interrupt frequency below 18 hz, but no operations are actually performed.

3: Experimental results

Figures 19 and 20 report some of the results obtained, starting from identical initial conditions imposing a 1.5 m/s velocity at the beam tip, by using the PWM control law and, for comparison, a control law which fires the air jet in the direction opposite to that of the velocity at the actuator position. The latter is also known as "Coulomb friction" modulator and can be shown to be dissipative (Appendix A). It is recalled that "active valve" simply indicates the direction of the force.

The use of a PWM controller is more flexible, since the sensors need not be co-located with the actuators. Comparing the PWM controller with the Coulomb friction controller, it appears that when the system is far from equilibrium (first 2.5 seconds approximately) the two responses are practically the same. This happens since the PWM controller is always saturated, so the duration of the pulses equals the sampling time and there is always one valve active, just as in the case of the Coulomb friction modulator. Slight differences come from the acceleration feedback term which advances the control signal compared to a pure velocity damping. When the controller does not saturate the system response becomes that of a linear system, so it is less damped, but this allows a much smoother approach to equilibrium. During this phase the duration of the pulses decreases continuously, and actually the valves, being driven in AC, may not even open if the command is too short so that the residual limit cycle is almost negligible even without providing a dead band. However the latter is in any case advisable to avoid a useless fuel consumption and must now be imposed directly on the duration of the pulses due to the nature of the modulator. A very narrow dead band appears enough to guarantee the complete rest, and the activity of the controller between 5 and 6 seconds is related to a high level of noise in the measurement that was difficult to suppress. It is remarked that the comparison has been made with a Coulomb friction modulator with the lowest possible dead band, and it appears evident that the PWM controller guarantees, for the same power expense, negligible residual vibrations. Moreover the velocity dead band equivalent to a 6 ms pulse dead band is of about 0.004 m/s, on the structure under test; once more it is confirmed that a PWM controller works well with narrower dead bands.

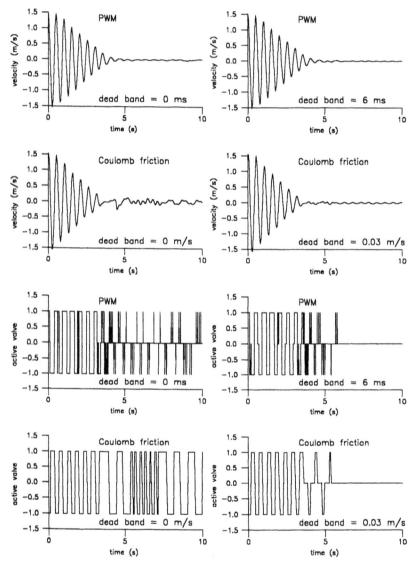

Figure 19: performances of the PWM controller (tip accelerometer)

Figure 20 shows the effect of an increase in the pulse dead band on the performances of the controller. It can be noted that the global control power consumption does not decrease sensibly, but the residual oscillations are much more evident. This happens because the PWM automatically decreases the pulse durations as the beam approaches its equilibrium in such a way that, if the dead band is not excessively wide, it will affect the controller only when it is almost inactive; the residual activity being all devoted to leading   the structure to a complete stop.

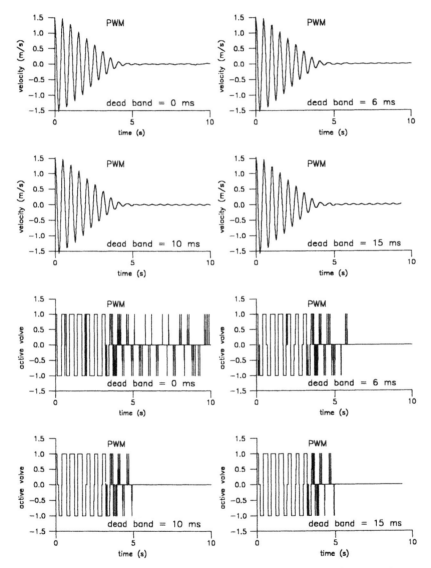

Figure 20: performances of the PWM controller (tip accelerometer)

Other results, not reported here, have demonstrated that the PWM controller is robust enough to be efficient also if the firing delay is almost double than the optimal one, i.e., when the optimal delay is added to the actuator's proper delay (see Fig. 18b). A possible explanation of this comes directly from the pulse duration analysis: when the errors are large there is no difference from the optimal delay, since the controller is saturated and active during the whole sampling period, and as the error decreases also the pulse duration decreases, thus alleviating also the possible error sources.

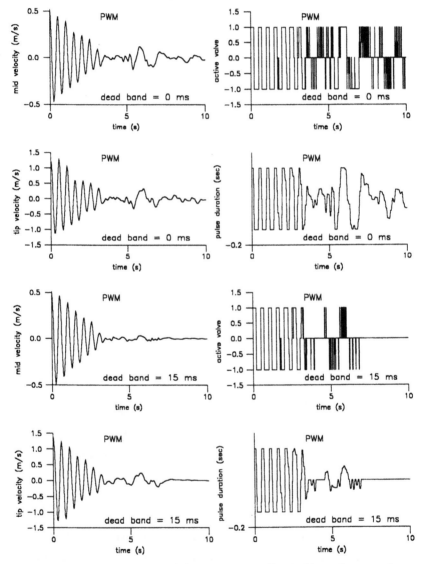

Figure 21: performances of the PWM controller (mid accelerometer)

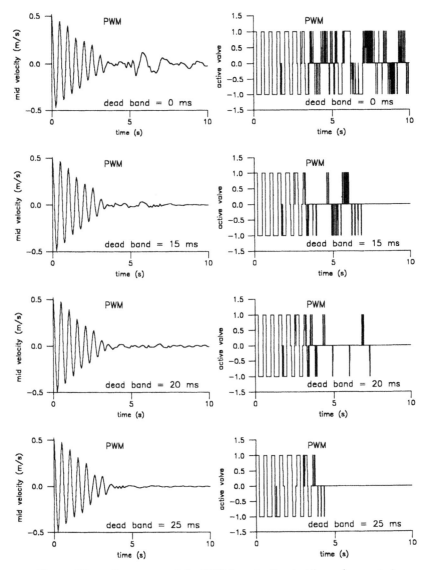

Figure 22: performances of the PWM controller (mid accelerometer)

The performances of the non co-located PWM controller are still satisfactory, and allow a considerable extension of the operational capabilities of this type of modulator. It is recalled that the feedback law uses non co-located velocity and acceleration signals. Figures 21 and 22 report the results obtained with this type of controller, and show also how the pulse duration is translated into the activity of the valves. The performances are not as good as in the previous case, and the arbitrary position of the sensor would require a further investigation on possible spillover effects, as well as an "ad hoc" tailoring of the hardware set-up. Nevertheless with the introduction of a dead band, although higher than the one required for the co-located control, the residual oscillations are still far smaller than those of a Coulomb friction damper having an equivalent velocity dead band (0.01 m/s approximately). Figure 22 displays the effects of increasing dead band widths, and show that in this case the optimal performances are not obtained with the narrowest dead band which switches the controller off; this is again regarded as an effect of the model uncertainties, which lead to unpredicted spurious switchings close to rest conditions unless the controller is forced to be turned off.

V. CONCLUDING REMARKS

On the basis of the results obtained it can be stated that the viability of using the linearization techniques for the PWM control of linear systems presented in this work has been proven on a variety of applications. The theoretical developments leading to the concept of equivalent pulse and optimal firing delay contain some assumptions which do not appear as a severe limitation. On the contrary the conversion technique from discrete linear to pulse width control maintains the dynamic performances of the equivalent parent discrete linear system. So all the computational advantages in the controller design related to the possibility of designing on a parent discrete linear system allows the designer to exploit the many techniques already available and the maximum freedom in choosing the controller structure. Furthermore the added cost of implementing the equivalent PWM control is minimal with almost no implementation burden added. In fact the conversion requires a simple equivalence of the total impulse applied during each sampling interval, which simply adds a further multiplication for each actuator, and the evaluation of a constant firing delay that can be computed off-line. The implementation of the latter is simple and affordable and can be easily carried out even on a personal computer by using low cost additional hardware.

The examples presented refer to a variety of systems which are traditionally driven by on/off actuators, such as power drives for DC motors, as well as non conventional vibration suppression applications. It should be remarked that the experimental results here shown can be regarded as a mere proof of concept, and show the direction for many other applications.

VI. REFERENCES

1. G.P. Sutton, "Rocket Propulsion Elements: An Introduction to the Engineering of Rockets", Wiley, 1986.

2. Y.Z. Tsypkin, "Relay Control Systems", Cambridge, Cambridge University Press, 1984.

3. A. Gelb, W.E. Vander Velde, "Multiple-Input Describing Functions and Nonlinear System Design", McGraw-Hill, 1968.

4. U. Itkis, "Control Systems of Variable Structure", John Wiley & Sons, 1976.

5. V.I. Utkin, "Variable Structure Systems with Sliding Modes", *IEEE Transactions on Automatic Control*, Vol. AC-22, n.2, pp. 212-222, 1977.

6. V.I. Utkin, K.D. Yang, "Methods for Constructing Discontinuity Planes in Multidimensional Variable Structure Systems", *Automation and Remote Control*, pp. 1466-1470, 1979.

7. H. Bühler, "Réglage par mode de glissement", Presses Polytechniques Romandes, 1986.

8. V.I. Utkin, "Sliding Modes in Control Optimization", Berlin, Springer Verlag, 1992.

9. R.A. Skoog, G.L. Blankenship, "Generalized Pulse-Modulated Feedback Systems: Norms, Gains, Lipschitz Constants, and Stability", *IEEE Transactions on Automatic Control*, Vol. AC-15, n.3, pp. 300-315, 1970.

10. B. Friedland, "Modeling Linear Systems for Pulsewidth Modulated Control", *IEEE Transactions on Automatic Control*, Vol. AC-21, n.5, pp. 739-746, 1976.

11. B. Wie, C.T. Plescia, "Attitude Stabilization of Flexible Spacecraft During Stationkeeping Maneuvers", *Journal of Guidance, Control and Dynamics*, Vol. 7, n.4, pp. 430-436, 1984.

12. T.C. Anthony, B. Wie, S. Carroll, "Pulse Modulated Control Synthesis for a Flexible Spacecraft", *Proceedings of the AIAA Guidance, Navigation and Control Conference - Boston, Massachusetts*, pp. 65-76, 1989.

13. S.B. Skaar, L. Tang, I. Yalda-Mooshabad, "On-Off Attitude Control of Flexible Satellites", *Journal of Guidance, Control and Dynamics*, Vol. 9, n.4, pp. 507-510, 1986.

14. H. Sira-Ramirez, M. Zribi, S. Ahmad, "Pulse Width Modulated Control of Robotic Manipulators", *Proceedings of the 29th IEEE Conference on Decision and Control - Honolulu, Hawaii*, pp.366-367, 1990.

15. H. Sira-Ramirez, T.A.W. Dwyer III, "Variable Structure Controller Design for Spacecraft Nutation Damping", *IEEE Transactions on Automatic Control*, Vol. AC-32, n.5, pp. 435-438, 1987.

16. T.A.W. Dwyer III, H. Sira-Ramirez, S. Monaco, S. Stornelli, "Variable Structure Control of Globally Feedback-Decoupled Deformable Vehicle Maneuvers", *Proceedings of the 26th IEEE Conference on Decision and Control - Los Angeles, California*, pp.1281-1287, 1987.

17. T.A.W. Dwyer III, H. Sira-Ramirez, "Variable Structure Control of Spacecraft Attitude Maneuvers", *Journal of Guidance, Control and Dynamics*, Vol. 11, n.3, pp. 262-270, 1988.

18. H. Sira-Ramirez," Sliding Regimes in Pulse-Width-Modulation Systems Design", *Proceedings of the 28th IEEE Conference on Decision and Control - Tampa, Florida*, pp.2199-2204, 1989.

19. S.R. Vadali, "Variable Structure Control of Spacecraft Large Angle Maneuvers", *Journal of Guidance, Control and Dynamics*, Vol. 9, n.2, pp. 235-239, 1986.

20. A.J. Calise, F.S. Kramer, "A Variable Structure Approach to Robust Control of VTOL Aircraft", *Journal of Guidance, Control and Dynamics*, Vol. 7, n.5, pp. 620-626, 1984.

21. S.K. Mudge, R.J. Patton, "Enhanced Assessment of Robustness for an Aircraft's Sliding Mode Controller", *Journal of Guidance, Control and Dynamics*, Vol. 11, n.6, pp. 500-507, 1988.

22. S.K. Spurgeon, R.J. Patton, "An Assessment of Robustness of Variable Structure Control Systems for Advanced Aircraft Manoeuvres", *Proceedings of the 29th IEEE Conference on Decision and Control - Honolulu, Hawaii*, pp.3588-3593, 1990.

23. S.K. Spurgeon, "Choice of Discontinuous Control Component for Robust Sliding Mode Performance", *International Journal of Control*, Vol. 53, n.1, pp. 163-179, 1991.

24. S. Yurkovich, Ü. Özgüner, F. Al-Abbass, "Model Reference, Sliding Mode Adaptive Control for Flexible Structures", *The Journal of the Astronautical Sciences*, Vol. 36, n.3, pp. 285-310, 1988.

25. Y.P. Chen, K.S. Yeung, "Sliding Mode Control of Multi Link Flexible Manipulators", *International Journal of Control*, Vol. 54, n.2, pp. 257-278, 1991.

26. S.J. Dodds, A.B. Walker, "Sliding Mode Control System for the Three Axis Attitude Control of Rigid Body Spacecraft with Unknown Dynamics Parameters", *International Journal of Control*, Vol. 54, n.4, pp. 737-761, 1991.

27. S.J. Dodds, M. Senior, "A Sliding Mode Approach to the Simultaneous Shape and Attitude Control of Flexible Space Structures with Uncertain Dynamics", *in* "Dynamics and Control of Structures in Space II", (C.L. Kirk and P.C. Hughes ed.), Computational Mechanics Publications, pp. 257-277, 1993.

28. H. Sira-Ramirez, "A Geometric Approach to Pulse-Width-Modulated Control Design", *Proceedings of the 26th IEEE Conference on Decision and Control - Los Angeles, California*, pp. 1771-1776, 1987.

29. H. Sira-Ramirez, "A Geometric Approach to Pulse-Width-Modulated Control in Nonlinear Dynamical Systems", *IEEE Transactions on Automatic Control*, Vol. AC-34, n.2, pp. 184-187, 1989.

30. F. Bernelli-Zazzera, P. Mantegazza, "Pulse Width Equivalent to Pulse Amplitude Discrete Control of Linear Systems", *Journal of Guidance Control and Dynamics*, Vol. 15, n.2, pp. 461-467, 1992.

31. Author(s) not shown, "DC Motors Speed Controls Servo Systems", Electro-Craft Corporation, 1980.

32. F. Bernelli-Zazzera, A. Ercoli-Finzi, P. Mantegazza, "Thermoelastic Behaviour of Large Space Structures: Modelling and Control", *ESA SP-289*, pp. 219-224, 1988.

33. F. Bernelli-Zazzera, A. Ercoli-Finzi, P. Mantegazza, "Control Strategies for Thermally Induced Vibrations of Space Structures", *Proceedings of the 7th*

VPI&SU Symposium on Dynamics and Control of Large Structures - Blacksburg, Virginia, pp. 539-551, 1989.

34.   F. Bernelli-Zazzera, P. Mantegazza, "Control of Flexible Structures by Means of Air Jet Thrusters: Experimental Results", *Proceedings of the 9th VPI&SU Symposium on Dynamics and Control of Large Structures - Blacksburg, Virginia*, pp. 231-242, 1993.

35.   F. Bernelli-Zazzera, P. Mantegazza, F. Ongaro "A Method to Design Structurally Constrained Discrete Suboptimal Control Laws for Actively Controlled Aircrafts", *Aerotecnica Missili e Spazio*, Vol.67, pp. 18-25, 1988.

36.   M.J. Balas, "Direct Velocity Feedback Control of Large Space Structures", *Journal of Guidance and Control*, Vol. 2, n.3, pp. 252-253, 1979.

APPENDIX A: COULOMB FRICTION MODULATORS

This is the simplest modulator that can be used to actively damp a mechanical system since it determines only the sign of the applied force. From a conceptual point of view it can be seen as a PWM with $\delta(.)=\Delta$ and an infinite sampling rate. When applied to a systems with a velocity sensor co-located with the actuator, this type of modulator acts as a pure Coulomb friction like damper. It is surely stable whatever the number of actuators; due to the importance of this fact, a brief demonstration will follow.

Let $\{q\}$ be the vector of generalized coordinates of the linear mechanical system to be controlled, [M], [C], [K] the positively defined mass, damping and stiffness matrices, $\{F\}$ the vector of the applied forces and [B] the input distribution matrix. The system dynamics and the total mechanical energy E are then determined by the following equations

$$[M]\{\ddot{q}\} + [C]\{\dot{q}\} + [K]\{q\} = [B]\{F\} \tag{A.1}$$

$$E = \tfrac{1}{2}\{\dot{q}\}^{T}[M]\{\dot{q}\} + \tfrac{1}{2}\{q\}^{T}[K]\{q\} \tag{A.2}$$

Calling $\{v\}$ the vector of the sensed velocities, obtained from $\{\dot{q}\}$ by setting to zero the elements corresponding to points where no sensors and actuators are located, the control forces are

$$\{F\} = -\{F_M\} \, \text{sgn}(\{v\})$$

(A.3)

with $\{v\} = [B]^T \{\dot{q}\}$ because of co-location. Choosing E as Lyapunov function we can prove the stability of this type of control. In fact, after differentiating Eq. (A.2) with respect to time, we obtain

$$\dot{E} = \{\dot{q}\}^T [M]\{\ddot{q}\} + \{\dot{q}\}^T [K]\{q\} =$$
$$= \{\dot{q}\}^T (\{F\} - [C]\{\dot{q}\}) = -\{F_M\}\{\dot{q}\}^T \text{sgn}\{v\} - \{\dot{q}\}^T [C]\{\dot{q}\}$$

(A.4)

For any non zero vector we can write

$$\{u\}^T \text{sgn}(\{u\}) = \sum_{i=1}^{N} |u_i| > 0$$

(A.5)

and since the contribution to $\dot{E}$ of the natural damping is surely dissipative, Eq. (A.4) gives, recalling the definition of the vector $\{v\}$

$$\dot{E} < -\{F_M\} \{v\}^T \text{sgn}(\{v\}) < 0$$

(A.6)

Hence, this type of modulator is surely stable with any number of actuators, provided that each of them works with a co-located velocity sensor. This result, well known for continuous time controllers under the name of Direct Velocity Feedback Control [36], is valid also for discrete time controllers until there are no sign changes in the velocities within a sampling period, i.e., until the velocities are quite large; as the structure approaches a rest condition, the controller must be modified in some way (see section I.A.5) to preserve stability.

Algorithms for Discretization and Continualization of MIMO State Space Representations

Stanoje Bingulac
Electrical and Computer Engineering Department
Kuwait University - P.O.Box 5969
13060 - Safat - KUWAIT

Hugh F. VanLandingham
The Bradley Department of Electrical Engineering
Virginia Polytechnic Institute and State University
Blacksburg, Virginia 24061-0111, U.S.A.

I. INTRODUCTION

With the widespread use of computers in control loops it is inevitable that control engineers will face problems associated with sampled-data systems. Such systems by their very definition contain a mixture of continuous-time (C-T) and discrete-time (D-T) signals. A common problem that arises with sampled-data control systems is to find the equivalent effect of C-T operations as seen by the computer in the loop. Typically, the modeling of the signal converters assumes an ideal uniform sampler for the analog-to-digital converter and a simple (zero-order) hold device synchronized with the samples for a digital-to-analog converter. With these assumptions one may find in many references the standard *zero-order hold* model, also known as the *step invariant (SI)* model which will be discussed subsequently.

In addition to simple plant modeling with SI equivalents there are occasions, such as in digital redesign, that demand more accuracy between a given C-T system and its D-T equivalent model. In these instances higher-order discrete models are required. Two models, [1-3], which have been introduced for this purpose the *bilinear transformation (BT)* (without prewarping), and a method which assumes a linearly interpolated input. This latter method is referred to as a *ramp invariant (RI)* model in contrast to the standard *ZOH* model's being a *step invariant (SI)* model. There are many other useful models, but this Chapter will focus on only these three methods of discretization as being the most useful in practice.

The reverse problem, called *continualization*, is that of reconstructing a C-T model from a given D-T model. This problem could arise, for instance, when measured discrete data are used to identify a C-T system, [3-5]. The particular method of continualization selected would depend on how the discrete data was derived (if known). The method of continualization is presented for each of the three discretization techniques, thereby offering the designer a great deal of flexibility in going between the continuous and the discrete domains. All algorithms described in this Chapter are implemented in the Computer-aided design package *L-A-S*, [6].

Both the forward, *discretization*, and reverse, *continualization*, problems may be viewed as functional transformations on a given matrix A, i.e. in calculating $exp(A_cT)$ for discretization or $ln(A_d)/T$ for continualization. If the matrix A is transformed into its Jordan canonical form, A_J, then

$$A = QA_JQ^{-1}$$

[7-11]. The *modal* matrix Q contains as columns the eigenvectors and/or generalized eigenvectors of A, depending on the eigenstructure of A. Then, relating the problem at hand, it is well known that

$$f(A) = Qf(A_J)Q^{-1}$$

when the scalar function $f(x)$ is analytic at the eigenvalues of A. This approach is convenient if A_J is diagonal because $f(A_J)$ is then itself diagonal. However, in the general case this approach is very restrictive in that it is not so straightforward to evaluate either the matrix Q or $f(A_J)$. Since it was desired to have robust algorithms to solve the continualization and discretization problems which are completely general, this method will not be pursued here.

II. PROBLEM FORMULATION

In the area of Systems and Controls, as well as related areas such as Signal Processing, it is useful to be able to discretize a given C-T system. This problem and its reverse problem of continualizing a D-T system are considered here. We assume a basic state variable representation for a C-T system as follows. A *state space realization* for a linear, C-T, constant parameter system consists of a 4-tuple of matrices; namely,

$$R_c = \{A_c, B_c, C_c, D_c\} \tag{1}$$

which defines the state model

$$\dot{x}(t) = A_c x(t) + B_c u(t)$$
$$y(t) = C_c x(t) + D_c u(t) \tag{2}$$

where $x(t)$, $u(t)$ and $y(t)$ are the state, input and output vectors with dimensions n, m and p, respectively, while the matrices A_c, B_c, C_c and D_c are constant matrices with compatible dimensions.

A. Discretization Procedures

In this Chapter some computational issues of the discretization and continualization procedures will be discussed with emphasis on explaining different algorithms which are easily implementable. The problem of discretization will be discussed first.

The familiar SI (ZOH) equivalent D-T model assumes that the input vector $u(t)$ in Eq.(2) is constant between (uniform) samples. The equivalent D-T model can be represented as

$$R_d = \{A_d, B_d, C_d, D_d\} \tag{3}$$

which implies the D-T state model

$$x(k+1) = A_d x(k) + B_d u(k)$$
$$y(k) = C_d x(k) + D_d u(k) \tag{4}$$

The matrices A_d and B_d in Eq.(3) are related to A_c and B_c in Eq.(2) by the well known relations, [2,4,7],

$$A_d = e^{A_c T} = \sum_{i=0}^{\infty} \frac{(A_c T)^i}{i!} \ , \quad B_d = \int_0^T e^{A_c s} B_c \, dt = \sum_{i=0}^{\infty} \frac{(A_c T)^i}{(i+1)!} B_c T \tag{5}$$

$$C_d = C_c \quad and \quad D_d = D_c \tag{6}$$

Also, if A_c is nonsingular, $\quad B_d = A_c^{-1}(e^{A_c T} - I) B_c \tag{7}$

In the following development three system discretization techniques are discussed.

1. Algorithm 1: $(E\text{-}A_d)$ Step Invariant (SI) Equivalent Model

This algorithm is a numerically robust procedure for calculating A_d and B_d described above. The standard general method for calculating A_d is to compute a truncated version of Eq.(5). The problem with this approach is that for matrices A_c and sampling intervals T satisfying that

$$|A_c T| > 1 \qquad (8)$$

a truncated version of Eq.(5) may either require large N, leading to considerable round-off errors, or may not converge at all, [11-13]. The concept of *norm* is used here to have a scalar measure of the relative size of the entries of a matrix, usually for the comparison of convergence errors after different numbers of steps of a particular algorithm. For this purpose the *Frobenius (F)* norm, defined as the square root of the sum of squares of all matrix elements, is used. Any other standard matrix norm could be used to measure the same relative effects.

It has been shown in [2] that the *SI* model can be calculated using an intermediate matrix E as follows:

$$A_d = I + EA_c T , \quad and \quad B_d = EB_c T \quad where \quad E = \sum_{i=0}^{\infty} \frac{(A_c T)^i}{(i+1)!} \qquad (9)$$

It is well known that to resolve the problem associated with Eq.(8), it is possible to utilize the property of the exponential function that

$$\exp(x) = e^x = \left(e^{(x/r)}\right)^r \qquad (10)$$

The present method extends this technique to permit calculation of both A_d and B_d under the condition of Eq.(8) as well as the condition that A_c may be singular.

In Appendix A1, Algorithm $E\text{-}A_d$, it is shown that the truncated version of E in Eq.(9) can be calculated by the following recursive process:

$$\begin{aligned} T_{k+1} &= 2T_k \\ E_{k+1} &= E_k(I + E_k A_c T_k/2) \end{aligned} \qquad (11)$$

for $k = 1, 2, 3, \cdots, j$ where

$$T_1 = \frac{T}{r} \quad and \quad E_1 = \sum_{i=0}^{N} \frac{(A_c T/r)^i}{(i+1)!} \qquad (12)$$

for $r = 2^j$ and
$$j = \left[\frac{\ln(|A_c T|)}{\ln(2)}\right]_{\substack{integer \\ part}} + 1 \tag{13}$$

The desired $E = E_{j+1}$. The series will obviously converge satisfactorily with the value of j given in Eq.(13) since $\|A_c T/r\| < 1$. Once E has been calculated, A_d can be obtained using Eq.(9).

2. Algorithm 2: $(F\text{-}E\text{-}A_d)$ Ramp Invariant (RI) Equivalent Model

This algorithm provides a robust method for the conversion from a C-T model Eq.(2) to a five matrix D-T state model, [14-16], represented by

$$R_{dr} = \{A_d, B_{d0}, B_{d1}, C_d, D_d\} \tag{14}$$

which, in turn, can be written as

$$\begin{aligned} x(k+1) &= A_d x(k) + B_{d0} u(k) + B_{d1} u(k+1) \\ y(k) &= C_d x(k) + D_d u(k) \end{aligned} \tag{15}$$

The matrices A_d, E, C_d and D_d have been described previously, see Eqs.(5), (6) and (9). To specify the remaining matrices, we define

$$F = \sum_{i=0}^{\infty} \frac{(A_c T)^i}{(i+2)!} \tag{16}$$

from which we obtain

$$B_{d0} = (E - F)B_c T, \quad \text{and} \quad B_{d1} = PB_{d0}, \quad \text{where} \quad P = F(E - F)^{-1} \tag{17}$$

Also, if A_c is nonsingular,

$$F = (A_d - I - A_c T)(A_c T)^{-2}$$

Following the guidelines of Algorithm $E\text{-}A_d$, it is desirable to create an algorithm which allows the condition of Eq.(8) and singular A_c matrices. The development for this algorithm, referred to as Algorithm $F\text{-}E\text{-}A_d$, is given in Appendix A2 and is summarized by the following recursive process:

$$\begin{aligned} T_{k+1} &= 2T_k \\ F_{k+1} &= 0.5 F_k + 0.25 (I + F_k A_c T_k)^2 \end{aligned} \tag{18}$$

for $k = 1, 2, 3, \cdots, j$ where (with j as in Eq.(13) and $r = 2^j$ as before)

$$T_1 = \frac{T}{r} , \qquad F_1 = \sum_{i=0}^{N} \frac{(A_c T/r)^i}{(i+2)!} \qquad (19)$$

and the desired $F = F_{j+1}$. Once F has been calculated, it follows that

$$E = (I + FA_c T) , \qquad A_d = I + EA_c T \qquad (20)$$

Comparing algorithms of Eqs.(11) and (18), it is clear that it is more convenient to use the algorithm of Eq.(18) since it may be used when either of the SI or RI equivalent models is required, as well as when only the transition matrix $A_d = \exp(A_c T)$ is sought.

3. Algorithm 3: (*BT-C-D*) Bilinear Transformation (BT) Model

This algorithm is better known in the transform domain as a conversion from the s-domain to the z-domain using the direct substitution:

$$s = \frac{2 (z - 1)}{T (z + 1)} \qquad (21)$$

The *BT-C-D* Algorithm provides a five matrix D-T model as in Eq.(15) where, in this case, (with $a = 2/T$)

$$A_d = (aI - A_c)^{-1}(aI + A_c)$$
$$B_{d0} = (aI - A_c)^{-1}B_c , \quad and \quad B_{d1} = PB_{d0} \quad where \quad P = I \qquad (22)$$

And, as in the previous results, $C_d = C_c$ and $D_d = D_c$.

4. Algorithm 4: (*R5R4*) Equivalent Standard State Model

Since both Algorithm 2, (*F-E-A_d*), and Algorithm 3, (*BT-C-D*), result in a non-standard five matrix model, it is useful to have a method of converting to a standard model as given in Eq.(4). Specifically, we describe the transformation from Eq.(15) or Eq.(22) to the following *equivalent* model:

$$x(k+1) = A_{de} x(k) + B_{de} u(k)$$
$$y(k) = C_{de} x(k) + D_{de} u(k) \qquad (23)$$

The simplest computational procedure for converting to a standard state model is derived using the identity of transfer function matrices, *i.e.*

$$C_d (zI - A_d)^{-1} (B_{d0} + z B_{d1}) + D_d$$
$$= C_{de} (zI - A_{de})^{-1} B_{de} + D_{de} \tag{24}$$

The detailed algorithm, referred to as Algorithm *R5R4*, is presented in Appendix A3.

B. Continualization Procedures

The reverse process of converting from a D-T model to an *equivalent* C-T model will now be considered, *i.e.* converting between the model in Eq.(4) and the model in Eq.(2), $R_d \rightarrow R_c$ in the *SI* case, or between Eq.(15) and Eq.(2), $R_{dr} \rightarrow R_c$ in the *RI* and *BT* sense. Of course, by itself R_d has no information regarding the signal values between samples and model conversion in this direction must be taken in context.

1. Algorithm 5: (*Ln*) SI to Continuous-Time Model

The algorithms for continualization require *logarithmic* operations instead of matrix exponentiation. When $(A_d - I)$ or A_c is non-singular, it is easily concluded that the matrices of R_c in Eq.(2) may be obtained from Eqs.(5) and (7) by:

$$A_c = \frac{1}{T} \ln (A_d) , \qquad B_c = (A_d - I)^{-1} A_c B_d \tag{25}$$

with the understanding that $C_c = C_d$ and $D_c = D_d$ as before. Appendix A5 contains the detailed algorithm for calculating A_c, but an outline of the method is given in the following. In a manner similar to the series definition of the exponential function in Eq.(5), the Taylor series expansion for the function *ln(x)* in the neighborhood of $x = 1$ leads to

$$A_c = \frac{1}{T} \sum_{i=1}^{\infty} \frac{(A_d - I)^i}{i} (-1)^{(i+1)} \tag{26}$$

The problem of using a truncated version of Eq.(26), [17], is that for matrices A_d with

$$|\lambda_{max}| > 0.5 \tag{27}$$

where λ_{max} is the maximum magnitude eigenvalue of $(A_d - I)$, the series may require large N, leading to considerable round-off errors if it converges at all. Algorithm *Ln* resolves this problem, [5,18], by using the following basic property of the logarithm function.

$$\ln(x) = r \ln\left[(x)^{1/r}\right] = r \sum_{i=1}^{\infty} \frac{(x^{1/r}-1)^i}{i}(-1)^{(i+1)} = -r \sum_{i=1}^{\infty} \frac{(1-x^{1/r})^i}{i} \quad (28)$$

With this approach the truncated series for calculation becomes

$$A_c = -\frac{r}{T}\sum_{i=1}^{N} \frac{(I-A_d^{1/r})^i}{i} \quad (29)$$

where the integer j satisfies that

$$\left|\lambda(A_d^{1/r}-I)\right|_{\max} < 0.5 \,, \quad \text{with} \quad r = 2^j \quad (30)$$

A computational algorithm, referred to as Algorithm Ln, calculating A_c according to Eq.(29) is given in Appendix 5. It has been experimentally verified that the accuracy of using Eq.(29) is satisfactory even for matrices A_d where some eigenvalues of $L = A_d - I$ have magnitude greater than one.

Having determined A_c, the remaining matrices in the C-T SI equivalent state space model of Eq.(2) could be calculated using Eq.(25) if A_c is nonsingular. If, however, A_c is singular, then the matrix E, appearing in Eq.(9) should be calculated using the procedure given in Eqs.(11)-(13). It follows that $C_c = C_d$, $D_c = D_d$ and

$$B_c = \frac{1}{T}E^{-1}B_d \quad (31)$$

2. Algorithm 6: RI to Continuous-Time Model

It is easily determined that the C-T model in Eq.(2) can be obtained from the five matrix model in Eq.(15) by using the Algorithm Ln to calculate A_c and from the availability of F in Eq.(16), i.e. Algorithm F-E-A_d in Eqs.(18)-(20), solving Eq.(17) for B_c,

$$B_c = \frac{1}{T}F^{-1}B_{d1} = \frac{1}{T}(E-F)^{-1}B_{d0} \quad (32)$$

$$\text{with} \quad B_{d1} = PB_{d0} \,, \quad \text{where} \quad P = F(E-F)^{-1}$$

The required five matrix D-T model in Eq.(15), containing B_{d0} and B_{d1}, could be obtained from a standard four matrix D-T model as in Eq.(23) by applying the Algorithm $R4R5$ presented in Appendix A4. Note that either of the expressions for B_c in Eq.(32) will give exactly the same result.

3. Algorithm 7: (BT-D-C) BT to Continuous-Time Model

The C-T model of Eq.(2) which corresponds to the BT D-T model specified in Eq.(22) could be obtained by a direct substitution of

$$z = \frac{a + s}{a - s}, \quad \text{where} \quad a = \frac{2}{T} \tag{33}$$

into the z-domain transfer function, thereby providing an s-domain transfer function from which R_c could be derived. Specifically, taking the z-transform of Eq.(4), introducing Eq.(33) and converting back to the time domain:

$$\begin{aligned}
\dot{x}(t) &= A_c x(t) + B_{c0} u(t) + B_{c1} \dot{u}(t) \\
y(t) &= C_c x(t) + D_c u(t)
\end{aligned} \tag{34}$$

where (with $a = 2/T$)

$$A_c = a(A_d+I)^{-1}(A_d-I), \quad B_{c0} = a(A_d+I)^{-1} B_d$$

$$B_{c1} = PB_{c0}, \quad \text{where} \quad P = -\frac{1}{a}I, \quad C_c = C_d, \quad \text{and} \quad D_c = D_d \tag{35}$$

As was discussed previously in terms of the five-matrix model of Eq.(15), if a four-matrix C-T model is required, Algorithm $R5R4$ can be applied to Eq.(34) to obtain an equivalent standard model of the form in Eq.(2).

III. NUMERICAL EXAMPLES

Three examples are presented in this section. They have been selected to illustrate the computational accuracy that can be achieved using the exponential and the logarithmic matrix calculations discussed previously. The first example demonstrates convergence rates when calculating A_d given a 5×5 singular matrix A_c, followed by a similar development in the second example in calculating A_c given A_d. The third example briefly illustrates all other discretization and continualization procedures mentioned in the paper. The calculations were performed using the L-A-S computer-aided design package, [6].

A. Example 1: Discretization

For this example the matrix A_c is given by

$$A_c = \begin{bmatrix} 0 & 1 & 0 & 0 & 0 \\ 0 & 0 & 1 & 0 & 0 \\ -4 & -4 & -3 & 1 & 4 \\ 0 & 0 & 0 & -1 & 0 \\ 0 & .5 & 0 & 0 & 0 \end{bmatrix} \tag{36}$$

The eigenvalues of A_c are

$$\lambda(A_c) = \{0, -1, -1, -1+j1, -1-j1\} \tag{37}$$

Note that A_c is singular and has multiple eigenvalues. In addition, the Jordan form, A_J, corresponding to A_c is not diagonal. The selection of this matrix was motivated by the fact that some widely used control system packages are not capable of calculating either the Jordan form or the natural logarithm of non-diagonalizable matrices. For example, the well known package *MATLAB*, [19], is a case in point. It is suggested that the reader repeat the calculations in these examples with another package at his or her disposal. The desired sampling interval for the discretization is $T = 2\ sec.$; and the (Frobenius) norm of $A_c T$ is calculated to be 15.65. As was suggested in Sec. 2, A_d was determined from (20) using the matrix F calculated from the algorithm in Eqs.(18)-(20). Equations (5) and (9) combined provide the following truncated summation which is similar to Algorithm $F\text{-}E\text{-}A_d$ for calculating the exponential matrix.

$$A_d = \left[\sum_{i=0}^{N} \frac{(A_c T/r)^i}{i!} \right]^r \tag{38}$$

As in $F\text{-}E\text{-}A_d$, $r = 2^j$ where j is given in Eq.(13). Both the truncation number N and the scaling parameter j are of key interest to this development. To emphasize the dependence of our calculated matrix A_d on these parameters, we will use the notation $A_d(N, j)$. Results will be presented for the following 36 parameter combinations:

$$j = 0,\ 1,\ 2,\ 3,\ 4,\ 5$$

and

$$N = 16,\ 14,\ 12,\ 10,\ 8,\ 6 \tag{39}$$

Each A_d is compared to the "exact" matrix A_{d_e} given by

$$
A_{de} = \begin{bmatrix}
-.09990 & .50637 & .19165 & .14761 & 1.0999 \\
-.76662 & -.31657 & -.06859 & .04404 & .76662 \\
.27438 & -.10893 & -.11078 & -.11264 & -.27438 \\
.00000 & .00000 & .00000 & .13534 & .00000 \\
-.54995 & .25318 & .09583 & .07381 & 1.5500
\end{bmatrix}
$$

The "exact" matrix A_{de} above was calculated by transforming A_c, [6,15], into its Jordan canonical form A_J and then using

$$
e^{A_c T} = Q e^{A_J T} Q^{-1} \tag{40}
$$

It may be verified that for A_c given in Eq.(36),

$$
A_J = \begin{bmatrix}
-1 & 1 & 0 & 0 & 0 \\
-1 & -1 & 0 & 0 & 0 \\
0 & 0 & -1 & 1 & 0 \\
0 & 0 & 0 & -1 & 0 \\
0 & 0 & 0 & 0 & 0
\end{bmatrix}, \quad
Q = \begin{bmatrix}
2 & 0 & 2 & 0 & 1 \\
-2 & 2 & -2 & 2 & 0 \\
0 & -4 & 2 & -4 & 0 \\
0 & 0 & 0 & 2 & 0 \\
1 & 0 & 1 & 0 & 1
\end{bmatrix}
$$

and for $T = 2$ seconds, $exp[A_J T]$ equals

$$
\begin{bmatrix}
e^{-T}\cos T & e^{-T}\sin T & 0 & 0 & 0 \\
-e^{-T}\sin T & e^{-T}\cos T & 0 & 0 & 0 \\
0 & 0 & e^{-T} & Te^{-T} & 0 \\
0 & 0 & 0 & e^{-T} & 0 \\
0 & 0 & 0 & 0 & 1
\end{bmatrix} =
\begin{bmatrix}
0.0563 & 0.1231 & 0 & 0 & 0 \\
-0.1231 & 0.0563 & 0 & 0 & 0 \\
0 & 0 & 0.1353 & 0.2707 & 0 \\
0 & 0 & 0 & 0.1353 & 0 \\
0 & 0 & 0 & 0 & 1
\end{bmatrix}
$$

Note that A_J is given in the *real number* Jordan form.

The log_{10} of the norm of the error matrix $E_d = A_{de} - A_d(N, j)$ is tabulated for each combination in Eq.(39) in Table I below. From Table I it can be seen that $N = 16$ terms is sufficient for A_d in Eq.(38) even for matrices $A_c T$ with relatively high norms. And, as we can see from Table I, N may be chosen as low as $N = 6$ with judicious choice of the scaling parameter j. The information is also shown graphically in Fig. 1.

TABLE I.						
Log$_{10}$($\|E_d\|$) vs. Truncation No. N						
and Scaling Parameter j						
N	j=0	j=1	j=2	j=3	j=4	j=5
16	-3.78	-10.07	-14.42	-14.45	-14.55	-14.55
14	-3.78	-8.08	-12.11	-14.61	-14.55	-14.55
12	-2.31	-6.04	-9.63	-13.10	-14.55	-14.55
10	-1.24	-4.34	-7.16	-9.94	-12.68	-14.44
8	-0.10	-2.60	-4.96	-7.20	-9.36	-11.50
6	0.64	-1.16	-2.78	-4.36	-5.91	-7.44

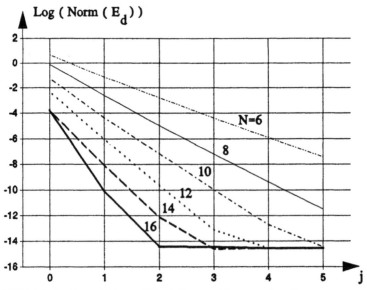

FIGURE 1. Log (Norm (E_d)) vs. Computation Parameters

B. Example 2: Continualization

In this example the matrix A_d is taken to be A_{d_e} given above. The calculation used to determine A_c is the truncated series in Eq.(29). We note that the eigenvalues of $L = I - A_d$, influencing the convergence of the series, can exceed unit magnitude. In particular, $\lambda(L)$ are:

$$\lambda(I - A_d) = \{0, -0.86, -0.86, -1.06 + j0.12, -1.06 - j0.12\} \qquad (41)$$

To illustrate the convergence properties, the power series Eq.(29) was evaluated for all combinations of the parameters N and j given by

$$j = 0, 1, 2, 3, 4, 5$$
and $\qquad\qquad\qquad\qquad\qquad\qquad\qquad\qquad\qquad\qquad\qquad$ (42)
$$N = 35, 30, 25, 20, 15, 10$$

As in Example 1, the error matrix is defined to be

$$E_c = A_c - A_c(N, j) \qquad (43)$$

where the explicit notation $A_c(N, j)$ is used to emphasize the dependence of the calculated matrix on the computation parameters N and j. The log_{10} of the (Frobenius) norm of the matrix E_c is tabulated for the combinations indicated in Eq.(42) in Table II. Figure 2 presents the tabulated data in graphical form.

We see from the results in Table II that the series of Eq.(29) can be truncated as high as $N = 35$ even when the maximum eigenvalue of L is greater than unity. It is also noted that the truncation may be as low as $N = 10$

TABLE II. $Log_{10}(\|E_c\|)$ vs. Truncation No. N and Scaling Parameter j						
N	j=0	j=1	j=2	j=3	j=4	j=5
35	1.02	-2.56	-9.25	-10.12	-10.12	-10.12
30	0.92	-2.00	-7.84	-10.12	-10.12	-10.12
25	0.75	-1.75	-6.40	-10.12	-10.12	-10.12
20	0.68	-1.21	-4.96	-9.76	-10.12	-10.12
15	0.78	-0.78	-3.52	-7.19	-10.10	-10.12
10	0.90	-0.37	-2.08	-4.49	-7.02	-9.59

provided that the scaling parameter j is appropriately selected. In practice, N can be fixed at a nominal value, say 20, and j can be varied over 3 or 4 values to ensure good convergence to the desired matrix. This is true whether the problem requires *discretization* or *continualization*.

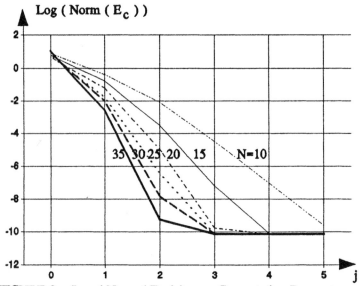

FIGURE 2.   Log (Norm (E_c)) vs. Computation Parameters

C. Example 3: C-T Identification

In this example the following C-T state space representation is considered:

$$
R_c = \begin{bmatrix} A_c & B_c \\ C_c & D_c \end{bmatrix} = \begin{bmatrix}
0 & 1 & 0 & 0 & 0 & | & 0 & 0 & 0 \\
0 & 0 & 1 & 0 & 0 & | & 0 & 0 & 0 \\
-4 & -4 & -3 & 1 & 4 & | & 0 & 0 & 1 \\
0 & 0 & 0 & -1 & 0 & | & 0 & 1 & 0 \\
0 & 0.5 & 0 & 0 & 0 & | & 1 & 0 & 0 \\
- & - & - & - & - & + & - & - & - \\
1 & 0 & 0 & 0 & 0 & | & 1 & 0 & 0 \\
0 & 1 & 0 & 0 & 0 & | & 0 & 0 & 0
\end{bmatrix}
\tag{44}
$$

Note that A_c is the same matrix used in Example 1, Eq.(36). The input signal $u(t)$ is specified over the interval $0 < t < 8T$, with $T = 2$ seconds by the components defined in Table III.

TABLE III. Input Signal for Example 3.			
Time Interval	$u_1(t)$	$u_2(t)$	$u_3(t)$
$0 \leq t \leq 4T$	$t/(4T)$	$\sin(\pi t/(4T))$	$\{\cos(\pi t/(4T))-1\}/2$
$4T \leq t \leq 8T$	$2 - t/(4T)$	0	$t/(4T) - 2$

1. Discretization

Using the sampling interval $T = 2$, the representation R_c is discretized into the following equivalents, each represented in the partitioned *system matrix* form of the given state-space model in Eq.(44):

(a) $R_{ds} = \{ A_{ds} , B_{ds} , C_{ds} , D_{ds} \}$, D-T step-invariant equivalent

(b) $R_{dr} = \{ A_{dr} , B_{dr} , C_{dr} , D_{dr} \}$, DT ramp-invariant equivalent

(c) $R_{db} = \{ A_{db} , B_{db} , C_{db} , D_{db} \}$, DT bilinear transform equivalent

As was pointed out in Sec. 2, in the cases of (b) and (c) the five-matrix D-T models of Eqs.(15) and (22) were first calculated. This was followed by a conversion to the standard four-matrix D-T model using algorithm R5R4 in Appendix 3. These three results are given below for the D-T models R_{ds}, R_{dr}, and R_{db}, respectively:

$$\begin{bmatrix} A_{ds} & B_{ds} \\ C_{ds} & D_{ds} \end{bmatrix} = \begin{bmatrix} -.100 & .506 & .192 & .148 & 1.100 & | & .787 & .127 & .275 \\ -.767 & -.317 & -.069 & .044 & .767 & | & 1.100 & .148 & .192 \\ .274 & -.109 & -.111 & -.113 & -.274 & | & .767 & .044 & -.069 \\ 0 & 0 & 0 & .135 & 0 & | & 0 & .865 & 0 \\ -.550 & .253 & .096 & .074 & 1.550 & | & 2.394 & .064 & .137 \\ - & - & - & - & - & + & - & - & - \\ 1 & 0 & 0 & 0 & 0 & | & 1 & 0 & 0 \\ 0 & 1 & 0 & 0 & 0 & | & 0 & 0 & 0 \end{bmatrix}$$

$$
\begin{bmatrix} A_{dr} & B_{dr} \\ C_{dr} & D_{dr} \end{bmatrix} =
\begin{bmatrix}
-.100 & .506 & .192 & .148 & 1.100 & | & 2.084 & .238 & .309 \\
-.767 & -.317 & -.069 & .044 & .767 & | & 1.235 & .070 & -.034 \\
.274 & -.109 & -.111 & -.113 & -.274 & | & -.135 & -.104 & -.177 \\
0 & 0 & 0 & .135 & 0 & | & 0 & .374 & 0 \\
-.550 & .253 & .096 & .074 & 1.550 & | & 3.042 & .119 & .154 \\
- & - & - & - & - & + & - & - & - \\
1 & 0 & 0 & 0 & 0 & | & 1.196 & .035 & .098 \\
0 & 1 & 0 & 0 & 0 & | & .394 & .064 & .137
\end{bmatrix}
$$

$$
\begin{bmatrix} A_{db} & B_{db} \\ C_{db} & D_{db} \end{bmatrix} =
\begin{bmatrix}
.200 & .800 & .200 & .100 & .800 & | & 1.840 & .180 & .260 \\
-.800 & -.200 & .200 & .100 & .800 & | & 1.040 & .080 & .060 \\
-.800 & -1.200 & -.800 & .100 & .800 & | & .240 & -.020 & -.140 \\
0 & 0 & 0 & 0 & 0 & | & 0 & .500 & 0 \\
-.400 & .400 & .100 & .050 & 1.400 & | & 2.920 & .090 & .130 \\
- & - & - & - & - & + & - & - & - \\
1 & 0 & 0 & 0 & 0 & | & 1.400 & .050 & .100 \\
0 & 1 & 0 & 0 & 0 & | & .400 & .050 & .100
\end{bmatrix}
$$

Responses of these models to samples of the input signal u(t) at $t_i = iT$ for $i =$ 0, 1, 2, 3, 4, are given in Table IV below. Also included in Table IV for comparison are the samples of the C-T system response. The norms of the differences between the C-T response, $y_c(t_i)$, and those of the three D-T models are:

$$\Delta_{ds} = \| \, y_c(t_i) - y_{ds}(t_i) \, \| = 1.6975$$

$$\Delta_{dr} = \| \, y_c(t_i) - y_{dr}(t_i) \, \| = 0.59173 \times 10^{-5}$$

$$\Delta_{db} = \| \, y_c(t_i) - y_{db}(t_i) \, \| = 0.87992 \times 10^{-1}$$

TABLE IV. Simulation Results for D-T Equivalents						
t_i (sec.)		0	2	4	6	8
$y_c(t_i)$	y_1	0	0.309	1.228	3.011	5.594
	y_2	0	0.123	0.555	0.966	1.374
$y_{ds}(t_i)$	y_1	0	0.250	0.747	2.107	4.290
	y_2	0	0.000	0.351	0.790	1.171
$y_{dr}(t_i)$	y_1	0	0.309	1.228	3.011	5.594
	y_2	0	0.123	0.555	0.966	1.374
$y_{db}(t_i)$	y_1	0	0.371	1.249	2.994	5.615
	y_2	0	0.121	0.508	0.987	1.383

2. Identification from Sampled Input/Output Data

As is well known, in order to perform a successful identification, the input signal selected should be sufficiently long and *sufficiently rich*, [3]. To this end the selected input vector $u^*(t)$ is defined by

$$u_1^*(t) = u_1(t) + u_1(t - 10T)$$

$$u_2^*(t) = u_2(t) + u_2(t - 12T)$$

$$u_3^*(t) = u_3(t) + u_3(t - 14T)$$

where $u_i(t)$ are given in Table III. Using $u^*(t)$, the response $y^*(t)$ of the system in Eq.(44) was calculated in the time interval $0 \leq t \leq 22T = 44$ seconds. The simulation of the C-T system in Eq.(44) was accomplished by solving the state-space differential equations at points 0.5 seconds apart. No measurement noise was added to the system response. The signals $\{u^*(t), y^*(t)\}$, shown in Figs.3 and 4, were then sampled at intervals of $T = 2$ seconds yielding the input-output samples $\{u^*(t_i), y^*(t_i)\}$ of a C-T system to be identified. The identification procedure presented in [4,5,16] and Appendix A9, has been applied to obtain the "identified D-T" model, given by the following D-T realization, R_d:

$$\begin{bmatrix} A_d & B_d \\ C_d & D_d \end{bmatrix} = \left[\begin{array}{ccccc|ccc} 0 & 0 & 1 & 0 & 0 & 2.084 & .238 & .309 \\ 0 & 0 & 0 & 1 & 0 & 1.235 & .070 & -.034 \\ .124 & .058 & .876 & 1.020 & 1.530 & 3.737 & .178 & .088 \\ 0 & 0 & 0 & 0 & 1 & .352 & -.090 & -.096 \\ -.030 & -.017 & .030 & -.086 & .281 & .015 & -.037 & -.008 \\ - & - & - & - & - & - & - & - \\ 1 & 0 & 0 & 0 & 0 & 1.196 & .035 & .098 \\ 0 & 1 & 0 & 0 & 0 & .394 & .064 & .137 \end{array}\right]$$

With the identification procedure used, the representation R_d is in the *Pseudo-Observable* form. As an admissible set of pseudo-observability indices, $\{n_i\}$, the following was selected, namely $\{n_i\} = \{2, 3\}$. Note that the unique set of observability indices of R_c is $\{n_i\} = \{3, 2\}$. More details on this identification procedure is to be found in Appendix A9.

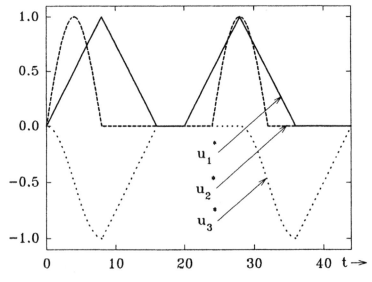

FIGURE 3. Excitations for Example 3

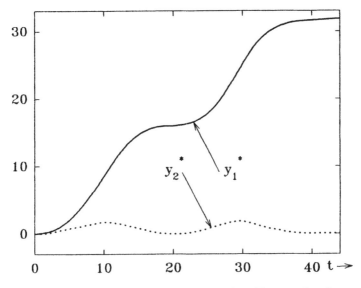

FIGURE 4.   Responses for Example 3.

3. Continualization

Using the sampling interval T = 2, the D-T representation R_d above is *continualized* into:

(a) R_{cs} - C-T step-invariant equivalent

(b) R_{cr} - C-T ramp-invariant equivalent

(c) R_{cb} - C-T bilinear transform equivalent

As was pointed out in Sec. 2, to determine R_{cr}, it was first necessary to convert R_d into an equivalent five-matrix representation, R_{ds}, using algorithm R4R5 in Appendix A4. Subsequently, using Eqs.(29), (30) and (32), the desired R_{cr} was obtained. To determine R_{cb}, the identified R_d was first converted to a five-matrix C-T representation using Eqs.(33)-(35). Following this, algorithm R5R4 was used to obtain the desired four-matrix model, R_{cb}. The C-T representations thus obtained are given below:

$$
\begin{bmatrix} A_{cs} & B_{cs} \\ C_{cs} & D_{cs} \end{bmatrix} =
\left[\begin{array}{ccccc|ccc}
-.001 & 1.00 & .001 & -.002 & -.010 & .394 & .064 & .137 \\
-4.07 & -3.71 & 4.07 & -6.70 & 5.47 & .552 & .076 & .096 \\
.000 & .000 & -.000 & 1.00 & -.004 & 1.63 & .134 & .104 \\
.344 & .147 & -.344 & -.033 & 1.06 & .415 & -.030 & -.081 \\
-.074 & -.038 & .074 & -.295 & -.260 & .033 & -.036 & -.015 \\
- & - & - & - & - & - & - & - \\
1 & 0 & 0 & 0 & 0 & 1.20 & .035 & .098 \\
0 & 1 & 0 & 0 & 0 & .394 & .064 & .137
\end{array}\right]
$$

$$
\begin{bmatrix} A_{cr} & B_{cr} \\ C_{cr} & D_{cr} \end{bmatrix} =
\left[\begin{array}{ccccc|ccc}
-.001 & 1.00 & .001 & -.002 & -.010 & -.003 & -.005 & .000 \\
-4.07 & -3.71 & 4.07 & -6.70 & 5.47 & .012 & .016 & -.003 \\
.000 & .000 & .000 & 1.00 & -.004 & 1.10 & .148 & .192 \\
.344 & .147 & -.344 & -.033 & 1.06 & .766 & .044 & -.069 \\
-.074 & -.038 & .074 & -.295 & -.260 & .121 & -.057 & -.044 \\
- & - & - & - & - & - & - & - \\
1 & 0 & 0 & 0 & 0 & 1.00 & .001 & .000 \\
0 & 1 & 0 & 0 & 0 & -.003 & -.004 & .001
\end{array}\right]
$$

$$
\begin{bmatrix} A_{cb} & B_{cb} \\ C_{cb} & D_{cb} \end{bmatrix} =
\left[\begin{array}{ccccc|ccc}
-1.16 & -.076 & 1.16 & -1.15 & -.487 & -.832 & -.030 & .285 \\
-.051 & -1.03 & .051 & 1.85 & -1.50 & 1.14 & .316 & .265 \\
.158 & .076 & -.158 & 1.15 & .487 & 1.51 & .222 & .163 \\
.051 & .028 & -.051 & -.847 & 1.50 & .651 & -.059 & -.158 \\
-.051 & -.028 & .051 & -.153 & -.502 & .026 & -.056 & -.017 \\
- & - & - & - & - & - & - & - \\
1 & 0 & 0 & 0 & 0 & .858 & -.061 & -.126 \\
0 & 1 & 0 & 0 & 0 & -.503 & -.064 & .084
\end{array}\right]
$$

It is worth mentioning that the eigenvalues of the matrix A_{cs} $(= A_{cr})$ obtained by the continualization of A_d are:

$$\{0 , -0.9844 , -1.0211 , -1.0004 + j1.0002 , -1.0004 - j1.0002\}$$

which are slightly different from those of A_c given by Eq.(37).

Having determined the C-T models above, the responses of these models to the four samples of the input signal u(t), Table III, were calculated, as was done for the D-T models, Table IV. In order to assess the accuracy of the proposed continualization procedures, only the samples of these C-T responses at the sampling instants are considered. Table V contains these results as well as the samples of the identified D-T model for comparison. The norms of the differences between $y_d(t_i)$ and the responses of the three derived C-T models are as follows:

$$\Delta_{cs} = \| y_d(t_i) - y_{cs}(t_i) \| = 2.0550$$

$$\Delta_{cr} = \| y_d(t_i) - y_{cr}(t_i) \| = 0.12935 \times 10^{-4}$$

$$\Delta_{cb} = \| y_d(t_i) - y_{cb}(t_i) \| = 0.73568 \times 10^{-1}$$

From the normed differences, both for the discretization and the continualization, it may be concluded that the RI transformation is superior to either of the other two, primarily because of the particular selection of $u^*(t)$ which does not contain step discontinuities. This should be expected since the SI transformation assumes constant values in input between samples, and the bilinear transformation is only satisfactory if $|p_iT| < 0.5$ for all poles p_i of the C-T system, [3], which is not the case for this example.

With the accuracy given in Tables IV and V it cannot be seen just how well the C-T *identified* system output $y_{cr}(t_i)$ matches that of the original C-T system, but the largest magnitude difference between the two, component by component, is 0.421×10^{-4}. The sequence of algorithm executions required in Example 3 is outlined in Appendix A7.

TABLE V. Simulation Results for C-T Equivalents						
t_i (sec.)		0	2	4	6	8
$y_d(t_i)$	y_1	0	0.309	1.228	3.011	5.594
	y_2	0	0.123	0.555	0.966	1.374
$y_{cs}(t_i)$	y_1	0	0.567	1.927	4.108	7.110
	y_2	0	0.322	0.765	1.161	1.585
$y_{cr}(t_i)$	y_1	0	0.309	1.228	3.011	5.594
	y_2	0	0.123	0.555	0.966	1.374
$y_{cb}(t_i)$	y_1	0	0.253	1.219	3.001	5.578
	y_2	0	0.164	0.561	0.956	1.374

IV. CONCLUSIONS

A newly developed set of numerically robust algorithms has been presented. These algorithms deal with the often encountered problems of *discretization* of C-T models as well as the inverse problem of recreating a C-T model from a given D-T model. This latter operation we have referred to as *continualization*. The seven algorithms described in the paper are further elaborated in the Appendix. They comprise, in addition to the standard *Step Invariant, (SI or ZOH)* procedures, two methods which are commonly referred to in the Signal Processing literature, namely the *Bilinear Transformation (BT)* and the *Ramp Invariant (RI)* method that is equivalent to the next higher order approximation than the *SI* approximation, representing a piecewise linear approximation to the input functions. With these algorithms the design engineer has complete flexibility to move between the continuous and discrete model domains.

V. ACKNOWLEDGEMENTS

The research presented in this chapter has been supported in part by the research administration, Kuwait University, under research grant EE 058.

References

1.   S. S. Haykin, "A Unified Treatment of Recursive Digital Filtering",
        IEEE Trans. on Automatic Control, **AC-17**, 104-108 (1972).

2.   H. F. VanLandingham, *Introduction to Digital Control Systems*,
        Macmillan Pub. Co., New York, NY (1985).

3.   N. K. Sinha and B. Kusta, *Modeling and Identification of Dynamic
        Systems*, Van Nostrand-Reinhold Publishers, New York, NY (1983).

4.   S. Bingulac and N. K. Sinha, "On the Identification of Continuous-
        Time Multivariable Systems", *Math. Comput. Modeling*, **14**, 203-208
        (1990).

5.   S. Bingulac and D. Cooper, "Use of Pseudo-Observability Indices in
        Identification of Continuous-Time Multivariable Models", *Identification
        of Continuous-Time Systems*, (Sinha, N. K. and G. P. Rao, editors),
        Kluver Academic Publishers, The Netherlands (1991).

6.   S. Bingulac and D. Cooper, *L-A-S User's Guide*, Bradley Department
        of Electrical Engineering, Virginia Polytechnic Institute and State
        University, (1990).

7.   W. L. Brogan, *Modern Control Theory*, Prentice-Hall Pub. Co.,
        Englewood Cliffs, NJ (1984).

8.   C. T. Chen, *Linear System Theory and Design*, Holt,Rinehart and
        Winston, Inc., New York, NY (1984).

9.   T. Kailath, *Linear Systems*, Prentice-Hall Pub. Co., Englewood Cliffs,
        NJ (1980).

10.   N. K. Sinha and G. J. Lastman, "Identification of Continuous-Time
        Multivariable Systems from Sampled Data", *International Journal of
        Control*, **35**, 117-126 (1982).

11.   G. Golub and C. F. Van Loan, *Matrix Computations*, The Johns-
        Hopkins University Press, Baltimore, MD (1991).

12.   A. J. Laub, "Numerical Linear Algebra Aspects of Control Design Computations", *IEEE Trans. on Automatic Control*, **AC-30**, 2, 97-108 (1985).

13.   C. B. Moler and C. F. Van Loan, "Nineteen Dubious Ways to Compute the Exponential of a Matrix", *SIAM Review*, **20**, 801-836 (1978).

14.   S. Strmcnik and F. Bremsak, "Some New Transformation Algorithms on the Identification of Continuous-Time Multivariable Systems Using Discrete Identification Methods", *Proc. 5th IFAC Symposium on Identification and System Parameter Estimation*, Darmstad, Germany, 397-405 (1979).

15.   S. Bingulac and D. Cooper, "Derivation of Discrete- and Continuous-Time Ramp Invariant Representations", *Electronics Letters*, **26**, 10, 664-666 (1990).

16.   S. Bingulac and D. Cooper, "Identification of First-Order Hold Continuous-Time Systems", *Proc. 9th IFAC Symposium on Identification and System Parameter Estimation*, Budapest, Hungary (1991).

17.   G. J. Lastman, S. C. Puthenpura and N. K. Sinha, "Algorithm for the Identification of Continuous-Time Multivariable Systems from Their Discrete-Time Models", *Electronics Letters*, **20**, 22, 918-919 (1984).

18.   D. Cooper and S. Bingulac, "Computational Improvement in the Calculation of the Natural Log of a Square Matrix", *Electronics Letters*, **26**, 13, 861-862 (1990).

19.   C. B. Moler, *The Student Edition of Matlab*, MathWorks, Inc. and Prentice-Hall, Inc., Englewood Cliffs, NJ (1992).

20.   L. Ljung, *System Identification: Theory for the User*, Prentice-Hall, Inc., Englewood Cliffs, NJ (1987).

21.   S. Bingulac and R. V. Krtolica, "On Admissibility of Pseudo-Observability Indices," *IEEE Trans. on Automatic Control*, **AC-32**, 920-922 (1987).

22.   B. M. Gorti, S. Binigulac and H. F. VanLandingham, "Deterministic Identification of Linear Multi-Variable Systems", *Proceedings of the 22nd Southeastern Symposium on System Theory*, Cookeville, TN, 126-131 (1990).

APPENDICES

For the most part these appendices expand and explain in more detail the various algorithms discussed earlier. A computational procedure, i.e. algorithm, using input data (arrays):

$$A_1, A_2, \cdots, A_i, \cdots, A_n$$

and producing the resulting arrays:

$$B_1, B_2, \cdots, B_j, \cdots, B_m$$

will be denoted symbolically by the following notation:

$$A_1, A_2, \cdots, A_n \ (Algorithm) \rightarrow B_1, B_2, \cdots, B_m$$

where "*Algorithm*" represents the specific algorithm name performing this calculation.

The symbolic expression above should be interpreted as a generic "block diagram" of either form shown in Fig. A-1 below.

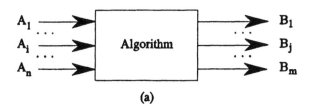

(a)

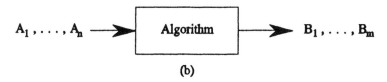

(b)

FIGURE A-1.   "Black Box" Notation for a Computational
Process (Algorithm)

Thus, the statement:

$$\text{Set} \quad A_1 , \dots , A_n \ (Algorithm) \Rightarrow B_1 , \dots , B_m$$

means:

Apply *Algorithm* to the input data A_i , $i=[1, n]$ to obtain the results B_j , $j=[1, m]$.

A1. Algorithm E-A_d {Calculation of matrices E and A_d, Eq.(9)}

It is easily shown that the matrix A_d given in Eq.(5) may be calculated using the following recursive process:

$$(A_d)_{k+1} = (A_d)_k^2 , \quad for \quad k=1, 2, \cdots, j \tag{A1.1}$$

where

$$(A_d)_1 = \sum_{i=0}^{N} \frac{(A_c T/r)^i}{i!} \tag{A1.2}$$

with j and r defined by Eq.(13). Also, from Eq.(9) we obtain the following recursions:

$$\begin{aligned} (A_d)_k &= I + E_k A_c T_k \\ (A_d)_{k+1} &= I + E_{k+1} A_c T_{k+1} \end{aligned} \tag{A1.3}$$

where

$$T_{k+1} = 2 \, T_k.$$

Introducing Eq.(A1.3) into Eq.(A1.1), we obtain the following:

$$\begin{aligned} (I + E_k A_c T_k)^2 &= I + E_{k+1} A_c T_{k+1} \\ I + 2 E_k A_c T_k + (E_k A_c T_k)^2 &= I + E_{k+1} A_c T_{k+1} \end{aligned} \tag{A1.4}$$

leading finally to the verification of Eq.(11), namely,

$$\begin{aligned} T_{k+1} &= 2 T_k \\ E_{k+1} &= E_k (I + E_k A_c T_k/2) \end{aligned} \tag{A1.5}$$

Thus, the following computational algorithm is obtained:

1. Define the matrix A_c, scalar T and integers j and N. The suggested value for N is 16, and the integer j should be chosen according to Eq.(13).
2. Calculate E_1 and T_1 using Eq.(12).
3. For $k = 1, 2, \cdots , j$ calculate E_{k+1} recursively from Eq.(A1.5).

4.   Set $E_{j+1} \Rightarrow E$ and $I + EA_cT \Rightarrow A_d$.

Symbolically, $T, A_c (E-A_d) \rightarrow A_d, E$

Some hints for computational saving in calculating truncated power series, such as Eq.(12), are given in Algorithm *C-H-T*.

A2. Algorithm *F-E-A$_d$* {Calculation of F, E, A_d in Eqs.(16) and (20)}

By comparing power series in Eqs.(9) and (16) it may be determined that the matrices E and F satisfy the following relation:

$$E = I + FA_cT \qquad (A2.1)$$

Using Eqs.(A2.1) and (A1.1), we can write, similarly to Eq.(A1.3), the recursion equations

$$E_k = I + F_kA_cT_k$$
$$E_{k+1} = I + F_{k+1}A_cT_{k+1} \qquad (A2.2)$$

where $T_{k+1} = 2\,T_k.$

Using now Eq.(A1.5) and eliminating E_k and E_{k+1} from Eq.(A2.2), the following relationship between F_k and F_{k+1} can be derived:

$$T_{k+1} = 2\,T_k$$
$$F_{k+1} = 0.5\,F_k + 0.25\,(I + F_kA_cT_k)^2 \qquad (A2.3)$$

which verifies Eq.(18). This development leads to the following computational procedure:

1.   Define the matrix A_c, scalar T and integers j and N. The integer j should satisfy Eq.(13) and a suggested value for N is 16.
2.   Calculate F_1 and T_1 using Eq.(19).
3.   For $k = 1, 2, \ldots, j$ calculate F_{k+1} recursively using Eq.(A2.3).
4.   Set $F_{j+1} \Rightarrow F$, $I + FA_cT \Rightarrow E$ and $I + EA_cT \Rightarrow A_d$.

Symbolically, $T, A_c (F-E-A_d) \rightarrow A_d, E, F$

(See computational saving hints given in Algorithm *C-H-T*.)

A3. Algorithm *R5R4* {Transformation from a non-standard five-matrix representation to an equivalent (standard) four-matrix representation.}

Consider two D-T state space representations:
$$R_5 = \{A, B_0, B_1, C, D\} \quad \text{and} \quad R_4 = \{A_e, B_e, C_e, D_e\} \quad \text{(A3.1)}$$

defining state models as in Eqs.(15) and (23), respectively, where the *d* notation has been dropped for convenience. Since these two models represent the same D-T system, the corresponding transfer function matrices should be the same. Thus, we obtain the following equality:

$$C(zI - A)^{-1}(B_0 + zB_1) + D = C_e(zI - A_e)^{-1}B_e + D_e \quad \text{(A3.2)}$$

Also the two transfer matrices should have identical characteristic polynomials. So, without loss of generality, it may be assumed that in both representations the system and output matrices are equal, i.e.

$$A_e = A \quad \text{and} \quad C_e = C \quad \text{(A3.3)}$$

In each of the five-matrix representations given in Eqs.(15), (22) and (34) there is a distinct relationship between matrices B_0 and B_1. It can be verified from Eqs.(17), (22) and (35), respectively, that this relationship is given as

$$B_1 = P B_0 \quad \text{(A3.4)}$$

where the $n \times n$ matrix P is expressible in each case by

$$P = F(E - F)^{-1} \quad , \quad P = I_n \quad , \quad P = -\frac{T}{2}I_n \quad \text{(A3.5)}$$

respectively. Using Eq.(A3.4) and the identity

$$(zI - A)^{-1}z = I + (zI - A)^{-1}A \quad \text{(A3.6)}$$

Equation (A3.2) may be written as

$$C(zI - A)^{-1}[(I + AP)B_0 - B_e] + (CPB_0 + D - D_e) = 0 \quad \text{(A3.7)}$$

Since Eq.(A3.7) should be satisfied for all z, it reduces to

$$C(zI - A)^{-1}[(I + AP)B_0 - B_e] = 0$$
$$D_e = CPB_0 + D \quad \text{(A3.8)}$$

We now introduce the following notation

$$C(zI - A)^{-1} = \frac{V(z)}{d(z)} \tag{A3.9}$$

where

$$V(z) = C\,Adj[zI - A] \quad , \quad d(z) = \det[zI - A] \tag{A3.10}$$

The $p \times n$ polynomial matrix $V(z) = \{v_{ij}(z)\}$, consisting generally of $(n-1)^{st}$ order polynomials, can also be represented as a matrix polynomial with real-number $p \times n$ matrices, i.e.

$$V(z) = \sum_{i=0}^{n-1} V_i z^i \tag{A3.11}$$

Using Eqs.(A3.10) and (A3.11) and defining the arrays

$$V = \begin{bmatrix} V_0 \\ V_1 \\ \vdots \\ V_{n-1} \end{bmatrix} \quad , \quad I(z) = \begin{bmatrix} I_p & zI_p & \cdots & z^{n-1}I_p \end{bmatrix} \tag{A3.12}$$

Eq.(A3.8) becomes

$$V[(I + AP)B_0 - B_e] = 0 \tag{A3.13}$$

It is easily shown that if the pair $\{A, C\}$ is observable, V is a full (column) rank matrix and that the unknown matrix B_e becomes

$$B_e = (I + AP)B_0 \tag{A3.14}$$

However, if $\{A, C\}$ is not observable, the general solution to Eq.(A3.13) may be written as

$$B_e = (I + AP)B_0 + NT \tag{A3.15}$$

where N is an $n \times h$ "null space" matrix satisfying that

$$VN = 0 \tag{A3.16}$$

and T is an arbitrary $h \times m$ matrix, which, if desired, may be chosen to be a zero matrix. If, however, T is selected as

$$T = -(N^T N)^{-1} N^T (I + AP)B_0$$

then B_e may be written as

$$B_e = (I - N(N^TN)^{-1}N^T)(I + AP)B_0 \qquad (A3.17)$$

which represents the minimum norm solution for B_e. It should be mentioned that even when $\{A, C\}$ is unobservable, the matrix B_e given in Eq.(A3.14) satisfies the transfer function matrix identity Eq.(A3.2).

The result of the previous development is the following computational procedure:

1. Define the matrices A, B_0, P, C and D.
2. If the pair $\{A, C\}$ is observable, calculate the unknown matrices B_e and D_e from Eqs.(A3.14) and (A3.8).
3. If $\{A, C\}$ is unobservable, Eq.(A3.14) might be substituted by Eq.(A3.17) which requires the evaluation of the polynomial matrix $V(z)$, Eq.(A3.10), building the $(pn \times n)$ matrix V, Eq.(A3.12), and calculation of the null space matrix N, Eq.(A3.16).

Matrices B_e and D_e are the results of applying Algorithm *R5R4* to the input data A, B_0, P, C and D, or symbolically:

$$A, B_0, P, C, D \;\; (R5R4) \;\; \rightarrow \;\; B_e, D_e \qquad (A3.18)$$

A4. Algorithm *R4R5* {Transformation from a standard four-matrix representation to an equivalent five-matrix representation.}

This is the reverse process of the previous algorithm, *R5R4*, and is used primarily as an intermediate step in the continualization procedure of Algorithm 6. The relation indicated in Eqs.(A3.4) and (32) will be used in this procedure to ensure that a unique four-matrix state space representation is obtained, [16]. Thus, assuming Eqs.(A3.3) and (A3.4), only B_0 and D are unknown. Following the same line of reasoning as in the previous algorithm, if the pair $\{A, C\}$ is observable, from Eq.(A3.13) we obtain

$$B_0 = (I + AP)^{-1} B_e \qquad (A4.1)$$

while from Eq.(A3.8)

$$D = D_e - CPB_0 \qquad (A4.2)$$

If $\{A, C\}$ is an unobservable pair, the minimum norm solution for B_0 can be obtained in a manner similar to the development of Eq.(A3.17) from

$$B_0 = (I - N(N^TN)^{-1}N^T)(I + AP)^{-1}B_e \qquad (A4.3)$$

where N was defined above, see Eq.(A3.16).

Thus, from Eq.(A3.2) we are led to the following procedure:

1. Define the matrices A, B_e, C, D_e and P.
2. If the pair $\{A, C\}$ is observable, calculate the unknown martices B_0 and D from Eqs.(A4.1) and (A4.2), respectively.
3. If $\{A, C\}$ is unobservable, Eq.(A4.3) may be substituted for Eq.(A4.1). This necessitates the evaluation of the polynomial matrix $V(z)$, Eq.(A3.11), building V, Eq.(A3.12), and calculation of the matrix N in Eq.(A4.16).

As in *R5R4* we use the symbolic notation

$$A, B_e, C, D_e, P \ (R4R5) \rightarrow B_0, D \qquad (A4.4)$$

to indicate that the matrices B_0 and D are the results of applying *R4R5* to the input data A, B_e, C, D_e and P.

A5. Algorithm *Ln* {Calculation of the natural log of an $n \times n$ matrix A_d.}

In order to calculate A_c defined by Eq.(26), it is first necessary to determine integers j and r satisfying Eq.(30). This is done as specified by the first five steps of the procedure below. Following the theoretical development, we will use the equivalent form Eq.(29) (as derived from Eq.(28)) instead of a truncated version of Eq.(26).

1. Define the matrix A_d, scalar T and integer N. A suggested value for N is 25.
2. Set $0 \Rightarrow j$ and $A_d \Rightarrow A_j$.
3. Set $I - A_j \Rightarrow L$.
4. If $|\lambda(L)|_{max} < 0.5$, go to 6, else go to 5.
5. Set $j+1 \Rightarrow j$; $(A_j)^{1/2} \Rightarrow A_j$ and go to 3.
6. Set $2^j \Rightarrow r$ and calculate A_c using $A_j = A_d^{1/r}$ in Eq.(29), i.e.

$$A_c = -\frac{r}{T} \sum_{i=1}^{N} \frac{(I - A_j)^i}{i}$$

(See computational saving hints given in Algorithm *C-H-T*.)
The square root of the matrix A_j required in step 5 above could be calculated by

the algorithm *SQM* given below. This algorithm is based on the standard recursive procedure:

$$x_{i+1} = 0.5(x_i + \frac{b}{x_i}) \qquad (A5.1)$$

used to determine the square root $x = (b)^{1/2}$ of a positive scalar b, [11].

Symbolically, $T, A_d \ (Ln) \rightarrow A_c$

Algorithm *SQM* {Square root of a positive definite matrix A, $X = (A)^{1/2}$.}

1.   Define the matrix A and a small scalar parameter $\epsilon << 1$.
2.   Set $X_0 = I$ and $i = 0$.
3.   Set $i = i+1$ and $X_{i+1} = 0.5 \ (X_i + A \ X_i^{-1})$.
4.   If $\| X_{i+1} - X_i \| > \epsilon$, go to 3, else *stop*.

Symbolically, $A, \epsilon \ (SQM) \rightarrow X$

A6. Algorithm *C-H-T* {Efficient calculation of truncated power series.}

In several developments presented earlier, e.g. Eqs.(12), (19) and (29), it was required that an $n \times n$ matrix, say A, be calculated using a truncated power series of the form

$$A = \sum_{i=0}^{N} c_i X^i \qquad (A6.1)$$

The series Eq.(A6.1) can be interpreted as an evaluation of the matrix polynomial $c(X)$ of the matrix X where the Nth order polynomial $c(s)$ is given by

$$c(s) = \sum_{i=0}^{N} c_i s^i \qquad (A6.2)$$

As a result of the Cayley-Hamilton Theorem, [7-9], the calculation of Eq.(A6.1) can be reduced to the evaluation of an $(n-1)^{st}$ order matrix polynomial $r(X)$ of the matrix X where the coefficients r_i of the scalar polynomial $r(s)$ satisfy the following n conditions:

$$r(\lambda_i) = c(\lambda_i), \quad for \ i = 1, 2, \cdots, n \qquad (A6.3)$$

where λ_i is the ith eigenvalue of the $n \times n$ matrix X. Using this approach, it can be verified that given the matrix X and the $N+1$ coefficients c_i of the polynomial $c(s)$, the n coefficients of the polynomial $r(s)$ can be obtained with the following procedure:

1. Set $c_i \Rightarrow r_i$, for $0 \le i \le N$.
2. Set $N+1 \Rightarrow k$ and $det(sI - X) \Rightarrow f(s)$.
3. Set $k-1 \Rightarrow k$.
4. Set $r_{k-n+j} - r_k f_j \Rightarrow r_{k-n+j}$, for $0 \le j \le n-1$.
5. If $k > n$, go to 3, else *stop*.

If the coefficients f_j, define the (monic) characteristic polynomial of X:

$$f(s) = \det(sI - X) \triangleq s^n + \sum_{i=0}^{n-1} f_j s^j \qquad (A6.4)$$

then the first n coefficients r_j, $0 \le j \le n-1$ define the $(n-1)^{st}$ order polynomial $r(s)$ satisfying Eq.(A6.3). Evaluating the matrix A in Eq.(A6.1) is then equivalent to evaluating

$$A = \sum_{i=0}^{n-1} r_i X^i \qquad (A6.5)$$

thereby considerably reducing the computational time and more importantly the accumulation of round-off errors. This method works well even if X is completely general with multiple eigenvalues. The C-H-T algorithm given above may be considered as a computational simplification of a standard procedure based on the Cayley-Hamilton Theorem and described in [8]. This standard procedure calculates coefficients r_i of the polynomial $r(s)$ from:

$$r^{(k)}(\lambda_i) = c^{(k)}(\lambda_i), \quad for \ i = 1, 2, \ldots, m \ \ and \ \ k = 1, 2, \ldots, n_i \qquad (A6.6)$$

where λ_i is an eigenvalue of X, n_i is its algebraic multiplicity and m is the number of distinct eigenvalues. Obviously, if all of the eigenvalues of X are distinct, then $m = n$ and $n_i = 1$ for all i. The notation of (A6.6) is defined by

$$r^{(k)}(\lambda_i) = \frac{d^k r(\lambda)}{d\lambda^k}\Big|_{\lambda = \lambda_i} \qquad\qquad (A6.7)$$

The C-H-T algorithm is useful when X has multiple eigenvalues. It is neither necessary to determine the algebraic multiplicities, nor to evaluate the derivatives in (A6.7). The C-H-T algorithm also works for matrices having a spectral radius greater than 0.5.

Symbolically, $X, c\ (C\text{-}H\text{-}T)\ \rightarrow A$

A7. Sequence of Algorithms Used for Example 3.

1.   Data definition: Define:
      C-T state space representation $R_c = \{A_c, B_c, C_c, D_c\}$,
      Sample interval T
      Number of samples N
      Samples $u_i(t_k)$ of input signal $u(t)$, $i=[1,m]$, $k=[0,N]$

2.   Discretization of R_c:
      Using A_c and T, define scaling and truncation parameters r_e and N_e
      Set: T, A_c, r_e, N_e (F-E-Ad) $\Rightarrow A_d, E, F$

     2.1.   Step invariant D-T equivalent model $R_{ds} = \{A_{ds}, B_{ds}, C_{ds}, D_{ds}\}$
           Set: $A_d \Rightarrow A_{ds}$; $E*B_c*T \Rightarrow B_{ds}$; $C_c \Rightarrow C_{ds}$; $D_c \Rightarrow D_{ds}$
     2.2.   Ramp invariant D-T equivalent model $R_{dr} = \{A_{dr}, B_{dr}, C_{dr}, D_{dr}\}$

          2.2.1.   Ramp invariant 5-matrix D-T model
                 Set: $A_d \Rightarrow A_{dr}$; $(E-F)*B_c*T \Rightarrow B_{dr0}$
                 Set: $C_c \Rightarrow C_{dr}$; $D_c \Rightarrow D_{drs}$; $F(E-F)^{-1} \Rightarrow P_r$

          2.2.2.   Ramp invariant 4-matrix D-T model
                 Set: $A_d, B_{dr0}, P_r, C_{dr}, D_{drs}$ (R5R4) $\Rightarrow B_{dr}, D_{dr}$

     2.3.   Bilinear D-T equivalent model

          2.3.1.   Bilinear 5-matrix D-T model
                 Set: A_c, B_c, T (BTCD) $\Rightarrow A_{db}, B_{db0}, B_{db1}, P_{cd}$
                 Set: $C_c \Rightarrow C_{db}$; $D_c \Rightarrow D_{dbs}$

2.3.2. Bilinear 4-matrix D-T model
    Set: A_{db}, B_{db0}, P_{cd}, C_{db}, D_{db5} (R5R4) $\Rightarrow$ B_{db}, D_{db}

3.   Responses of C-T model R_c and D-T models R_{ds}, R_{dr} and R_{db}
Set: A_c, B_c, C_c, D_c, x_0, $u(t_k)$, $T*N$ (RCS) $\Rightarrow$ $y_c(t_k)$
Set: A_{ds}, B_{ds}, C_{ds}, D_{ds}, x_0, $u(t_k)$ (RDS) $\Rightarrow$ $y_{ds}(t_k)$
Set: A_{dr}, B_{dr}, C_{dr}, D_{dr}, x_0, $u(t_k)$ (RDS) $\Rightarrow$ $y_{dr}(t_k)$
Set: A_{db}, B_{db}, C_{db}, D_{db}, x_0, $u(t_k)$ (RDS) $\Rightarrow$ $y_{db}(t_k)$

4.   Input\output identification, i.e. definition of D-T model
$R_d = \{A_d, B_d, C_d, D_d\}$ to be continualized into SI, RI and BT models

$$n_o = \{ n_{oi} \}, \quad i = [1, p]$$

Set: $u(t_k)$, $y_c(t_k)$ (IDMV) $\Rightarrow$ A_d, B_d, C_d, D_d, n_o, $x(0)$

5.   Continualization of R_d
Using A_d, define scaling and truncation parameters r_L and N_L
Set: T, A_d, r_L, N_L (Ln) $\Rightarrow$ A_{cs}
Set: T, A_{cs}, r_e, N_e (F-E-Ad) $\Rightarrow$ A_d, E, F

    5.1.   Step invariant C-T equivalent model $R_{cs} = \{A_{cs}, B_{cs}, C_{cs}, D_{cs}\}$
        Set: $E^{-1}*B_d/T \Rightarrow B_{cs}$; $C_d \Rightarrow C_{cs}$; $D_d \Rightarrow D_{cs}$

    5.2.   Ramp invariant C-T equivalent model $R_{cr} = \{A_{cr}, B_{cr}, C_{cr}, D_{cr}\}$

        5.2.1.   Ramp invariant 5-matrix C-T model
            Set: $F*(E-F)^{-1} \Rightarrow P_r$
            Set: A_d, B_d, P_r, C_d, D_d (R4R5) $\Rightarrow$ B_{dr0}, D_{dr}

        5.2.2.   Ramp invariant 4-matrix C-T model
            Set: $A_{cs} \Rightarrow A_{cr}$; $(E-F)^{-1}*B_{dr0}/T \Rightarrow B_{cr}$; $C_d \Rightarrow C_{cr}$; $D_d \Rightarrow D_{cr}$

    5.3.   Bilinear C-T equivalent model $R_{cb} = \{A_{cb}, B_{cb}, C_{cb}, D_{cb}\}$

        5.3.1.   Bilinear 5-matrix C-T model
            Set: A_d, B_d, T (BTDC) $\Rightarrow$ A_{cb}, B_{cb0}, B_{cb1}, P_{dc}

        5.3.2.   Bilinear 4-matrix C-T model
            Set: A_{cb}, B_{cb0}, P_{dc}, C_d, D_d (R5R4) $\Rightarrow$ B_{cb}, D_{cb}

6.   Responses of D-T model R_d and C-T models R_{cs}, R_{cr} and R_{cb}

Set: A_d, B_d, C_d, D_d, x_0, $u(t_k)$ *(RDS)* $\Rightarrow y_d(t_k)$
Set: A_{cs}, B_{cs}, C_{cs}, D_{cs}, x_0, $u(t_k)$, $T*N$ *(RCS)* $\Rightarrow y_{cs}(t_k)$
Set: A_{cr}, B_{cr}, C_{cr}, D_{cr}, x_0, $u(t_k)$, $T*N$ *(RCS)* $\Rightarrow y_{cr}(t_k)$
Set: A_{cb}, B_{cb}, C_{cb}, D_{cb}, x_0, $u(t_k)$, $T*N$ *(RCS)* $\Rightarrow y_{cb}(t_k)$

Explanation of indices:

- c and d denote *continuous* and *discrete* models
- s, r and b denote *step*-invariant, *ramp*-invariant and *bilinear* models
- *5* indicates a 5-matrix model
- *0* and *1* are used to separate the two input matrices in 5-matrix models

Algorithms used and not explained in the text:

A, B, C, D, x_0, $u(t)$, T *(RCS)* $\Rightarrow y(t)$
A, B, C, D, x_0, $u(t_k)$ *(RDS)* $\Rightarrow y(t_k)$
U, Y, *(IDMV)* $\Rightarrow A$, B, C, D, n_o, $x(0)$

These algorithms are available in the CAD software package *L-A-S*.

RCS: Response $y(t)$ of a C-T system specified in state space, $\{A,B,C,D\}$, having an initial state x_0 and input signal $u(t)$ defined over the time interval $0 < t < T$.

RDS: Response $y(t_k)$ of a D-T system specified in state space, $\{A,B,C,D\}$, having an initial state x_0 and samples, $u(t_k)$, of the input $u(t)$.

IDMV   Input/output deterministic identification of a D-T MIMO system. U and Y contain m and p sequences of input/output data, m and p being dimensions of input and output vectors, respectively, $n_o = \{n_{oi}\}$, $0 \leq i \leq p$, is a set of admissible pseudo-observability indices. $R = \{A,B,C,D\}$ is the identified state space representation of a D-T system corresponding to the given input/output sequences U and Y. The pair $\{A,C\}$ is in a pseudo-observable form specified by the selected set n_o and $x(0)$ is the initial condition vector.

Note that the input/output identification performed in Step 4 could be avoided by taking for R_d any of the D-T models R_{dc}, R_{dr} or R_{db} obtained in Step 2 by discretizing the given C-T model R_c. Algorithm *IDMV* is explained in Appendix A9.

To facilitate better understanding of Example 3, the following Table indicates equation numbers from the text and algorithm names used in the above calculations. The complete listing of Example 3, implemented using the CAD software package *L-A-S*, [6], is given in Appendix A8, following the table.

TABLE A-I   Algorithm Summary for Example 3

Step Numbers	Equations and Algorithms Used
2	Eqs. (13), (20) and algorithm *F-E-Ad*
2.1	Eq. (9)
2.2.1	Eq. (17)
2.2.2	Algorithm *R5R4*
2.3.1	Eq. (22) and algorithm *BT-C-D*
2.3.2	Algorithm *R5R4*
5	Eq. (30) and algorithms *Ln* and *F-E-Ad*
5.1	Eq. (31)
5.2.1	Eq. (32) and algorithm *R4R5*
5.2.2	Eq. (32)
5.3.1	Eq. (35) and algorithm *BT-D-C*
5.3.2	Algorithm *R5R4*

A8. *L-A-S* Software Implementation of Example 3

```
L - A - S   print file  LASR
created  2/21/1992  at   7:57

1   _Example_#_3
2   _1._DATA_Definition
```

```
 3    _Reading_Rc_from_disk_
 4    (rbf)=Ac,Bc,Cc,Dc,{r532}
 5    1e-5(dma),1(atg),4(s*),Ac(cdi)(mcp)=eps,pi,n
 6    n,1(dzm)=xo
 7    2,.5,18,.3,30,5(dma)=x
 8    x(tvc)=T,nrm,Ne,Egm,NL,N
 9    Ac(nrr),T(*,t)=nAcT
10    _Definition_of_input_vector_u(tk)
11    T,N(dec)(*,t)=TT
12    0,1,N(dma,t)=v
13    v(gts),N(dec)(s/,t)=t
14    t,pi(*)(sin)=sin
15    t,pi(*)(cos)(dec),2(s/,t)=cos
16    t,sin,cos(cti)=u
17    u,TT(dis)=
18    (sto)=
19    _2._Discretization_of_Rc
20    T,Ac,nrm,Ne(FEAd,t)=Ad,E,F
21    _2.1_SI-D-T_Model
22    E,Bc(*),T(s*,t)=Bds
23    _2.2_RI-D-T_Model
24    E,F(-),Bc(*),T(s*)=Bdro
25    F,E,F(-)(-1)(*,t)=Pr
26    Ad,Bdro,Pr,Cc,Dc,eps(R5R4,SBR)=Bdr,Ddr
27    _2.3_BT-C-D_Model
28    Ac,Bc,T(btcd,sub)=Adb,Bdbo,Bdb1,Pcd
29    Adb,Bdbo,Pcd,Cc,Dc,eps(R5R4,SBR)=Bdb,Ddb
30    _3._Responses_of_Rc_,_Rds_,_Rdr_and_Rdb
31    Ac,Bc,Cc,Dc,xo,u,TT(Rcs,sub)=yc
32    Ad,Bds,Cc,Dc,xo,u(Rds,sub)=yds
33    Ad,Bdr,Cc,Ddr,xo,u(Rds,sub)=ydr
34    Adb,Bdb,Cc,Ddb,xo,u(Rds,sub)=ydb
35    _Response_Plotting
36    yc,TT(dis)=
37    yds,TT(dis)=
38    ydr,TT(dis)=
39    ydb,TT(dis)=
40    _Norms_of_response_differences
41    yc,yds(-),yc,ydr(-),yc,ydb(-)(mcp)=dds,ddr,ddb
42    dds(nrr),ddr(nrr),ddb(nrr)(cti,e)=difd
43    _Building_Representations_Rc_,_Rds_,_Rdr_and_Rdb
44    Ac,Bc,Cc,Dc(Sysm,sub)=Rc
```

```
45    Ad,Bds,Cc,Dc(Sysm,sub)=Rds
46    Ad,Bdr,Cc,Ddr(Sysm,sub)=Rdr
47    Adb,Bdb,Cc,Ddb(Sysm,sub)=Rdb
48    _4._Identification;_Definition_of_Rd
49    _Reading_identified_model_Rd_from_disk
50    (rbf)=Rd,{rd4}
51    Rd,n,n(m14,sub)=Ad,Bd,Cd,Dd
52    (sto)=
53    _5._Continualization_of_Rd
54    T,Ad,Egm,NL(Ln,t)=Acs
55    T,Acs,nrm,Ne(FEAd,t)=Adn,E,F
56    (sto)=
57    _5.1_SI_C-T_Model
58    E(-1),Bd(*),T(s/,t)=Bcs
59    _5.2_RI_C-T_Model
60    F,E,F(-)(-1)(*,t)=Pr
61    Ad,Bd,Pr,Cd,Dd,eps(R4R5,SBR)=Bdro,Ddr
62    E,F(-)(-1),Bdro(*),T(s/,t)=Bcr
63    _5.3_BT_C-T_Model
64    Ad,Bd,T(Btdc,sub)=Acb,Bbco,Bbc1,Pdc
65    Acb,Bbco,Pdc,Cd,Dd,eps(R5R4,SBR)=Bcb,Dcb
66    _Responses_of_Rd_,_Rcs_,_Rcr_and_Rcb
67    Ad,Bd,Cd,Dd,xo,u(Rds,sub)=yd
68    Acs,Bcs,Cd,Dd,xo,u,TT(Rcs,sub)=ycs
69    Acs,Bcr,Cd,Ddr,xo,u,TT(Rcs,sub)=ycr
70    Acb,Bcb,Cd,Dcb,xo,u,TT(Rcs,sub)=ycb
71    _Response_Plotting
72    yd,TT(dis)=
73    ycs,TT(dis)=
74    ycr,TT(dis)=
75    ycb,TT(dis)=
76    _Norms_of_response_differences
77    yd,ycs(-),yd,ycr(-),yd,ycb(-)(mcp)=dcs,dcr,dcb
78    dcs(nrr),dcr(nrr),dcb(nrr)(cti,e)=difc
79    _Building_Representations_Rd_,_Rcs_,_Rcr_and_Rcb
80    Ad,Bd,Cd,Dd(Sysm,sub)=Rd
81    Acs,Bcs,Cd,Dd(Sysm,sub)=Rcs
82    Acs,Bcr,Cd,Ddr(Sysm,sub)=Rcr
83    Acb,Bcb,Cd,Dcb(Sysm,sub)=Rcb
84    (sto)=
```

Comparing Appendix A7 with the *L-A-S* implementation given above, a striking similarity between the required sequence of calculation and corresponding commands of the *L-A-S* software may be noted.

Explanation of Algorithms *SYSM* and *M14* used in the *L-A-S* implementation:

- The input/output arguments: *A, B, C, D* and *R* in

$$A,B,C,D \ (SYSM,sub) \ \Rightarrow \ R \quad \text{and} \quad R,n,n \ (M14,sub) \ \Rightarrow \ A,B,C,D$$

are related to each other by:

$$R = \left[\begin{array}{c|c} A & B \\ \hline C & D \end{array} \right] \qquad (A8.1)$$

where *A, B, C* and *D* are *(n×n), (n×m), (p×n)* and *(p×m)* martices, respectively, defining the MIMO model, while *R* is an *[(n+p)×(n+m)]* martix given by Eq.(A8.1).

A9.   Deterministic Identification of D-T Models from Input/Output Data

The identification procedure used in Example 3 for building the D-T representation R_d is described here. We will assume that sampled data from the unknown C-T system has been made available. In particular, suppose that the following input/output vector data pairs are given for a system with *m*-inputs and *p*-outputs:

$$\{ \mathbf{u}(k), \mathbf{y}(k) \} \quad \text{for} \quad k = [0, N] \qquad (A9.1)$$

where *N* is sufficiently large for identification purposes. The input and output vectors are of dimensions $m \times 1$ and $p \times 1$, respectively, where $m, p \geq 1$.

A9.1 MIMO Structure Determination

An initial step is used to re-organize a part of the data as follows, typically with $l < q < N$:

$$\mathbf{Z}_l = \left[\begin{array}{c} \mathbf{U}_l \\ ---- \\ \mathbf{Y}_l \end{array}\right] \tag{A9.2}$$

where

$$\mathbf{U}_l = \begin{bmatrix} \mathbf{u}(0) & \mathbf{u}(1) & \cdots & \mathbf{u}(q) \\ \mathbf{u}(1) & \mathbf{u}(2) & \cdots & \mathbf{u}(1+q) \\ \vdots & \vdots & \cdots & \vdots \\ \mathbf{u}(l) & \mathbf{u}(l+1) & \cdots & \mathbf{u}(l+q) \end{bmatrix} \tag{A9.3}$$

and

$$\mathbf{Y}_l = \begin{bmatrix} \mathbf{y}(0) & \mathbf{y}(1) & \cdots & \mathbf{y}(q) \\ \mathbf{y}(1) & \mathbf{y}(2) & \cdots & \mathbf{y}(1+q) \\ \vdots & \vdots & \cdots & \vdots \\ \mathbf{y}(l) & \mathbf{y}(l+1) & \cdots & \mathbf{y}(l+q) \end{bmatrix} \tag{A9.4}$$

An effort is made to select q large enough so that as additional rows are added, the number of columns will continue to exceed the number of rows of $\mathbf{Z}$, as will be explained in the following.

Starting with a "small" value of l, the procedure calls for the rank of $\mathbf{Z}_l$ to be checked, followed by an augmentation of $\mathbf{Z}_l$ to $\mathbf{Z}_{l+1}$, i.e. with rows [$\mathbf{u}(l+1) \ldots \mathbf{u}(l+q+1)$] and [$\mathbf{y}(l+1) \ldots \mathbf{y}(l+q+1)$] appended to $\mathbf{U}_l$ and $\mathbf{Y}_l$, respectively, and a subsequent check of the rank of $\mathbf{Z}_{l+1}$. More specifically, suppose that

$$\text{rank}\,\mathbf{Z}_l = r_l \quad \text{and} \quad \text{rank}\,\mathbf{Z}_{l+1} = r_{l+1} \tag{A9.5}$$

then if

$$d_l = d_{l+1} \quad \text{where} \quad d_l = r_l - m(l+1) \tag{A9.6}$$

the system order is $n = d_l$, and if (A9.6) is not satisfied, the augmentation step is repeated. In this manner, starting from the first input/output pair and sequentially augmenting additional pairs until the d_l ceases to increase, the system order is determined. Note that the effect of rank due to the input vectors is subtracted out. For a "sufficiently rich" input the rank of $\mathbf{Z}$ would continue to increase with additional augmentation. Once the system order, n, has been determined, the *observability index*, n_x, defined by

$$n_x = \max_{i=[1,p]} \{ n_i \} \qquad\qquad (A9.7)$$

can also be determined. In fact, as it will be shown later, n_x is given by $n_x = l+1$, where l is the smallest integer satisfying (A9.6). In other words, it is equal to the number of blocks of outputs $y(i)$, $i=[0, n_x-1]$, containing at least one linearly independent row, $y_j(i)$, $j=[1,p]$. In (A9.7) n_i is the number of linearly independent rows of Z arising from the ith output. The set $\{n_i\}$ is referred to as the unique set of observability indices or, as will be explained later, as a set of admissible pseudo-observability indices. A specific example will help to illustrate the procedure.

Consider a system with order $n=7$, $m=2$ inputs and $p=3$ outputs. A typical instance of the above augmentation process might result in the array Z_3 being constructed after Z_2 with the determination that $d_2 = d_3$ from (A9.6). It is found that the rank of Z_3 is *15* which, after subtracting $m(l+1)=8$, corresponding to the number of linearly independent rows of input vectors, gives for the system order $n=7$. Two rows among the those beginning with $y(1)$ and $y(2)$, are found to be linearly dependent; and all $p=3$ rows beginning with $y(3)$ are linearly dependent.

If we are interested in the unique set of observability indices, we should determine which two rows among those beginning with $y(1)$ and $y(2)$ are linearly dependent. Assuming that the particular rows beginning with $y_2(1)$ and $y_2(2)$ are, in fact, the dependent rows, then, according to the definition of observability indices,[8,9], it may be concluded that this case leads to the set of observability indices given by

$$\{ n_i \} = \{ 3, 1, 3 \}$$

It has recently been shown, [20], that the use of this unique set of observability indices does not necessarily lead to the most convenient system representation, and that the use of so called admissible sets of pseudo-observability indices, [21], offers more flexibility in choosing the appropriate model. For these reasons in the sequel we will pursue the selection of the most convenient set of (pseudo) observability indices.

Knowing that the system order is *7* from the rank calculations and that the observability index is *3*, corresponding to the minimum number of output vectors needed to achieve that rank, there are several possible observable form structures that may be considered. These pseudo-observable indices are given by

Case	1	2	3	4	5	6
Pseudo-Observability Indices	3,2,2	2,3,2	2,2,3	3,3,1	3,1,3	1,3,3

In all six cases $d_3 = d_2 = 7$, but in the first three cases $d_1 = 6$, whereas in the last three cases $d_1 = 5$. Note that in each case the "observability indices" sum to $n = 7$. Let us further assume that, in fact, $d_1 = 5$, i.e. cases 4, 5 and 6. We can use a *crate* diagram to represent each of these three cases, [9].

{3,3,1}		
1	1	1
1	1	0
1	1	
0	0	

{3,1,3}		
1	1	1
1	0	1
1		1
0		0

{1,3,3}		
1	1	1
0	1	1
	1	1
	0	0

Crate diagrams are simply a graphical method of visualizing the selection of linearly independent rows from the given output data. For example, with the columns of the crate being associated with particular output strings, the center crate above indicates that the independent elements are rows beginning with $y_1(0)$, $y_2(0)$, $y_3(0)$, $y_1(1)$, $y_3(1)$, $y_1(2)$, $y_3(2)$.

From the crate diagrams several related "selection vectors" are generated:

- By omitting the first row of, say the center diagram, corresponding to the indices {3,1,3}, the vector v_i is created by selecting the non-blank elements row-wise:

$$v_i = [\,1 \quad 0 \quad 1 \quad 1 \quad 1 \quad 0 \quad 0\,]^T \qquad (A9.8)$$

- From v_i the binary complement is formed, and denoted as v_a:

$$v_a = [\,0 \quad 1 \quad 0 \quad 0 \quad 0 \quad 1 \quad 1\,]^T \qquad (A9.9)$$

- By considering the blank elements to be zeros, v_{ii} is formed in like manner, but with row 1 included:

$$\mathbf{v}_{li} = [\,1 \quad 1 \quad 1 \quad 1 \quad 0 \quad 1 \quad 1 \quad 0 \quad 1 \quad 0 \quad 0 \quad 0\,]^T \quad (A9.10)$$

- Finally, $\mathbf{v}_{ld}$ is formed by again including the first row, but now taking the blank elements of the diagram to be unit valued, and finally taking the binary complement, leading to:

$$\mathbf{v}_{ld} = [\,0 \quad 0 \quad 0 \quad 0 \quad 1 \quad 0 \quad 0 \quad 0 \quad 0 \quad 1 \quad 0 \quad 1\,]^T \quad (A9.11)$$

By appropriately using the above "selector vectors", an observable form state space model can be obtained. In addition, corresponding "selector matrices", [5,22], which carry the same information as the selector vectors, serve to simplify the subsequent discussion. Thus,

$$S_i = \begin{bmatrix} 1 & 0 & 0 & 0 & 0 & 0 & 0 \\ 0 & 0 & 1 & 0 & 0 & 0 & 0 \\ 0 & 0 & 0 & 1 & 0 & 0 & 0 \\ 0 & 0 & 0 & 0 & 1 & 0 & 0 \end{bmatrix}^T , \quad S_{li} = \begin{bmatrix} 1 & 0 & 0 & 0 & 0 & 0 & 0 & 0 & 0 & 0 & 0 & 0 \\ 0 & 1 & 0 & 0 & 0 & 0 & 0 & 0 & 0 & 0 & 0 & 0 \\ 0 & 0 & 1 & 0 & 0 & 0 & 0 & 0 & 0 & 0 & 0 & 0 \\ 0 & 0 & 0 & 1 & 0 & 0 & 0 & 0 & 0 & 0 & 0 & 0 \\ 0 & 0 & 0 & 0 & 0 & 1 & 0 & 0 & 0 & 0 & 0 & 0 \\ 0 & 0 & 0 & 0 & 0 & 0 & 1 & 0 & 0 & 0 & 0 & 0 \\ 0 & 0 & 0 & 0 & 0 & 0 & 0 & 1 & 0 & 0 & 0 \end{bmatrix}^T \quad (A9.12)$$

$$S_a = \begin{bmatrix} 0 & 1 & 0 & 0 & 0 & 0 & 0 \\ 0 & 0 & 0 & 0 & 0 & 1 & 0 \\ 0 & 0 & 0 & 0 & 0 & 0 & 1 \end{bmatrix}^T , \quad S_{ld} = \begin{bmatrix} 0 & 0 & 0 & 0 & 1 & 0 & 0 & 0 & 0 & 0 & 0 & 0 \\ 0 & 0 & 0 & 0 & 0 & 0 & 0 & 0 & 0 & 1 & 0 & 0 \\ 0 & 0 & 0 & 0 & 0 & 0 & 0 & 0 & 0 & 0 & 0 & 1 \end{bmatrix}^T$$

The relationship between selector vectors and selector matrices is relatively straightforward, having to do only with the selection of rows from an identity matrix corresponding to the unities in the associated selector vector.

A9.2 MIMO System Parameterization

In this Section the structural details described in the previous Section will be used in the process of obtaining the system parameters. The eventual representation of the identified system is a state space observable form $R_o =$

$\{A_o, B_o, C_o, D_o, x(0)\}$ where C_o and A_o have the following structure, (continuing to use the $\{3,1,3\}$ example from Section A9.1):

$$A_o = \begin{bmatrix} 0 & 0 & 0 & 1 & 0 & 0 & 0 \\ x & x & x & x & x & x & x \\ 0 & 0 & 0 & 0 & 1 & 0 & 0 \\ 0 & 0 & 0 & 0 & 0 & 1 & 0 \\ 0 & 0 & 0 & 0 & 0 & 0 & 1 \\ x & x & x & x & x & x & x \\ x & x & x & x & x & x & x \end{bmatrix} \qquad C_o = \begin{bmatrix} 1 & 0 & 0 & 0 & 0 & 0 & 0 \\ 0 & 1 & 0 & 0 & 0 & 0 & 0 \\ 0 & 0 & 1 & 0 & 0 & 0 & 0 \end{bmatrix} \qquad \text{(A9.13)}$$

The structure of the pair $\{A_o, C_o\}$ is characterized by the following points:

- C_o consists of the first $p=3$ rows of the $(n \times n)$ identity matrix I_n.
- At locations specified by the unities in the selector vector v_l, the matrix A_o contains the last $n-p = 4$ rows of I_n.
- At locations specified by the $p=3$ unities in the selector vector v_a, the matrix A_o contains rows of elements which are not necessarily of zero or unit value.
- The observability matrix Q_{oo} of the pair $\{A_o, C_o\}$, i.e.

$$Q_{oo} = \begin{bmatrix} C_o^T & (C_oA_o)^T & \cdots & (C_oA_o^{n_x})^T \end{bmatrix}^T \qquad \text{(A9.14)}$$

contains all n rows of I_n at locations specified by the $n=7$ unities in the selector vector v_u.
- The $p=3$ rows of A_o containing not necessarily zero or unit elements appear in Q_{oo} at locations specified by the unities in the selector vector v_{ld}.

The results of (A9.13) derive from the basic similarity transformation, or change of state,

$$\begin{aligned} A_o &= TAT^{-1}, & B_o &= TB \\ C_o &= CT^{-1}, & D_o &= D \end{aligned} \qquad \text{(A9.15)}$$

where $R = \{A, B, C, D\}$ is an arbitrary n^{th} order observable state space representation. In order to obtain A_o and C_o given by (A9.13), the transformation matrix T in (A9.15), corresponding to the observability indices $\{3,1,3\}$, is given by

$$\mathbf{T} = \left[\begin{array}{ccccccc} \mathbf{c_1}^T & \mathbf{c_2}^T & \mathbf{c_3}^T & (\mathbf{c_1 A})^T & (\mathbf{c_3 A})^T & (\mathbf{c_1 A^2})^T & (\mathbf{c_3 A^2})^T \end{array}\right]^T \quad \text{(A9.16)}$$

It may be verified that all $n=7$ rows of $\mathbf{T}$ are located in the observability matrix $\mathbf{Q_o}$ of the pair $\{\mathbf{A}, \mathbf{C}\}$, i.e.

$$\mathbf{Q_o} = \left[\begin{array}{cccc} \mathbf{C}^T & (\mathbf{CA})^T & \cdots & (\mathbf{CA}^{n_z})^T \end{array}\right]^T$$

at locations specified by $n=7$ unities in the selector vector $\mathbf{v_u}$.

A9.3   MIMO System Parameter Determination

In Section A9.2 the existence of an observable state space representation having the structure (A9.13) has been established. In other words, by a similarity transformation $\mathbf{T}$, given by (A9.16), any observable realization $R = \{\mathbf{A}, \mathbf{B}, \mathbf{C}, \mathbf{D}\}$ can be transformed into the realization $R_o = \{\mathbf{A_o}, \mathbf{B_o}, \mathbf{C_o}, \mathbf{D_o}\}$ as described in (A9.13). In this Section we describe the *Identification Identity* which relates the input-output data (A9.1) to the matrices of R_o. Since $\mathbf{C_o}$ is completely specified, as is the structure of $\mathbf{A_o}$, it remains to relate the data to matrices $\mathbf{B_o}$, $\mathbf{D_o}$ and the unspecified p rows of $\mathbf{A_o}$ and $\mathbf{x}(0)$.

To this end, consider the order-n system with m-inputs and p-outputs:

$$\begin{aligned} \mathbf{x}(k+1) &= \mathbf{A_o x}(k) + \mathbf{B_o u}(k), \quad \mathbf{x}(0) \\ \mathbf{y}(k) &= \mathbf{C_o x}(k) + \mathbf{D_o u}(k) \end{aligned} \quad \text{(A9.17)}$$

where $\{\mathbf{A_o}, \mathbf{B_o}, \mathbf{C_o}, \mathbf{D_o}\}$ is in the observable form corresponding to a set of admissible (pseudo) observability indices $n_o = \{n_{ij}\}$, $i=[1,p]$. From (A9.17) we may write

$$\begin{bmatrix} \mathbf{y}(k) \\ \mathbf{y}(k+1) \\ \vdots \\ \mathbf{y}(k+r) \end{bmatrix} = \begin{bmatrix} \mathbf{C_o} \\ \mathbf{C_o A_o} \\ \vdots \\ \mathbf{C_o A_o}^r \end{bmatrix} \mathbf{x}(k) + \begin{bmatrix} \mathbf{D_o} & 0 & \cdots & 0 & 0 \\ \mathbf{C_o B_o} & \mathbf{D_o} & \cdots & 0 & 0 \\ & & \cdots & & \\ \mathbf{C_o A_o}^{r-1}\mathbf{B_o} & \cdots & \mathbf{C_o A_o B_o} & \mathbf{C_o B_o} & \mathbf{D_o} \end{bmatrix} \begin{bmatrix} \mathbf{u}(k) \\ \mathbf{u}(k+1) \\ \vdots \\ \mathbf{u}(k+r) \end{bmatrix}$$

$$\text{(A9.18)}$$

Now we let $r = n_x$. Clearly, (A9.18) holds for any $k = [0, N-r]$ and can be rewritten as

$$\mathbf{y}_k = \mathbf{Q_{oo} x}(k) + \mathbf{H u}_k \quad \text{(A9.19)}$$

where y_k and u_k are $(n_x+1)p$ and $(n_x+1)m$ dimensional columns containing output and input vectors $y(k+j)$ and $u(k+j)$, $j = [0, n_x]$. The matrix Q_{oo} is the observability matrix of the pair $\{A_o, C_o\}$, while H is the $(r+1)p \times (r+1)m$ lower block triangular matrix containing along the main diagonal the $(p \times m)$ blocks D_o. The other nonzero blocks of H are the $p \times m$ dimensional *Markov parameters*:

$$C_o A_o^j B_o \, , \quad \text{for } j = [0, n_x-1] \tag{A9.20}$$

Our goal is to eliminate from (A9.18) the $x(k)$ terms, thereby obtaining the *Identification Identity*, which relates the available sampled data to the unknown elements in R_o.

Equation (A9.18) can be considered to represent $(n_x+1)p$ scalar equations in the samples

$$y_{ij} = y_i(k+j) \tag{A9.21}$$

i.e. the ith element of the output vector $y(k+j)$, $i=[1, p]$, $j=[0, n_x]$. In Section A9.2 it was shown that Q_{oo} has n rows of an identity matrix and p rows that correspond to the unknown rows of A_o. Furthermore, the location of these rows are determined by the selector vectors v_{li} and v_{ld}, respectively.

Premultiplying (A9.19) by the selector matrices S_{li}^T and S_{ld}^T defined by (A9.12), we obtain, respectively,

$$y_{1k} = x(k) + H_1 u_k \, , \quad \text{and} \quad y_{2k} = A_r x(k) + H_2 u_k \tag{A9.22}$$

where

$$y_{1k} = S_{li}^T y_k \, , \quad y_{2k} = S_{ld}^T y_k \quad \text{with} \quad H_1 = S_{li}^T H \, , \quad H_2 = S_{ld}^T H$$

Eliminating $x(k)$ from (A9.22),

$$y_{2k} = \left[\, (H_2 - A_r H_1) \quad A_r \, \right] \begin{bmatrix} u_k \\ y_{1k} \end{bmatrix} \tag{A9.23}$$

The matrix A_r in (A9.22) and (A9.23) is a $(p \times n)$ matrix containing the unknown rows of A_o, whose locations in A_o are specified by the selector vector v_a. Equation (A9.23) may be written in a more concise form given by

$$y_{2k} = \left[\, B_r \quad A_r \, \right] z_k \tag{A9.24}$$

where $B_r = H_2 - A_r H_1$ is a $p \times (n_x+1)m$ matrix and z_k is an h-dimensional vector of data where $h = (n_x+1)m + n$. Equation (A9.24) is referred to as the *Identification Identity* since it relates input-output data samples arranged into columns y_{2k} and z_k to the unknown parameters of the state space representation R_o, i.e. in the matrices A_o, B_o and D_o.

Concatenating vectors y_{2k} and z_k corresponding to samples $k = 0, 1, 2, ..., q-1$, assuming that $(n_x+1)m + n \leq q < N$, into $p \times q$ and $h \times q$ matrices Y_{2k} and Z_k, respectively, yields:

$$Y_{2k} = \begin{bmatrix} B_r & A_r \end{bmatrix} Z_k \qquad (A9.25)$$

where
$$Z_k = \begin{bmatrix} U_k \\ Y_{1k} \end{bmatrix} \begin{matrix} \} & (n_x+1)m \\ \} & n \end{matrix} \qquad (A9.26)$$

Note that the matrix U_l, for $l=n_x$, given in (A9.3) is equal to U_k in (A9.26). Also, Y_{1k} and Y_{2k} in (A9.26) could be obtained from Y_l in (A9.3) by premultiplying it with selector matrices S_{ll}^T and S_{ld}^T, respectively. In other words, Y_{1k} and Y_{2k} are obtained from Y_l in (3) "according to" the selector vectors v_{ll} and v_{ld}, respectively.

One quickly concludes that the input sequence used to generate the response is "sufficiently rich" if and only if the matrix U_k is full rank, i.e.

$$\text{rank } U_k = (n_x + 1) m \qquad (A9.27)$$

and that the set of (pseudo) observability indices $n_o = \{n_i\}$ is admissible if Z_k is of full (row) rank, i.e. if

$$\text{rank } Z_k = h \qquad (A9.28)$$

If the *condition number* of Z_k is relatively large, it indicates that it might be advisable to try another set of (pseudo) observability indices which, through a different set of selector vectors and matrices, could lead to a better conditioning of Z_k. The (D-T) identification procedure is summarized in the following section.

Finally, the solution of (A9.24), containing the parameter information for R_o, is

$$\begin{bmatrix} B_r & A_r \end{bmatrix} = Y_{2k} Z_k^T (Z_k Z_k^T)^{-1} \qquad (A9.29)$$

In the next section a step-by-step procedure is suggested for extracting the required arrays.

A9.4 Algorithm

After determining the system order, n, and the observability index, n_x, one must select an admissible set of (pseudo) observability indices, $n_o = \{n_i\}$;

see Section A9.2. Using some identification method, such as a least-squared error procedure, [3,20], estimate the matrices A_r and B_r defined in (A9.25). This process can be symbolized as

$$U, Y_1, Y_2 \rightarrow A_r, B_r \qquad (A9.30)$$

where the k subscript notation has been dropped for convenience. Once A_r and B_r have been determined, R_o may be found, as formalized in the following steps:

1.   Set $I_n \Rightarrow \begin{bmatrix} C_o \\ L_2 \end{bmatrix} \begin{matrix} \} \ p \\ \} \ n-p \end{matrix}$

2.   Set $S_{ll} \, I_2 + S_a \, A_r \Rightarrow A_o$

3.a   Partition B_r into $p \times m$ blocks:

$$B_r \Rightarrow \begin{bmatrix} X_0 & X_1 & \cdots & X_{nx} \end{bmatrix} \qquad \text{where each block has } m\text{-}$$

columns, and similarly,

3.b   Partition A_r into two blocks:

$A_r \Rightarrow \begin{bmatrix} A_{r0} & A_{r1} \end{bmatrix}$ where the block A_{r0} has dimensions $p \times p$.

4.   Concatenate $\begin{bmatrix} X_0 \\ \vdots \\ X_{nx} \end{bmatrix} \Rightarrow B_c$, a $(n_x+1)p \times m$ matrix

5.   Calculate the $n \times (n_x+1)m$ controllability matrix Q_c of the pair $\{A_o, S_a\}$ having n_x+1 blocks, i.e.

$$Q_c = \begin{bmatrix} S_a & A_o S_a & \cdots & (A_o)^{n_x} S_a \end{bmatrix}$$

6.   Set $Q_c \, B_c \Rightarrow B_o$
7.   Set $C_o \, A_o^{-1} \, B_o - A_{r0}^{-1} \, X_o \Rightarrow D_o$
8.   Set $y_{10} - H_1 \, u_0 \Rightarrow x(0)$ (from (A9.22)).

As was stated in Appendix A7, this algorithm is given symbolically by:

$$u(k), \ y(k) \ (IDMV) \Rightarrow A_o, \ B_o, \ C_o, \ D_o, \ n_o, \ x(0)$$

Any steps that seem obscure can be said to have been arrived at by "divine inspiration".

Discrete-Time Control Systems Design via Loop Transfer Recovery

Tadashi Ishihara

Graduate School of Information Sciences
Tohoku University, Sendai 980-77, Japan

I. INTRODUCTION

Loop transfer recovery (LTR) techniques found by Kwakernaak [1] and Doyle and Stein [2] provide systematic procedures for determining design parameters of LQG controllers. Although LQG controllers have defects, which have stimulated the emergence of H∞ control theory, LQG/LTR methods are still attractive for control system designers. Stein and Athans [3] have given an excellent exposition of the flexibility of LQG/LTR methods in shaping feedback properties.

Most theoretical results on the LTR techniques have been given in continuous-time formulation. On the other hand, recent advances in digital technology have promoted use of digital controllers. Direct discretization of a continuous-time controller is often used to implement a digital controller. However, this approach overlooks inherent limitations of digital control systems; digital controllers usually include computation delays; the discretization of a continuous-time plant model with a zeroth order holder often yields a non-minimum phase discrete-time model even if the continuous-time model is minimum phase [4]. Discrete-time LTR techniques taking account of these issues have been discussed [5]-[9].

The purpose of this contribution is to provide a tutorial discussion on discrete-time LTR techniques together with some new results. We discuss feedback properties achieved at the input side of a non-minimum phase plant model with a controller using a prediction type observer to admit full sampling delay. Two LTR techniques for non-minimum phase plants are considered: one is the formal application of the conventional LTR technique for minimum phase plants and the other is a discrete-time version of the partial LTR technique proposed by Moore and Xia [10] for the continuous-time case. Since integral controllers have been widely used in practical applications, we discuss the application to an integral controller design as well as a standard LQG design.

In Section II, we discuss discrete-time LTR techniques for standard LQG controllers. We show that the feedback property achieved by the formal application of the conventional LTR technique can be achieved by the partial LTR technique. In addition, we show that the partial LTR technique provides more design freedom than the enforcement of the conventional LTR technique. The application to an integral controller design is discussed in Section III. The techniques used in Section II are applied to an efficient integral controller design [11]. In Section IV, we give numerical examples which make use of the design freedom provided by the partial LTR technique. Concluding remarks are given in Section V.

II. DISCRETE-TIME LQG/LTR METHOD

A. DISCRETE-TIME LQG CONTROLLER

Consider a discrete-time plant model given by

$$x(t+1) = Ax(t) + Bu(t) + w(t), \quad y(t) = Cx(t) + v(t), \tag{1}$$

where $x(t) \in R^n$ is the state vector, $u(t) \in R^m$ is the control vector and $y(t) \in R^m$ is the output vector. We assume that the pair (A, B) is controllable and (C, A) is observable. In addition, the disturbance $w(t)$ and the sensor noise $v(t)$ are assumed to be zero mean white noise processes with covariance matrices Q and R, respectively.

For the model described by (1), we introduce a quadratic performance index

$$V^\infty = \lim_{T \to \infty} E\{ \frac{1}{T} \sum_{t=0}^{T-1} [y'(t)y(t) + \rho u'(t)u(t)] \}, \tag{2}$$

where ρ is a non-negative scalar. It is well-known that the optimal controller minimizing the performance index (2) is given by

$$u(t) = -F \hat{x}(t|t-1), \tag{3}$$

where F is the optimal feedback gain matrix for the LQ problem and $\hat{x}(t|t-1)$ is the optimal 1-step ahead prediction of the state generated by the prediction type Kalman filter. The optimal feedback gain matrix F is given by

$$F = (\rho I + B'PB)^{-1} B' PA, \tag{4}$$

where P is the non-negative definite solution of the algebraic Riccati equation

$$P = A'PA - A'PB(\rho I + B'PB)^{-1} B'PA + C'C. \tag{5}$$

The algorithm of the prediction type Kalman filter is given by

$$\hat{x}(t+1|t) = A\hat{x}(t|t-1) + Bu(t) + K[y(t) - C\hat{x}(t|t-1)], \tag{6}$$

where K is the Kalman filter gain matrix given by

$$K = AM C'(CMC' + R)^{-1}, \tag{7}$$

with M being a non-negative definite solution of the algebraic Riccati equation

$$M = AMA' - AM C'(CMC' + R)^{-1} CMA' + Q. \tag{8}$$

To discuss LTR techniques, we use sensitivity matrices instead of loop transfer function matrices. For the LQG control system, we can easily obtain the following expression for the sensitivity matrix at the plant input.

Lemma 1: Consider the control system consisting of the plant described by (1) and the LQG controller given by (3). Then the sensitivity matrix $\Sigma(z)$ at the plant input can be decomposed as

$$\Sigma(z) = S(z)[I + F(zI - A + KC)^{-1} B], \tag{9}$$

where

$$S(z) = [I + F(zI - A)^{-1} B]^{-1}, \tag{10}$$

which is the sensitivity matrix for the state feedback case.

B. FORMAL APPLICATION OF THE LTR PROCEDURE

Maciejowski [5] has shown that the perfect recovery of a target property is possible for discrete-time LQG controllers using a filtering type Kalman filter provided the discrete-time plant model is minimum phase. For non-minimum phase plants, he has pointed out that the discrete-time LTR techniques are still useful to shape feedback properties in the frequency region where the effect of the unstable zeros is insignificant. Also he has pointed out that the perfect recovery is impossible for discrete-time LQG controllers using a prediction type Kalman filter. Ishihara and Takeda [6] have given a simple system-theoretic interpretation for this case; for minimum phase plant models, the feedback property achieved at the plant input side can be interpreted as that achieved by a predictor-based LQ controller [12]; the dual result holds for the plant output side. Recently, Zheng et al. [8] have used this interpretation to clarify the feedback property achieved by discrete-time LTR techniques using a prediction type reduced order observer.

For the LQG controller given by (3)-(8), we first discuss sensitivity property obtained by applying the conventional LTR procedure for the plant input side. Namely, we are interested in the asymptotic behavior of the sensitivity matrix (9) obtained by letting $\sigma \to \infty$ for the covariance matrices

$$Q = \sigma BB', \quad R = I. \tag{11}$$

In the case that the plant transfer function matrix $C(zI - A)^{-1}B$ is non-minimum phase, we need to decompose the transfer function matrix into all-pass and minimum phase parts. As a discrete-time version of Enns' result [13] for the continuous-time case, Zhang and Freudenberg [7] have given an expression for general multiple unstable zeros. However, their expression is somewhat complicated. Assuming for simplicity that the plant model has only one real unstable zero, we have the following result.

Lemma 2: Assume that the transfer function matrix $G(z) = C(zI - A)^{-1}B$ has a single real unstable zero q ($|q| > 1$). Let η and ξ denote the unit zero direction input vector and the zero direction state vector for the unstable zero q, respectively, i.e.,

$$\begin{bmatrix} qI - A & -B \\ -C & 0 \end{bmatrix} \begin{bmatrix} \xi \\ \eta \end{bmatrix} = 0, \quad \eta'\eta = 1. \tag{12}$$

Then the transfer function matrix $G(z)$ can be decomposed as

$$G(z) = C(zI - A)^{-1} B_m G_a(z),$$ (13)

where $C(zI - A)^{-1} B_m$ is minimum phase part and $G_a(z)$ is all-pass part satisfying $G_a(z)G_a'(z^{-1}) = I$. The matrix B_m is defined by

$$B_m = B_q J_q$$ (14)

where

$$B_q = B - (q - \frac{1}{q})\xi\eta', \quad J_q = I - (q+1)\eta\eta'.$$ (15)

The all-pass part $G_a(z)$ is defined by

$$G_a(z) = (I - \eta\eta') + a(z)\eta\eta',$$ (16)

where

$$a(z) = \frac{q-z}{qz-1}.$$ (17)

Proof: See Appendix. ◆

The following result for the asymptotic behavior of the optimal filter gain matrix is well-known [14] and is indispensable for our discussion.

Lemma 3: Consider the optimal filter gain matrix K defined in (7) for the covariance matrices (11). As $\sigma \to \infty$, the optimal filter gain approaches

$$K^* = AB_m(CB_m)^{-1}.$$ (18)

Remark: It can easily be check by use of Lemma 2 that the non-singularity of the matrix CB_m in (18) is guaranteed if CB is non-singular.

From Lemmas 2 and 3, we have the following expression for the sensitivity matrix obtained by formal application of the LTR procedure.

Proposition 1: Consider the control system consisting of the non-minimum phase plant (1) and the LQG controller (3) for the performance index (2) and the covariance matrices (11). As $\sigma \to \infty$, the sensitivity matrix (9) approaches

$$\Sigma^*(z) = S(z)\{I + F(zI - A)^{-1}[B - B_m G_a(z)] + z^{-1}FB_m G_a(z)\} . \tag{19}$$

Proof: The following matrix identity holds by the well-known matrix inversion lemma.

$$F(zI - A + K^*C)^{-1}B = F(zI - A)^{-1}B - F(zI - A)^{-1}K^* \tag{20}$$
$$[I + C(zI - A)^{-1}K^*]^{-1}C(zI - A)^{-1}B$$

Substituting (18) into (20), we obtain

$$F(zI - A + K^*C)^{-1}B = F(zI - A)^{-1}B - F(zI - A)^{-1}AB_m \tag{21}$$
$$[zC(zI - A)^{-1}B_m]^{-1}C(zI - A)^{-1}B.$$

Noting that

$$[C(zI - A)^{-1}B_m]^{-1}C(zI - A)^{-1}B = G_a(z), \tag{22}$$

we have

$$F(zI - A + K^*C)^{-1}B = z^{-1}FB_m G_a(z) + F(zI - A)^{-1}[B - B_m G_a(z)]. \tag{23}$$

The expression (19) readily follows from (9) and (23). ◆

The expression for the minimum phase can easily be obtained as a special case of (19) by assuming that $B_m = B$ and $G_a(z) = I$.

Corollary 1: If the plant (1) is minimum phase, the expression (19) for the input sensitivity matrix is reduced to

$$\Sigma^P(z) = S(z)(I + z^{-1}FB), \tag{24}$$

which corresponds to the sensitivity matrix achieved by the predictor-based LQ regulator [6].

As a directional property of the achievable sensitivity matrix (19), we have the following result.

Proposition 2: The sensitivity matrices (19) and (24) satisfy the relation

$$\Sigma^*(z)(I - \eta\eta') = \Sigma^P(z)(I - \eta\eta'). \tag{25}$$

Proof: It follows from (15)-(17) that

$$J_q G_a(z) = (I - \eta\eta') - \frac{q(q-z)}{qz-1}\eta\eta'. \tag{26}$$

Using (12), (14) and (26), we obtain

$$B_m G_a(z) = B - \frac{q^2-1}{qz-1}[B\eta - (q-z)\xi]\eta'. \tag{27}$$

Noting that $\eta'(I - \eta\eta') = 0$, we have

$$B_m G_a(z)(I - \eta\eta') = B(I - \eta\eta'). \tag{28}$$

Using (19), (24) and (28), we can readily check that the relation (25) holds. ◆

The above result implies that, for the input direction orthogonal to the zero input direction, the target feedback property can be recovered.

C. PARTIAL LTR TECHNIQUE

Ishihara et al.[15] have applied the partial loop transfer recovery technique for discrete-time plants under general feedback delays. Here we give the discrete-time result simplified to the unit feedback delay case.

Let $\{A_a, B_a, C_a, D_a\}$ denote a state-space representation of the all-pass part $G_a(z)$ in the decomposition (13). Define the new state vector and the disturbance vector as

$$\chi(t) = [\, x_m'(t) \quad x_a'(t)]', \quad \omega(t) = [\, w_f'(t) \quad w'(t)]', \tag{29}$$

where $x_m(t)$ and $x_a(t)$ are state vectors corresponding the minimum phase and all-pass part, respectively, and $w_f(t)$ is a new disturbance vector. The covariance matrix of $\omega(t)$ is denoted by Ω. Then we can construct the following state-space model.

$$\chi(t+1) = \Phi\chi(t) + \Gamma u(t) + \Theta\omega(t),$$
$$y(t) = H\chi(t) + v(t), \qquad (30)$$

where

$$\Phi = \begin{bmatrix} A & B_m C_a \\ 0 & A_a \end{bmatrix}, \quad \Gamma = \begin{bmatrix} B_m D_a \\ B_a \end{bmatrix}, \quad \Theta = \begin{bmatrix} B_m & B_m D_a \\ 0 & B_a \end{bmatrix}, \quad H = \begin{bmatrix} C & 0 \end{bmatrix}. \qquad (31)$$

1. Minimum Phase State Feedback Controller

Assuming the perfect state observation and the existence of the unit feedback delay, we can construct a LQ controller for the model (30) and (the deterministic version of) the performance index (2). Then the optimal control input is given by

$$u(t+1) = -F_m \bar{x}_m (t+1|t) - F_a \bar{x}_a (t+1|t), \qquad (32)$$

where $\bar{x}_m(t+1|t)$ and $\bar{x}_a(t+1|t)$ are the optimal 1-step ahead prediction of the minimum phase state and all-pass state, respectively, and

$$\Psi = \begin{bmatrix} F_m & F_a \end{bmatrix} = (\rho I + \Gamma'\Pi\Gamma)^{-1}\Gamma'\Pi\Phi, \qquad (33)$$

with Π being a non-negative definite solution of the algebraic Riccati equation

$$\Pi = \Phi'\Pi\Phi - \Phi'\Pi\Gamma(\rho I + \Gamma'\Pi\Gamma)^{-1}\Gamma'\Pi\Phi + H'H. \qquad (34)$$

Since the pair (Φ, Γ) is stabilizable and (H, Φ) is detectable provided (A, B, C) is a minimal realization, the existence of the non-negative definite solution of (34) and the asymptotic stability of the matrix $(\Phi - \Gamma\Psi)$ are guaranteed by the well-known result for the Riccati equation.

Note that $\bar{x}_m(t+1|t)$ and $\bar{x}_a(t+1|t)$ are given by

$$\bar{x}_m(t+1|t) = A_m x_m(t) + B_m C_a x_a(t) + B_m D_a u(t),$$
$$\bar{x}_a(t+1|t) = A_a x_a(t) + B_a u(t), \qquad (35)$$

respectively, and that the z-transform of $x_a(t)$ can be expressed as

$$x_a(z) = (zI - A_a)^{-1} B_a u(z). \qquad (36)$$

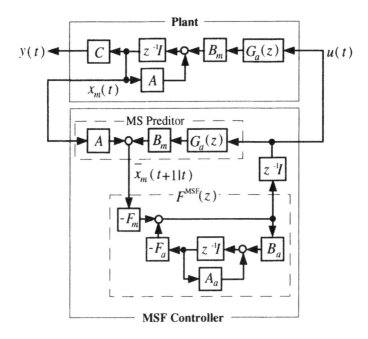

Fig. 1. Structure of MSF controller.

Using (36) in (35), we can express the z-transform of (32) as

$$u(z) = -F^{MSF}(z)\overline{x}_m(z),\qquad(37)$$

where $u(z)$ and $\overline{x}_m(z)$ are the z-transform of $u(t)$ and that of $\overline{x}_m(t|t-1)$, respectively, and

$$F^{MSF}(z) = [I + F_a(zI - A_a)^{-1}B_a]^{-1}F_m.\qquad(38)$$

Since the controller (37) feeds back only the minimum phase state prediction with the frequency-shaped feedback gain matrix (38), we call (37) the minimum phase state feedback (MSF) controller. The structure of the MSF controller is shown in Fig. 1. The controller consists of the minimum phase state (MS) predictor, the frequency-shaped feedback gain matrix $F^{MSF}(z)$ and the unit feedback delay.

The input sensitivity matrix achieved by the MSF controller is expressed as follows.

Lemma 4: Assume that the minimum phase state $x_m(t)$ in (29) is perfectly measurable. Then the input sensitivity matrix of the control system consisting of the plant model (30) and the MSF controller (37) is given by

$$\Sigma^{MSF}(z) = \Sigma_0^{MSF}(z)[I + z^{-1}F^{MSF}(z)B_m G_a(z)], \qquad (39)$$

where

$$\Sigma_0^{MSF}(z) = [I + F^{MSF}(z)(zI - A)^{-1}B_m G_a(z)]^{-1} \qquad (40)$$

is the sensitivity matrix achieved by the minimum phase state prediction feedback without the feedback delay. In terms of the optimal feedback gain matrices F_m and F_a, the expression (39) can be rewritten as

$$\Sigma^{MSF}(z) = \Sigma_0(z)[I + F_a(zI - A_a)^{-1}B_a + z^{-1}F_m B_m G_a(z)], \qquad (41)$$

where

$$\begin{aligned}
\Sigma_0(z) &= [I + \Psi(zI - \Phi)^{-1}\Gamma]^{-1} \\
&= [I + F_a(zI - A_a)^{-1}B_a + F_m(zI - A)^{-1}B_m G_a(z)]^{-1}
\end{aligned} \qquad (42)$$

is the sensitivity matrix achieved by the perfect state observation without the feedback delay.

Proof: The expression (39) can easily be obtained by the direct matrix calculation by noting that the controller transfer function matrix from $x_m(t)$ to $u(t)$ is given by

$$C^{MSF}(z) = -z^{-1}[I + F_a(zI - A_a)^{-1}B_a + z^{-1}F_m B_m G_a(z)]F_m A. \qquad (43)$$

An easier way to check (39) is to note that the right side of (39) can be obtained by replacing F and B in (24) with the frequency-shaped matrices $F^{MSF}(z)$ and $B_m G_a(z)$, respectively. The expression (41) readily follows from (38), (39) and (40). ◆

Remark: The sensitivity matrix achieved by the full state prediction feedback (32) is given by

$$\Sigma_1(z) = \Sigma_0(z)(I + z^{-1}\Psi\Gamma), \qquad (44)$$

which is different form (41).

2. Minimum Phase Estimator Feedback Controller

For the output feedback case, the minimum phase state feedback can be realized by use of an observer. For the non-minimal stochastic model (30), we can construct a prediction type Kalman filter as

$$\hat{\chi}(t+1|t) = \Phi\hat{\chi}(t|t-1) + \Gamma u(t) + K[y(t) - H\hat{\chi}(t|t-1)], \tag{45}$$

where K is the Kalman filter gain matrix defined by

$$K = \Phi\Xi H'(H\Xi H' + R)^{-1} \tag{46}$$

with Ξ being non-negative definite solution of the algebraic Riccati equation

$$\Xi = \Phi\Xi\Phi' - \Phi\Xi H'(H\Xi H' + R)^{-1}H\Xi\Phi' + \Theta\Omega\Theta'. \tag{47}$$

Define the partitions of the matrices K and Ξ as

$$K = \begin{bmatrix} K_m \\ K_a \end{bmatrix}, \qquad \Xi = \begin{bmatrix} \Xi_m & \Xi_{ma} \\ \Xi_{am} & \Xi_a \end{bmatrix}. \tag{48}$$

Using (31), (46) and (48), we can express the filter gain matrices as

$$K_m = (A\Xi_m C' + B_m C_a \Xi_{am} C')(C\Xi_m C' + R)^{-1}, \tag{49}$$
$$K_a = A_a \Xi_{am} C'(C\Xi_m C' + R)^{-1}. \tag{50}$$

For a special choice of the covariance matrix Ω, we have the following result which can easily be obtained as a discrete-time version of Lemma 1 in [10].

Lemma 5: For the covariance matrix Ω of the disturbance vector $\omega(t)$ given by

$$\Omega = \text{diag}[\mu_m I \quad \mu_a I], \tag{51}$$

where μ_m and μ_a are non-negative scalars, the Riccati equation (47) reduces to

$$\Xi_m = A\Xi_m A' - A\Xi_m C'(C\Xi_m C' + R)^{-1}C\Xi_m A' \\ \qquad + B_m(\mu_m I + \mu_a D_a D_a' + C_a \Xi_a C_a')B_m', \tag{52}$$

$$\Xi_a = A_a \Xi_a A_a' + \mu_a B_a B_a', \tag{53}$$

and $\Xi_{ma} = \Xi_{am}' = 0$. Consequently, the filter gain matrices are given by

$$K_m = A \Xi_m C' (C \Xi_m C' + R)^{-1}, \quad K_a = 0. \tag{54}$$

Remark: It is worth noting that the optimal filter gain matrix K_m depends on the state space matrices A_a, B_a for the all-pass part. This is not the case for the continuous-time case.

Applying the above result to (45), we can construct an optimal estimator for the minimum phase part as

$$\hat{x}_m(z) = (zI - A + K_m C)^{-1} [K_m y(z) + B_m G_a(z) u(z)], \tag{55}$$

where $\hat{x}_m(z)$, $u(z)$ and $y(z)$ is the z-transforms of $\hat{x}_m(t|t-1)$, $u(t)$ and $y(t)$, respectively. Using the frequency-shaped feedback gain matrix (38) and the estimator (55), we can construct an output feedback version of (37) as

$$u(z) = -F^{\text{MSF}}(z) \hat{x}_m(z), \tag{56}$$

which we call the minimum phase estimate feedback (MEF) controller. The structure of the MEF controller is shown in Fig. 2.

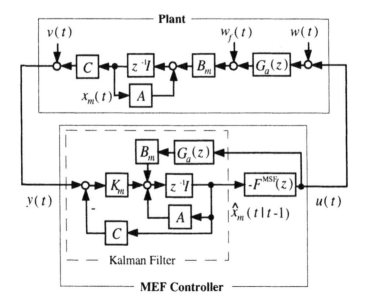

Fig. 2. Structure of MEF controller.

For the MEF controller, we have the following result for the sensitivity matrix at the plant input.

Lemma 6: The sensitivity matrix at the plant input for the control system consisting of the plant model (30) and the MEF controller (56) can be expressed as

$$\Sigma^{\text{MEF}}(z) = \Sigma_0^{\text{MSF}}(z)[I + F^{\text{MSF}}(z)(zI - A + K_m C)^{-1} B_m G_a(z)], \qquad (57)$$

where $F^{\text{MSF}}(z)$ and $\Sigma_0^{\text{MSF}}(z)$ are defined in (38) and (40), respectively.

Proof: The expression (57) can be obtained by direct matrix calculation. An easier way to check (57) is to note that the right side of (57) can be obtained by replacing F and B in the expression (9) in Lemma 1 with the frequency-shaped matrices $F^{\text{MSF}}(z)$ and $B_m G_a(z)$, respectively. ◆

3. Partial Loop Transfer Recovery

The following result for the MEF controller shows that the feedback property achieved by the MSF controller can be recovered.

Proposition 3: Consider the control system consisting of the plant model (30) and the MEF controller designed by the covariance matrix (51). Then the sensitivity matrix achieved by the MSF controller can be recovered by increasing the covariance matrix of the fictitious disturbance inserted into the input of the minimum phase part, i.e.,

$$\lim_{\mu_m \to \infty} \Sigma^{\text{MEF}}(z) = \Sigma^{\text{MSF}}(z), \qquad (58)$$

where $\Sigma^{\text{MSF}}(z)$ and $\Sigma^{\text{MEF}}(z)$ are defined by (39) and (57), respectively.

Proof: Applying the well known result for discrete-time Riccati equations to (52) in Lemma 5, we can explicitly express the asymptotic value of the filter gain matrix for the minimum phase part as

$$\lim_{\mu_m \to \infty} K_m = K^* = AB_m(CB_m)^{-1}. \qquad (59)$$

Note that

$$(A - K^* C)B_m = [A - AB_m(CB_m)^{-1}C]B_m = 0. \qquad (60)$$

In addition, we can write (57) as

$$\Sigma^{MEF}(z) = \Sigma_0^{MSF}(z)[I + z^{-1} F^{MSF}(z) B_m G_a(z)$$
$$+ z^{-1}(zI - A + K_m C)^{-1}(A - K_m C) B_m G_a(z)]. \tag{61}$$

Using (60) in (61), we have (58).   ♦

D. RELATION BETWEEN THE TWO LTR TECHNIQUES

We discuss the relation between the formal application of the conventional LTR and the partial LTR techniques. For the case that the plant has a single real unstable zero q, we have the following result for the two Riccati equations (5) and (34).

Lemma 7: Assume that the plant has a single real unstable zero q. As a realization of the all-pass transfer function matrix $G_a(z)$ of the decomposition (13), we take

$$A_a = \frac{1}{q}, \quad B_a = \frac{1}{q}(q - \frac{1}{q})\eta', \quad C_a = \eta, \quad D_a = I - (1 + \frac{1}{q})\eta\eta', \tag{62}$$

where η is the zero direction vector defined by (12). Define

$$\Pi_P = \begin{bmatrix} P & qP\xi \\ q\xi'P & q^2\xi'P\xi \end{bmatrix}, \tag{63}$$

where P is the positive definite solution of the Riccati equation (5). Then the matrix Π_P is non-negative definite and satisfies the Riccati equation (34) with the realization (62).

Proof: It can easily be checked that the state space parameters in (62) is a realization of the all-pass transfer function matrix (16). Note that the matrix defined by (63) can be expressed as

$$\Pi_P = N'PN, \tag{64}$$

where N is $n \times (n+1)$ matrix defined as

$$N = [I \quad q\xi]. \tag{65}$$

The non-negative definiteness of Π_P readily follows from (64). From (14), (15) and (62), we have

$$B_m D_a = B_q [I - (q+1)\eta\eta'][I - (1+\frac{1}{q})\eta\eta'] = B_q .$$
(66)

Using (66), we can write the matrix Γ in (31) as

$$\Gamma = \begin{bmatrix} B - (q-\frac{1}{q})\xi\eta' \\ \frac{1}{q}(q-\frac{1}{q})\eta' \end{bmatrix} .$$
(67)

From (65) and (67), we have

$$N\Gamma = B,$$
(68)

and therefore

$$\Gamma'\Pi_p\Gamma = \Gamma'N'PN\Gamma = B'PB .$$
(69)

Using (14), (15) and (62), we can show that

$$B_m C_a = B_q J_q C_a = -q[B\eta - (q-\frac{1}{q})\xi] .$$
(70)

It follows from (31), (62),(65) and (70) that

$$N\Phi = [A \quad q(q\xi - B\eta)] = [A \quad qA\xi] = AN .$$
(71)

In addition, multiplying C both sides of (65) and noting that $C\xi = 0$ holds by (12), we have

$$H = CN .$$
(72)

Using (64), (68), (69), (71) and (72), we finally have

$$\Phi'\Pi_p\Phi - \Phi'\Pi_p\Gamma(\rho I + \Gamma'\Pi_p\Gamma)^{-1}\Gamma'\Pi_p\Phi + H'H$$
$$= N'[A'PA - A'PB(\rho I + B'PB)^{-1}B'PA + C'C]N \qquad (73)$$
$$= N'PN = \Pi_p,$$

where we have used the fact that P satisfies the Riccati equation (5). Consequently, the matrix Π_p is a non-negative definite solution of the Riccati equation (34).   ♦

Using the above lemma, we can find explicit relation between the feedback gain matrices F and Ψ.

Lemma 8: Assume that the plant has a single real unstable zero q and that the optimal feedback gain matrix $\Psi = [\, F_m \quad F_a \,]$ defined by (33) is obtained from the Riccati equation (34) where the realization (62) for the all-pass part is used. Then the feedback gain matrices F_m and F_a can simply be expressed as

$$F_m = F, \quad F_a = qF\xi, \tag{74}$$

where ξ is a vector defined by (12) and F is the optimal feedback gain matrix defined by (4) with the Riccati equation (5).
Proof: Noting that the solution of the Riccati equation (34) can be expressed as (63) or (64) by Lemma 7, we have

$$\begin{aligned}
\Psi &= (\rho I + \Gamma' \Pi_p \Gamma)^{-1} \Gamma' \Pi_p \Phi \\
&= (\rho I + B'PB)^{-1} \Gamma' N' PN\Phi = (\rho I + B'PB)^{-1} B'PAN, \\
&= FN,
\end{aligned} \tag{75}$$

where we have used the identities (68), (69) and (71). Since the matrix N is defined by (65), the relation (74) readily follows from (75).   ♦

Now, we have the following result which clarifies the relation of the two LTR techniques for non-minimum phase plants.

Proposition 4: Under the quadratic performance index (2), the sensitivity matrix (19) achieved by the formal application of the conventional LTR procedure coincides with that achieved by the MEF controller, i.e.,

$$\Sigma^*(z) = \Sigma^{MSF}(z), \tag{76}$$

where $\Sigma^{MSF}(z)$ is given by (39).
Proof: To show (76), we first note that the following identity holds for the realization (62) and the feedback gain matrices (74) .

$$F_a(zI - A_a)^{-1} B_a = \frac{q^2 - 1}{qz - 1} F\xi\eta'. \tag{77}$$

On the other hand, noting that $(qI - A)\xi = B\eta$ by (12), we have

$$B_m G_a(z) = B - \frac{q^2 - 1}{qz - 1}(zI - A)\xi\eta', \qquad (78)$$

which is a rewritten version of the relation (27). Multiplying the both sides of (78) by $F(zI - A)^{-1}$, we obtain

$$F(zI - A)^{-1} B_m G_a(z) = F(zI - A)^{-1} B - \frac{q^2 - 1}{qz - 1}F\xi\eta'. \qquad (79)$$

Using (77) in (79), we have

$$F_a(zI - A_a)^{-1} B_a + F_m(zI - A)^{-1} B_m G_a(z) = F(zI - A)^{-1} B. \qquad (80)$$

It follows from (10), (42) and (80) that

$$\Sigma_0(z) = S(z). \qquad (81)$$

Noting that $F = F_m$ holds by Lemma 8, we can rewrite (80) as

$$F_a(zI - A_a)^{-1} B_a = F(zI - A)^{-1}[B - B_m G_a(z)]. \qquad (82)$$

Substitution of (81) and (82) into (41) and (42) yields (76) where $\Sigma^*(z)$ is defined by (19). ♦

The above result shows that the feedback property achieved by the formal application of the LTR technique can be achieved by the partial LTR technique. A meaning for the formal application of the conventional LTR procedure is obscure from the expression (19) for the sensitivity matrix $\Sigma^*(z)$ while the sensitivity matrix $\Sigma^{MSF}(z)$ has a clear system-theoretic meaning. By the above result, we can justify the enforcement of the conventional LTR procedure for non-minimum phase plants.

In addition, the above result suggests that, to recover the feedback property achieved by the MSF controller for the performance index (2), we need not use the partial LTR procedure which requires a higher order controller than the conventional LTR procedure. Note that this fact does not make the partial LTR technique useless for practical control systems design. The partial LTR procedure provides more design freedom than the conventional procedure. An example making use of this freedom will be given in Section IV.

III. INTEGRAL CONTROLLER DESIGN

A. OBSERVER-BASED INTEGRAL CONTROLLER DESIGN

Based on the novel state-feedback design proposed by Mita [12], Ishihara *et al.* [11] have proposed an efficient design of observer-based integral controllers accounting general feedback delays. This design does not require the solution of a regulator problem higher than the plant order. The application of the LTR technique for the minimum phase case has been discussed in [11].

In this section, we apply the LTR techniques for the non-minimum phase case discussed in the previous section to the integral controller design. We proceed almost parallel to the discussions in the previous section.

To guarantee the existence of a stabilizing integral controller, we assume that the plant model (1) has no zero at $z=1$. This assumption is equivalent to the non-singularity of the matrix

$$E = \begin{bmatrix} A - I & B \\ C & 0 \end{bmatrix}. \tag{83}$$

We also assume that the integral controller includes full sampling delay. For the plant model (1), the algorithm of the integral controller is given by

$$u(t+1) = s(t+1) - L\hat{x}(t+1|t), \quad s(t+1) = s(t) + T[r(t) - y(t)], \tag{84}$$

where $r(t)$ is a step reference input, $s(t)$ is the state of the integrator and $\hat{x}(t+1|t)$ is the estimate generated by the prediction type Kalman filter (6). The matrices L and T in (84) are determined by the linear matrix equation

$$[L \quad T]E = [FA \quad I + FB], \tag{85}$$

where F is the state feedback gain matrix of a regulator problem for the plant (1) and E is the matrix defined by (83). The structure of the integral controller is shown in Fig. 3.

The transfer function matrix from $r(t)$ to $y(t)$ is given by

$$G_{ry}(z) = z^{-1}C(zI - A + BF)^{-1}B[C(I - A + BF)^{-1}B]^{-1}, \tag{86}$$

which explicitly includes F.

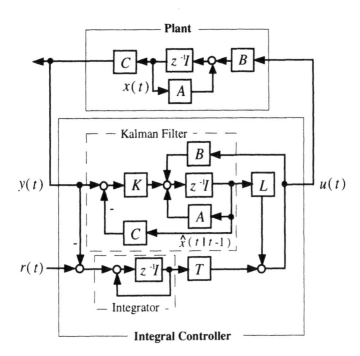

Fig. 3. Observer-based integral controller.

For the control system shown in Fig. 3, we can easily obtain the following expression for the input sensitivity matrix.

Lemma 9: Consider the control system consisting of the plant described by (1) and the integral controller given by (84). Then the sensitivity matrix $\Sigma_{INT}(z)$ at the plant input can be decomposed as

$$\Sigma_{INT}(z) = z^{-1}(z-1)S(z)[I + L(zI - A + KC)^{-1}B],   (87)$$

where $S(z)$ is the sensitivity matrix for the state feedback regulator defined by (10).
Proof: Omitted.   ◆

Note that the expression (87) includes the sensitivity matrix of the regulator used for the design and explicitly shows the effect of the integrators, i.e., $\Sigma_{INT}(1) = 0$.

B. FORMAL APPLICATION OF THE LTR PROCEDURE

For the observer-based integral controller, a target for the LTR at the plant input is a feedback property achieved by the integral controller using the state predictor [11]. The target feedback property can not be recovered for the non-minimum case. However, we have the following expression for the achievable sensitivity matrix using Lemmas 3 and 9.

Proposition 5: Consider the control system consisting of the non-minimum phase plant (1) and the integral controller (84) with the Kalman filter designed for the covariance matrices (11). As $\sigma \to \infty$, the sensitivity matrix (87) approaches

$$\Sigma^*_{\text{INT}}(z) = z^{-1}(z-1)S(z)\{I + L(zI - A)^{-1}[B - B_m G_a(z)] + z^{-1}LB_m G_a(z)\}, \quad (88)$$

where B_m and $G_a(z)$ are defined by the decomposition (13).

Proof: Noting that the identity (23) in the proof for Proposition 1 holds for arbitrary F, we have

$$L(zI - A + K^*C)^{-1}B = z^{-1}LB_m G_a(z) + L(zI - A)^{-1}[B - B_m G_a(z)]. \quad (89)$$

From (87) and (89), we have (88). ◆

The expression for the minimum phase case given in [11] can easily be obtained as a special case of (88) by assuming that $B_m = B$ and $G_a(z) = I$.

Corollary 2: If the plant model (1) is minimum phase, the expression (88) for the input sensitivity matrix is reduced to

$$\Sigma^p_{\text{INT}}(z) = z^{-1}(z-1)S(z)(I + z^{-1}LB), \quad (90)$$

which corresponds to the sensitivity matrix achieved by the target, i.e., the integral controller using the state-predictor [11].

As a directional property of the achievable sensitivity matrix (88), we have the result similar to Proposition 2 for the LQG case.

Proposition 6: The sensitivity matrices (88) and (90) satisfy the relation

$$\Sigma^*_{\text{INT}}(z)(I - \eta\eta') = \Sigma^p_{\text{INT}}(z)(I - \eta\eta'). \quad (91)$$

Proof: Omitted.   ◆

C. APPLICATION OF PARTIAL LTR TECHNIQUE

First, we construct an integral controller for the extended state space model (30) assuming the perfect state observation with full sampling delay. Then the controller is modified such that only the minimum phase state is used for the feedback. The feedback property of the resulting controller is a target of the partial LTR technique and is shown to be recoverable as in LQG case.

1. Minimum Phase State Feedback Integral Controller

Consider the state feedback integral controller given by

$$
\begin{aligned}
u(t+1) &= \tau(t+1) - \Lambda[\Phi\chi(t) + \Gamma u(t)], \\
\tau(t+1) &= \tau(t) + M[r(t) - y(t)],
\end{aligned}
\tag{92}
$$

where $\tau(t)$ is the state of the integrators. We determine Λ and M by

$$
[\Lambda \quad M]\mathcal{E} = [\Psi\Phi \quad I + \Psi\Gamma],
\tag{93}
$$

where Ψ is the feedback gain matrix for a regulator problem for the extended plant model (30) and

$$
\mathcal{E} = \begin{bmatrix} \Phi - I & \Gamma \\ H & 0 \end{bmatrix}.
\tag{94}
$$

Note that the non-singularity of the matrix (94) is guaranteed if the matrix E defined in (83) is non-singular.

Define the partition the state feedback matrix Λ as

$$
\Lambda = [L_m \quad L_a],
\tag{95}
$$

where L_m and L_a are feedback gain matrix for the minimum phase state and that for the all-pass state, respectively. Using (31) and (95) in (92), we obtain

$$
\begin{aligned}
u(t+1) &= \tau(t+1) - L_m\{Ax_m(t) + B_m[C_a x_a(t) + D_a u(t)]\} \\
&\quad - L_a[A_a x_a(t) + B_a u(t)].
\end{aligned}
\tag{96}
$$

Neglecting the disturbance and the sensor noise in (30), we have the following z-

transform relations

$$x_a(z) = (zI - A_a)^{-1} B_a u(z), \quad C_a x_a(z) + D_a u(z) = G_a(z)u(z). \tag{97}$$

From the second equation in (92), we can write the z-transform of $\tau(t)$ as

$$\tau(z) = z(z-1)^{-1} M[r(z) - y(z)], \tag{98}$$

where $r(z)$ and $y(z)$ are z-transforms of $r(t)$ and $y(t)$, respectively. Using (97) and (98), we can express the z-transform of (96) as

$$zu(z) = z(z-1)^{-1} M[r(z) - y(z)] - L_m[Ax_m(z) + B_m G_a(z)u(z)] \\ - zL_a(zI - A_a)^{-1} B_a u(z). \tag{99}$$

Define the transfer function matrices

$$T(z) = \Delta_a(z)M, \quad L(z) = \Delta_a(z)L_m, \tag{100}$$

where

$$\Delta_a(z) = [\, I + L_a(zI - A_a)^{-1} B_a\,]^{-1}. \tag{101}$$

Then we can rewrite (99) as

$$u(z) = (z-1)^{-1} T(z)[r(z) - y(z)] - z^{-1} L(z)[Ax_m(z) + B_m G_a(z)u(z)]. \tag{102}$$

Note that the term $Ax_m(z) + B_m G_a(z)u(z)$ can be regarded as the 1-step ahead prediction of the minimum phase state. Since the controller (102) feeds back only the minimum phase state prediction with the frequency-shaped feedback gain matrix (100), we call (102) the minimum phase state feedback (MSF) integral controller. The structure of this controller is shown in Fig. 4.

For the MSF integral controller, we have the following result corresponding to Lemma 4.

Lemma 9: Assume that the minimum phase state $x_m(t)$ in (30) is perfectly measurable. Then the input sensitivity matrix of the control system consisting of the plant model (30) and the MSF integral controller (102) is given by

$$\Sigma_{INT}^{MSF}(z) = z^{-1}(z-1)\Sigma_0(z)[I + L_a(zI - A_a)^{-1} B_a + z^{-1} L_m B_m G_a(z)], \tag{103}$$

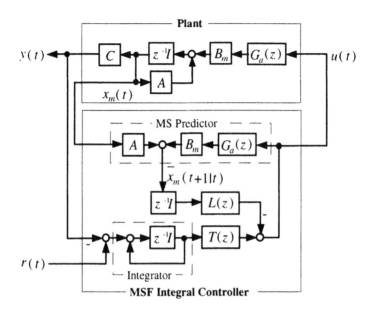

Fig. 4. MSF integral controller.

where $\Sigma_0(z) = [I + \Psi(zI - \Phi)^{-1}\Gamma]^{-1}$ is the input sensitivity matrix for the regulator with the state feedback gain matrix Ψ under the perfect state observation without the feedback delay.

Proof: Assuming that $r(z)=0$ in (102), we can easily show that the controller transfer function matrix from $x_m(t)$ to $u(t)$ is given by

$$C_{\text{INT}}^{\text{MSF}}(z) = -[I + z^{-1}L(z)B_mG_a(z)]^{-1}[(z-1)^{-1}T(z)C + z^{-1}L(z)A]. \quad (104)$$

Noting that the loop transfer function matrix at the plant input is given by $C_{\text{INT}}^{\text{MSF}}(z)(zI - A)^{-1}B_m$, we can write the sensitivity matrix as

$$\Sigma_{\text{INT}}^{\text{MSF}}(z) = [I + L(z)(zI - A)^{-1}B_mG_a(z) + (z-1)^{-1}T(z)G(z)]^{-1} \\ [I + z^{-1}L(z)B_mG_a(z)]. \quad (105)$$

Substituting (95) and (100) into (105), we can rewrite (105) as

$$\Sigma_{\text{INT}}^{\text{MSF}}(z) = D_{\text{MSF}}^{-1}(z)N_{\text{MSF}}(z), \quad (106)$$

where

$$D_{MSF}(z) = I + L_a(zI - A_a)^{-1}B_a + L_m(zI - A)^{-1}B_m G_a(z)$$
$$+ (z-1)^{-1}MG(z),$$

(107)

$$N_{MSF}(z) = I + L_a(zI - A_a)^{-1}B_a + z^{-1}L_m B_m G_a(z).$$

(108)

Using the state space matrices defined in (31), we can rewrite (107) as

$$D_{MSF}(z) = I + \Lambda(zI - \Phi)^{-1}\Gamma + (z-1)^{-1}MH(zI - \Phi)^{-1}\Gamma.$$

(109)

Notice that the solution of the matrix equation (93) can explicitly be given by

$$\Lambda = \Psi + (I - \Phi + \Gamma\Psi)^{-1}\Gamma[H(I - \Phi + \Gamma\Psi)^{-1}\Gamma]^{-1},$$
$$M = [H(I - \Phi + \Gamma\Psi)^{-1}\Gamma]^{-1}.$$

(110)

Using the explicit solutions (110) in (109), we have

$$D_{MSF}(z) = z(z-1)^{-1}[I + \Psi(zI - \Phi)^{-1}\Gamma].$$

(111)

The expression (103) follows from (106), (108) and (111).   ◆

2. Minimum Phase Estimator Feedback Integral Controller

By replacing $z^{-1}[Ax_m(z) + B_m G_a(z)u(z)]$ in (102) with the estimate $\hat{x}_m(z)$ generated by the prediction type Kalman filter (55) for the minimum phase state, we can construct the output feedback integral controller

$$u(z) = (z-1)^{-1}T(z)[r(z) - y(z)] - L(z)\hat{x}_m(z).$$

(112)

We call the controller (112) the minimum phase estimator feedback (MEF) integral controller.

The input sensitivity matrix for the MEF integral controller can be expressed as follows.

Lemma 10: The sensitivity matrix at the plant input for the control system consisting of the plant model (30) and the MEF integral controller (112) can be expressed as

$$\Sigma_{INT}^{MEF}(z) = z^{-1}(z-1)\Sigma_0(z)\{I + L_a(zI - A_a)B_a(z)$$
$$+ z^{-1}L_m[I + (zI - A + K_mC)^{-1}(A - K_mC)]B_mG_a(z)\}, \tag{113}$$

where $\Sigma_0(z) = [I + \Psi(zI - \Phi)^{-1}\Gamma]^{-1}$.

Proof: It follows from (55) and (112) that the transfer function matrix from $y(t)$ to $u(t)$ is given by

$$C_{INT}^{MEF}(z) = -[I + L(z)(zI - A + K_mC)^{-1}B_mG_a(z)]^{-1}$$
$$[(z-1)^{-1}T(z) + L(z)(zI - A + K_mC)^{-1}K_m]. \tag{114}$$

Using (95), (100) and (114), we can rewrite the input sensitivity matrix as

$$\Sigma_{INT}^{MEF}(z) = D_{MEF}^{-1}(z)N_{MEF}(z), \tag{115}$$

where

$$D_{MEF}(z) = I + L_a(zI - A_a)^{-1}B_a + L_m(zI - A + K_mC)^{-1}$$
$$[B_m + K_mG_m(z)]G_a(z) + (z-1)^{-1}MG(z), \tag{116}$$

$$N_{MEF}(z) = I + L_a(zI - A_a)^{-1}B_a + L_m(zI - A + K_mC)^{-1}B_mG_a(z). \tag{117}$$

Noting that $K_mG_m(z) = K_mC(zI - A)^{-1}B_m$, we can rewrite (116) as

$$D_{MEF}(z) = I + L_a(zI - A_a)^{-1}B_a + L_m(zI - A)^{-1}B_mG_a(z)$$
$$+ (z-1)^{-1}MG(z), \tag{118}$$

which is equal to $D_{MSF}(z)$ defined by (107). From (42) and (111), we have

$$D_{MEF}^{-1}(z) = z^{-1}(z-1)\Sigma_0(z). \tag{119}$$

On the other hand, noting that the matrix identity

$$(zI - A + K_mC)^{-1} = z^{-1}I + z^{-1}(zI - A + K_mC)^{-1}(A - K_mC) \tag{120}$$

holds, we can rewrite (117) as

$$N_{\text{MEF}}(z) = I + L_a(zI - A_a)^{-1}B_a$$
$$+ z^{-1}L_m[I + (zI - A + K_mC)^{-1}(A - K_mC)]B_mG_a(z). \quad (121)$$

Substitution of (119) and (121) into (115) yields (113).   ◆

3. Partial Loop Transfer Recovery

Using the filter gain matrix given by (59) and the above lemmas, we can easily show the result corresponding to Proposition 3 as follows.

Proposition 7: Consider the control system consisting of the plant model (30) and the MEF integral controller designed by use of the covariance matrix (51). Then the sensitivity matrix achieved by the MSF integral controller can be recovered by increasing the covariance matrix of the fictitious disturbance inserted into the input of the minimum phase part, i.e.,

$$\lim_{\mu_m \to \infty} \Sigma_{\text{INT}}^{\text{MEF}}(z) = \Sigma_{\text{INT}}^{\text{MSF}}(z), \quad (122)$$

where $\Sigma_{\text{INT}}^{\text{MSF}}(z)$ and $\Sigma_{\text{INT}}^{\text{MEF}}(z)$ are defined by (103) and (113), respectively.

D. RELATION BETWEEN THE TWO LTR TECHNIQUES

To discuss the relation between the two LTR techniques for the integral controller design, we first give the following result corresponds to Proposition 4 for the LQG case.

Lemma 11: Assume that the plant has a single real unstable zero q. Choose a feedback gain matrix F for the plant model (1). Consider the linear matrix equation (93) with $\Psi = [F_m \ F_a] = [F \ qF\xi]$. Then the solution $\Lambda = [L_m \ L_a]$ and M can be expressed as

$$L_m = L, \quad L_a = qL\xi, \quad M = T, \quad (123)$$

where the matrices L and T satisfy the linear matrix equation (85) with the feedback gain matrix F.
Proof: Using the submatrices in Λ and Ψ, we can rewrite (93) as

$$L_m(A - I) + MC = F_mA, \quad (124)$$
$$L_mB_mC_a + L_a(A_a - I) = F_mB_mC_a + F_aA_a, \quad (125)$$

$$L_m B_m D_a + L_a B_a = I + F_m B_m D_a + F_a B_a. \tag{126}$$

Noting that $(qI - A)\xi = B\eta$ holds by (12), we can rewrite the relation (70) as

$$B_m C_a = -q[B\eta - (q - \frac{1}{q})\xi] = (qA - I)\xi. \tag{127}$$

Using (66) and (127), we can rewrite (124)-(126) as

$$L_m(A - I) + MC = FA, \tag{128}$$

$$L_m(qA - I)\xi + L_a(\frac{1}{q} - 1) = qFA\xi, \tag{129}$$

$$L_m[B - (q - \frac{1}{q})\xi\eta'] + L_a \frac{1}{q}(q - \frac{1}{q})\eta' = I + FB. \tag{130}$$

Noting that L and T satisfy (85), we can easily check that the matrices given by (123) satisfy (128) and (130). To show that (123) also satisfies (129), we substitute (123) into the left side of (129). Then the left side of (129) is given by

$$L(qA - I)\xi + (1 - q)L\xi = qL(A - I)\xi = q(FA - TC)\xi, \tag{131}$$

which is equal to $qFA\xi$, the right side of (129), since $C\xi = 0$ by (2). Hence, the matrices given by (123) satisfy the matrix equation (93) which has the unique solution by the assumption on the matrix (94).   ◆

Using the above lemma, we have the following result corresponding to Proposition 4 for the LQG case.

Proposition 8: Assume that the plant has a single real unstable zero q. Choose a feedback gain matrix F for the plant model (1). Consider the observer-based integral controller (84) where the state feedback matrices L and T are determined by the linear matrix equation (85) with the feedback gain matrix F. Then the sensitivity matrix (88) achieved by the formal application of the conventional LTR procedure coincides with that achieved by the MSF integral controller with the feedback gain matrices Λ and M determined by the matrix equation (93) with $\Psi = [F_m \quad F_a] = [F \quad qF\xi]$, i.e.,

$$\Sigma^*_{\text{INT}}(z) = \Sigma^{\text{MSF}}_{\text{INT}}(z), \tag{132}$$

where $\Sigma^{\text{MSF}}_{\text{INT}}(z)$ is given by (103).

Proof: It follows from Lemma 11 that the feedback gain matrices for the MSF integral controller are given by (123) under the assumptions. Note that the matrix identity (81) holds. In addition, as in the proof for the Proposition 4, we can easily show that

$$L_a(zI - A_a)^{-1} B_a = L(zI - A)^{-1}[B - B_m G_a(z)]. \tag{133}$$

Substituting (81) and (133) into (103) and comparing with (88), we can readily show (132).   ♦

IV. NUMERICAL EXAMPLES

Consider the single-input single-output minimum phase continuous-time system

$$G(s) = \frac{s+3}{(s+1)(s+0.5\pm1.2j)}. \tag{134}$$

Discretizing with zeroth order holder and the sampling period T = 0.1 (sec.), we have a non-minimum phase discrete-time model of (134) with the single unstable zero $z = -1.034$. A state-space representation for the decomposed transfer function corresponding to (13) is given by

$$A = \begin{bmatrix} 2.7936 & -2.6139 & 0.8187 \\ 1 & 0 & 0 \\ 0 & 1 & 0 \end{bmatrix}, \quad B_m = \begin{bmatrix} 1 \\ 0 \\ 0 \end{bmatrix},$$

$$C = \begin{bmatrix} 0.0053 & 0.0012 & -0.0038 \end{bmatrix}, \tag{135}$$

$$A_a = -0.9676, \quad B_a = 1, \quad C_a = 0.0637, \quad D_a = 0.9676.$$

For the performance index (2), the gain characteristics of the sensitivity function (39) and the complementary sensitivity function for the MSF controller are shown in Fig. 5 for various values of the weighting coefficient ρ. In Fig. 6, the corresponding gain characteristics of (103) for the MSF integral controller are shown. The step response of the MSF integral controller is shown in Fig. 7. Comparison of Fig. 5 and 6 clearly reveals the effect of introducing an integrator. The sensitivity at low frequency region is reduced considerably with the large increase at high frequency region. The sensitivity properties shown in Fig. 5 and

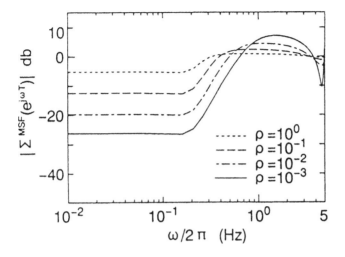

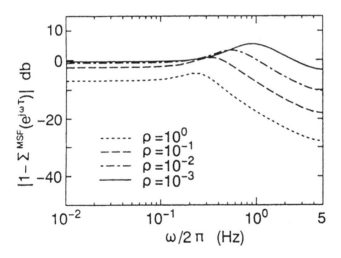

Fig. 5. Sensitivity and complementary sensitivity for the MSF controller.

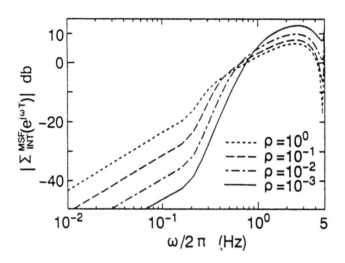

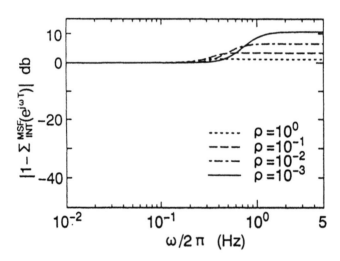

Fig. 6. Sensitivity and complementary sensitivity for
        the MSF integral controller.

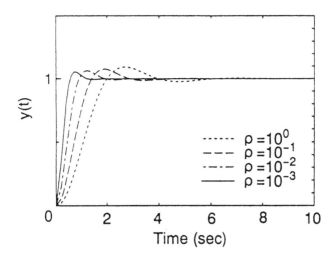

Fig. 7. Step response of the MSF integral controller.

6 can be recovered by the MEF controller and the MEF integral controller, respectively, using the partial LTR technique. As shown in SectionsII-D and III-D, these sensitivity properties can also be recovered by the formal applications of the conventional LTR technique.

To make use of design freedom provided by the partial LTR techniques, we determine the feedback gain matrix by the performance index

$$V_a^\infty = \sum_{t=0}^{\infty} [y^2(t) + \alpha x_a^2(t) + \rho u^2(t)], \tag{136}$$

where α is a non-negative weighting coefficient for the all-pass state. Denoting A_a in (135) by a, we can write $x_a(z)$ as

$$x_a(z) = \frac{1}{z-a} u(z). \tag{137}$$

Using (137), we can rewrite (136) by use of Parseval's theorem as

$$V_a^\sim = \frac{1}{2\pi j} \oint_{|z|=1} [y(z^{-1})y(z) + \rho u(z^{-1})r_a(z^{-1})r_a(z)u(z)]\frac{dz}{z}, \tag{138}$$

where $r_a(z)$ satisfies the spectral factorization

$$r_a(z^{-1})r_a(z) = 1 + \frac{\alpha}{\rho(z^{-1}-a)(z-a)}. \tag{139}$$

It can easily be checked that, for any $\alpha \geq 0$ and $\rho > 0$, there exists $r_a(z)$, which is stable and minimum phase, satisfying (139). Apparently, $\rho r_a(z^{-1})r_a(z)$ can be regarded as a frequency-shaped weighting coefficient on the control input. This interpretation is useful to determine the coefficients α and ρ. For an efficient determination, we restrict ourselves to the weighting coefficients defined by

$$\rho = \frac{\gamma}{r_a^2(1)}, \quad \alpha = \lambda\rho, \tag{140}$$

where γ and λ are new parameters for tuning. Apparently, the parameter γ fixes the dc gain of the frequency-shaped weighting $\rho r_a(z^{-1})r_a(z)$ and the parameter λ determines the shape of $r_a(z)$.

Gain characteristics of $r_a(z)$ are shown in Fig. 8. As λ increases, the gain at high frequency region near the Nyquist frequency increases significantly while the gain below 1 Hz is almost unity. For $\gamma = 0.01$, the gain characteristics of the

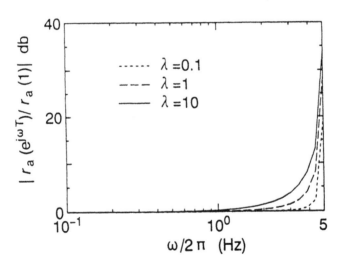

Fig. 8. Gain characteristics of $r_a(z)$.

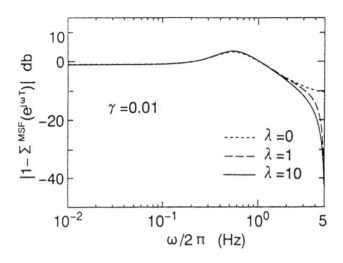

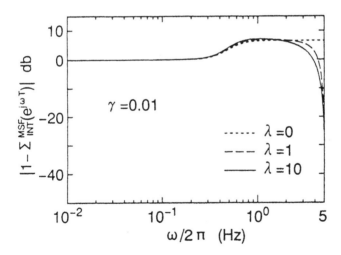

Fig. 9. Complementary sensitivity functions.

complementary sensitivity functions for the MSF controller and the MSF integral controller are shown in Fig. 9. The figures clearly show that the parameter λ affects the high frequency characteristics significantly with minor changes in the low frequency characteristics. The effect of the parameter λ on the sensitivity characteristics is minor compared with that for the complementary sensitivity characteristics. This property suggests that the performance index (138) with the weighting coefficients (140) is particularly useful to improve the robustness for high frequency uncertainties.

The technique used here is a simple example for making use of the design freedom provided by the partial LTR technique. More general frequency-shaped weighting matrices can be used to manipulate a target property for the partial LTR. In this case, however, the order of a controller increases and numerous design parameters in the weighting matrices have to be determined.

V. CONCLUSIONS

We have discussed applications of the LTR techniques for designing discrete-time controllers taking account of the inherent limitations of digital control systems. In addition to the standard LQG design, we have discussed applications to the integral controller design which provides transparent design perspective as well as computational advantage.

By clarifying the relation between the two LTR techniques, we have given a theoretical justification for enforcing the conventional LTR technique for non-minimum phase plants. We have also pointed out that the partial LTR technique provides more freedom in shaping a target property. Numerical examples have been presented to show that the freedom of the partial LTR technique can be used to enhance the robustness against the uncertainties near the Nyquist frequency.

Although we have assumed that the discrete-time model possesses a single unstable zero, our results can easily be generalized to the case of multiple unstable zeros.

ACKNOWLEDGMENTS

The author wishes to express his thanks to Prof. Hiroshi Takeda and Prof. Hikaru Inooka of Tohoku University for their encouragements. He also thanks Dr. Hai-Jiao Guo for stimulating discussions and for the assistance in preparing numerical examples.

VI. REFERENCES

1. H. Kwakernaak, "Optimal Low Sensitivity Linear Feedback Systems," *Automatica* 5, pp. 279-286 (1969).
2. J. C. Doyle and G. Stein, "Robustness with Observers", *IEEE Transactions on Automatic Control* AC-24, pp. 607-611 (1979).
3. G. Stein and M. Athans, "The LQG/LTR Procedure for Multivariable Feedback Control Design," *IEEE Transactions on Automatic Control* AC-32, pp. 105-114 (1987).
4. K. J. Astrom, P. Hagander and J. Sternby, "Zeros of Sampled Systems," *Automatica* 20, pp.31-38 (1984).
5. J. M. Maciejowski, "Asymptotic Recovery for Discrete-Time Systems," *IEEE Transactions on Automatic Control* AC-30, pp. 602-605 (1985).
6. T. Ishihara and H. Takeda, "Loop Transfer Recovery Techniques for Discrete-Time Optimal Regulators Using Prediction Estimators," *IEEE Transactions on Automatic Control* AC-31, pp. 1149-1151 (1986).
7. Z. Zhang and J. S. Freudenberg, "On Discrete-Time Loop Transfer Recovery," *Proceedings of the 1991 American Control Conference,* pp. 2214-2219 (1991).
8. L.-A. Zheng, T. Ishihara and H. Inooka, "On Prediction-Type Reduced-Order Observers for Discrete-Time Systems," *International Journal of Control* 57, pp. 55-73 (1993).
9. A. Saberi, P. Sannuti and B.M. Chen, "Loop Transfer Recovery," Springer-Verlag, New York (1993).
10. J. B. Moore and L. Xia, "Loop Recovery and Robust State Feedback Designs," *IEEE Transactions on Automatic Control* AC-32, pp. 512-517 (1979).
11. T. Ishihara, H.-J. Guo and H. Takeda, "A Design of Discrete-Time Integral Controllers with Computation Delays via Loop Transfer Recovery," *Automatica* 28, pp. 599-603 (1992).
12. T. Mita, "Optimal Digital Control Systems Counting Computation Time of Control Laws, *IEEE Transactions on Automatic Control* AC-30, pp. 542-548 (1985).
13. D. Enns, "Model Reduction for Control System Design," *Ph.D. dissertation, Stanford Univ.* (1984).
14. U. Shaked, "Explicit Solution to the Singular Discrete-Time Stationary Filtering Problem, *IEEE Transactions on Automatic Control* AC-30, pp. 34-47 (1985).
15. T. Ishihara, T. Watanabe and H. Inooka, "A Partial Loop Recovery for Discrete-Time Systems under Feedback Delays," *Recent Advances in Mathematical Theory of Systems, Networks and Signal Processing* I, pp. 523-528, Mita Press, Tokyo (1992).

Appendix: Proof of Lemma 2

Noting that the transfer function (17) is all-pass, i.e., $a(z)a(z^{-1}) = 1$, we can easily show that the transfer function matrix (16) is also all-pass and

$$\det G_a(q) = \det (I - \eta\eta') = 1 - \eta'\eta = 0. \tag{A.1}$$

Hence $z=q$ is a zero of $G_a(z)$. Using (12), we can rewrite (27) as

$$B_m G_a(z) = B - \frac{q^2 - 1}{qz - 1}(zI - A)\xi\eta'. \tag{A.2}$$

Multiplying the both sides of (A.2) by $C(zI - A)^{-1}$ and using $C\xi = 0$, we can easily check that the decomposition (13) holds.

To show that $C(zI - A)^{-1} B_m$ is minimum phase, we note that

$$\begin{bmatrix} q^{-1}I - A & -B_q \\ -C & 0 \end{bmatrix} \begin{bmatrix} \xi \\ \eta \end{bmatrix} = 0 \tag{A.3}$$

and

$$\det J_q = 1 - (q+1)\eta'\eta = -q \tag{A.4}$$

hold from (12) and (15). The relation (A.3) implies that $z = 1/q$ is a zero of $C(zI - A)^{-1} B_q$. The matrix J_q is non-singular from (A.4) under the assumption $|q| > 1$, which implies that the zeros of $C(zI - A)^{-1} B_q$ and those of $C(zI - A)^{-1} B_m$ coincide. Since $G(z)$ has a single unstable zero which is included in $G_a(z)$ in the decomposition (13), we can conclude that all the finite zeros of $C(zI - A)^{-1} B_m$ are in the open unit circle. ◆

The Study of Discrete Singularly Perturbed Linear-Quadratic Control Systems

Zoran Gajic

Myotaeg Lim

Department of Electrical and Computer Engineering
Rutgers University
Piscataway, NJ 08855–0909

Xuemin Shen

Department of Electrical Engineering
University of Alberta
Edmonton, T6G 2G7, Canada

I. INTRODUCTION

Theory of singular perturbations has been very fruitful control engineering research area in the last twenty five years, [1]-[3]. The discrete-time control systems have been the subject of research since early eighties. Several control researchers have produced important results on different aspects of control problems of deterministic singularly perturbed discrete systems such as Phillips, Blankenship, Mahmoud, Sawan, Khorasani, Naidu and their coworkers. Particularly important are the fundamental results of Khalil and Litkouhi, [4]-[6]. Along the lines of research of Khalil and Litkouhi in the papers of Gajic and Shen [7]-[8] an extension of the linear-quadratic control problem of [5] and the formulation and solution of the linear-quadratic Gaussian stochastic control problem are obtained.

Two main structures of singularly perturbed linear discrete systems have been considered: the fast time scale version [4]-[5], [9]-[14] and the slow time scale version [15]-[16]. Discrete-time models of singularly perturbed linear systems, similar to [15]-[16], were studied also in [17]-[18]. Since the slow time scale version presupposes the asymptotic stability of the fast modes, it seems that in the design procedure of stabilizing feedback controllers, the fast time scale version is much more appropriate [4].

In the first part of this chapter we study the reduced-order parallel algorithms for solving the algebraic Lyapunov and Riccati equations of discrete-time singularly perturbed systems. The algebraic Riccati equation is solved efficiently by using a bilinear transformation. The obtained results are extended to the discrete-time linear regulator problem. An F-8 aircraft model is used to demonstrate the presented technique. Also, we study the discrete-time Kalman filtering and the corresponding linear optimal stochastic control problem.

In the second part of this chapter the algebraic regulator and filter Riccati equations of singularly perturbed discrete-time control systems are completely and exactly decomposed into reduced-order continuous-time algebraic Riccati equations corresponding to the slow and fast time scales. That is, the exact solution of the global discrete algebraic Riccati equation is obtained in terms of the reduced-order pure-slow and pure-fast nonsymmetric continuous-time algebraic Riccati equations. In addition, the optimal global Kalman filter is decomposed into pure-slow and pure-fast local optimal filters both driven by the system measurements and the system optimal control input. It is shown that these two filters can be implemented independently in the different time scales. As a result, the optimal linear-quadratic Gaussian control problem for singularly perturbed linear discrete systems takes the complete decomposition and parallelism between pure-slow and pure-fast filters and controllers.

II. RECURSIVE METHODS FOR SINGULARLY PERTURBED LINEAR DISCRETE SYSTEMS

In this section, we adopt the structure of singularly perturbed discrete linear systems defined by Litkouhi and Khalil, and study the corresponding linear-quadratic discrete control problems. We will take the approach based on a bilinear transformation [19]. The main equations of the optimal linear control theory — the Lyapunov and Riccati equations — are solved by recursive, reduced-order parallel algorithms in the most general case when the system matrices are functions of a small perturbation parameter. Since the Riccati equation has quite complicated form in the discrete-time domain, partitioning this equation, in the spirit of singular perturbation methodology, will produce a lot of terms and make the corresponding problem numerically inefficient, even though the problem order-reduction is achieved. By applying a bilinear transformation, the solution of the discrete-time algebraic Riccati equation of singularly perturbed systems is obtained by using already known results for the corresponding continuous-time algebraic Riccati equation. The presented methods produce the reduced-order near-optimal solutions, up to an arbitrary order of accuracy, for both the Lyapunov and Riccati equations, that is $O(\epsilon^k)$, where ϵ is a small perturbation parameter. In addition, they reduce the size of required computations. The methods are very suitable for parallel and distributed computations. A real world example, an F-8 aircraft demonstrates the efficiency of the presented methods.

A. PARALLEL ALGORITHM FOR SOLVING DISCRETE ALGEBRAIC LYAPUNOV EQUATION

A discrete-time constant linear system with

$$x(k + 1) = Ax(k) \tag{1}$$

is asymptotically stable if and only if the solution of the algebraic discrete-time Lyapunov equation

$$A^T P A - P = -Q \tag{2}$$

is positive definite, where Q is any positive definite symmetric matrix. Equation (2) represents also a variance equation of a linear stochastic system driven by zero-mean stationary Gaussian white noise $\omega(k)$ with the intensity matrix Q

$$x(k + 1) = Ax(k) + \omega(k) \tag{3}$$

Consider the algebraic discrete Lyapunov equation of the singularly perturbed linear discrete system represented by the matrix partitions, [5]

$$A = \begin{bmatrix} I + \epsilon A_1 & \epsilon A_2 \\ A_3 & A_4 \end{bmatrix}, \quad Q = \begin{bmatrix} Q_1 & Q_2 \\ Q_2^T & Q_3 \end{bmatrix}, \quad P = \begin{bmatrix} P_1/\epsilon & P_2 \\ P_2^T & P_3 \end{bmatrix} \tag{4}$$

A_i, $i = 1, 2, 3, 4$, and Q_j, $j = 1, 2, 3$, are assumed to be continuous functions of ϵ. Matrices P_1 and P_3 are of dimensions $n \times n$ and $m \times m$, respectively. Remaining matrices are of compatible dimensions.

The partitioned form of (2) subject to (4) is

$$\begin{aligned} P_1 A_1 + A_1^T P_1 + P_2 A_3 + A_3^T P_2^T + A_3^T P_3 A_3 + Q_1 \\ + \epsilon (A_1^T P_1 A_1 + A_1^T P_2 A_3 + A_3^T P_2^T A_1) = 0 \end{aligned} \tag{5}$$

$$\begin{aligned} P_1 A_2 + P_2 A_4 + A_3^T P_3 A_4 - P_2 + Q_2 \\ + \epsilon (A_1^T P_2 A_4 + A_3^T P_2^T A_2) = 0 \end{aligned} \tag{6}$$

$$A_4^T P_3 A_4 - P_3 + Q_3 + \epsilon (A_2^T P_1 A_2 + A_2^T P_2 A_4 + A_4^T P_2^T A_2) = 0 \tag{7}$$

Define $O(\epsilon)$ perturbations of (5)–(7) by

$$\bar{P}_1 A_1 + A_1^T \bar{P}_1 + \bar{P}_2 A_3 + A_3^T \bar{P}_2^T + A_3^T \bar{P}_3 A_3 + Q_1 = 0 \tag{8}$$

$$\bar{P}_1 A_2 + \bar{P}_2 A_4 + A_3^T \bar{P}_3 A_4 - \bar{P}_2 + Q_2 = 0 \tag{9}$$

$$A_4^T \bar{P}_3 A_4 - \bar{P}_3 + Q_3 = 0 \tag{10}$$

Note that we did not set $\epsilon = 0$ in $A_i's$ and $Q_j's$. From Eq. (9) the matrix $\bar{P}_2$ can be expressed in terms of $\bar{P}_1$ and $\bar{P}_3$ as

$$\bar{P}_2 = L_1 + \bar{P}_1 L_2 \tag{11}$$

where

$$\begin{aligned} L_1 &= \left(A_3^T \bar{P}_3 A_4 - \bar{P}_3 + Q_2 \right)(I - A_4)^{-1} \\ L_2 &= A_2 (I - A_4)^{-1} \end{aligned} \tag{12}$$

The invertibility of the matrix $(I - A_4)$ follows from the stability assumption that $| \lambda(A_4) | < 1$.

After doing some algebraic calculations we get

$$\bar{P}_1 \mathbf{A} + \mathbf{A}^T \bar{P}_1 + \mathbf{Q} = 0 \tag{13}$$

where

$$\begin{aligned} \mathbf{A} &= A_1 + L_2 A_3 = A_1 + A_2 (I - A_4)^{-1} A_3 \\ \mathbf{Q} &= L_1 A_3 + A_3^T L_1^T + A_3^T \bar{P}_3 A_3 - Q_1 \end{aligned} \tag{14}$$

Thus, we can get solutions for $\bar{P}_1$, $\bar{P}_2$, and $\bar{P}_3$ by solving one lower order continuous-time Lyapunov equation and lower order discrete-time Lyapunov equation. It is assumed that $\mathbf{A}$ and A_4 are stable matrices so that solutions of (10) and (13) exist. These are standard assumptions in the theory of singularly perturbed linear discrete-time systems, [4].

Assumption 1 The matrix $\mathbf{A}$ is stable in the continuous-time domain and the matrix A_4 is stable in the discrete-time domain.

Define errors as

$$\begin{aligned} P_1 &= \bar{P}_1 + \epsilon E_1 \\ P_2 &= \bar{P}_2 + \epsilon E_2 \\ P_3 &= \bar{P}_3 + \epsilon E_3 \end{aligned} \tag{15}$$

Subtracting (8)-(10) from (5)-(7) and doing some algebra, the following set of equations is obtained

$$E_1 A + \mathbf{A}^T E_1 = -DA_3 - A_3^T D^T - A_1^T P_1 A_1 - A_1^T P_2 A_3$$
$$- A_3^T P_2^T A_1 - A_3^T E_3 A_4 - A_4^T E_3 A_3 - A_3^T E_3 A_3$$

$$A_4^T E_3 A_4 - E_3 = -A_2^T P_1 A_2 - A_2^T P_2 A_4 - A_4^T P_2^T A_2 \tag{16}$$

$$E_2 = E_1 L_2 + D + A_3^T E_3 A_4$$

where

$$D = \left(A_3^T E_3 A_4 + A_1^T P_1 A_2 + A_1^T P_2 A_4 + A_3^T P_2^T A_2 \right) (I - A_4)^{-1} \tag{17}$$

The solution of (16) of the given accuracy will produce the same accuracy for the solution of the Lyapunov equation (2). The proposed parallel synchronous algorithm for the numerical solution of (16) is as following, [20].

Algorithm 1:

$$E_1^{(i+1)} \mathbf{A} + \mathbf{A}^T E_1^{(i+1)} = -D^{(i)} A_3 - A_3^T D^{T^{(i)}} - A_1^T P_1^{(i)} A_1$$
$$- A_1^T P_2^{(i)} A_3 - A_3^T P_2^{T^{(i)}} A_1 - A_3^T E_3^{(i)} A_4 - A_4^T E_3^{(i)} A_3 - A_3^T E_3^{(i)} A_3$$

$$A_4^T E_3^{(i+1)} A_4 - E_3^{(i+1)} = -A_2^T P_1^{(i)} A_2 - A_2^T P_2^{(i)} A_4 - A_4^T P_2^{T^{(i)}} A_2 \tag{18}$$

$$E_2^{(i+1)} = E_1^{(i+1)} L_2 + D^{(i)} + A_3^T E_3^{(i+1)} A_4$$

with starting points $E_1^{(0)} = E_2^{(0)} = E_3^{(0)} = 0$ and

$$P_j^{(i)} = \bar{P}_j + \epsilon E_j^{(i)}, \quad j = 1, 2, 3; \quad i = 0, 1, 2... \tag{19}$$

The main feature of Algorithm 1 is given in the following theorem.

Theorem 1 *Based on the stability assumptions imposed on* **A** *and* A_4 *the algorithm (18)-(19) converges to the exact solutions for* $E_j's$ *with the rate of convergence of* $O(\epsilon)$.

The proof of this theorem can be found in [21].

1. CASE STUDY: AN F-8 AIRCRAFT

A numerical example for a linearized model of an F-8 aircraft with the small positive parameter $\epsilon = 0.03333$, [6], demonstrates the efficiency of the proposed method. The problem matrices A and Q are given by

$$A = 10^{-3} \begin{bmatrix} 998.51 & -8.044 & -0.10886 & -0.018697 \\ 0.15659 & 1000 & -0.76232 & 3.2272 \\ -213.94 & 0.88081 & 897.21 & 92.826 \\ 110.17 & -0.37821 & -445.56 & 929.68 \end{bmatrix}$$

$$Q = diag[0.1,\ 0.1,\ 0.1,\ 0.1]$$

Simulation results are presented in Table I. It can be seen that obtained numerical results are consistent with the established theoretical statements.

Table I: Recursive solution of the singularly perturbed discrete Lyapunov equation

iteration	$P_1^{(i)}$	$P_2^{(i)}$	$P_3^{(i)}$
0	7.26070 0.17779 0.17779 8.02970	-1.4252 -0.81171 -7.3907 1.29590	2.10690 -0.25231 -0.25231 0.54983
1	8.15990 0.20340 0.20340 8.95420	-1.6239 -0.89268 -8.2950 1.47450	2.33520 -0.29196 -0.29196 0.55729
2	8.27660 0.20656 0.20656 9.07410	-1.6494 -0.90328 -8.4104 1.49760	2.36500 -0.29714 -0.29714 0.55884
3	8.28940 0.20696 0.20696 9.08960	-1.6526 -0.90465 -8.4253 1.50060	2.36880 -0.29781 -0.29781 0.55884
4	8.29130 0.20701 0.20701 9.09160	-1.6531 -0.90483 -8.4272 1.50100	2.36930 -0.29790 -0.29790 0.55887
5 = exact	8.29160 0.20702 0.20702 9.09180	-1.6531 -0.90485 -8.4275 1.50100	2.36940 -0.29791 -0.29791 0.55887

B. PARALLEL ALGORITHM FOR SOLVING DISCRETE ALGEBRAIC RICCATI EQUATION

The algebraic Riccati equation of singularly perturbed linear discrete systems is given by

$$P = A^T P A + Q - A^T P B (B^T P B + R)^{-1} B^T P A \quad R > 0, \quad Q \geq 0 \quad (20)$$

where [4]-[6]

$$A = \begin{bmatrix} I + \epsilon A_1 & \epsilon A_2 \\ A_3 & A_4 \end{bmatrix}, \quad B = \begin{bmatrix} \epsilon B_1 \\ B_2 \end{bmatrix}, \quad Q = \begin{bmatrix} Q_1 & Q_2 \\ Q_2^T & Q_3 \end{bmatrix} \qquad (21)$$

and ϵ is a small positive singular perturbation parameter. In addition, the following condition is satisfied [4]

$$det(I - A_4) \neq 0 \qquad (22)$$

Due to the special structure of the problem matrices and its representation in the fast time scale, the required solution P has the form [5]

$$P = \begin{bmatrix} P_1/\epsilon & P_2 \\ P_2^T & P_3 \end{bmatrix} \qquad (23)$$

The main goal in the theory of singular perturbations is to obtain the required solution in terms of the reduced-order problems, namely subsystems. In the case of the algebraic singularly perturbed discrete-time Riccati equation, the expansion of the partitioned form of (20) will produce a lot of terms and make corresponding approach computationally very involved even though one is faced with the reduced-order numerical problems. In order to overcome this problem, we have used a bilinear transformation introduced in [19] to transform the discrete-time Riccati equation (20) into the continuous-time algebraic Riccati equation of the form

$$A_c^T P_c + P_c A_c + Q_c - P_c S_c P_c = 0, \quad S_c = B_c R_c^{-1} B_c^T \qquad (24)$$

such that the solution of (20) is equal to the solution of (24).

It will be shown that Eq. (24) preserves the structure of singularly perturbed systems. Equation (24) will be solved in terms of the reduced-order problems very efficiently by using the recursive method that converges with the rate of convergence of $O(\epsilon)$.

The bilinear transformation states that Eqs. (20) and (24) have the same solution if the following hold [19]

$$A_c = I - 2D^{-T}$$
$$S_c = 2(I + A)^{-1} S_d D^{-1}, \quad S_d = BR^{-1}B^T$$
$$Q_c = 2D^T Q(I + A)^{-1}$$
$$D = (I + A)^T + Q(I + A)^{-1} S_d \qquad (25)$$

assuming that $(I + A)^{-1}$ exists. It can be easily seen that the matrix

$$I + A = \begin{bmatrix} 2I + \epsilon A_1 & \epsilon A_2 \\ A_3 & I + A_4 \end{bmatrix} \tag{26}$$

is invertible for small values of ϵ if and only if the matrix $I + A_4$ is invertible. Using the standard result from [22] the invertibility is assured if the matrix A_4 has no eigenvalues at -1. Thus, the method proposed in this section will be applicable under the following assumption.

Assumption 2 The fast subsystem matrix has no eigenvalues located at -1.

It is important to point out that, under given Assumption 2, the matrix D defined in (25) is nonsingular [23].

Let us show that applying the bilinear transformation, the system still preserves the singularly perturbed structure, namely, matrices defined in (25) should correspond to the linear-quadratic (LQ) singularly perturbed continuous-time control problem.

Using the formula for an inversion of block partitioned matrices, the following can be obtained from (21) and (25)

$$(I + A)^{-1} = \begin{bmatrix} I + O(\epsilon) & O(\epsilon) \\ O(1) & (1) \end{bmatrix}, \quad S_d^f = \begin{bmatrix} O(\epsilon^2) & O(\epsilon) \\ O(\epsilon) & O(1) \end{bmatrix}$$

$$D^f = \begin{bmatrix} I + O(\epsilon) & O(1) \\ O(\epsilon) & O(1) \end{bmatrix}, \quad D^{f-T} = \begin{bmatrix} I + O(\epsilon) & O(\epsilon) \\ O(1) & O(1) \end{bmatrix} \tag{27}$$

so that

$$A_c^f = \begin{bmatrix} O(\epsilon) & O(\epsilon) \\ O(1) & (1) \end{bmatrix}, \quad Q_c^f = \begin{bmatrix} O(1) & O(1) \\ O(1) & O(1) \end{bmatrix}$$

$$S_c^f = \begin{bmatrix} O(\epsilon^2) & O(\epsilon) \\ O(\epsilon) & O(1) \end{bmatrix} \Rightarrow B_c^f = \begin{bmatrix} O(\epsilon) \\ O(1) \end{bmatrix} \tag{28}$$

where f indicates the fast time scale version quantities.

It is the well-known fact that the structure of matrices obtained in (28) corresponds to the fast time scale representation of the continuous-time singularly perturbed LQ control problem [1]-[2], [5].

Since there is no difference in the use of either the slow or fast time scale representation for the continuous-time LQ control problem of singularly perturbed systems, we will adopt the slow time scale version for this problem. It is customary to represent continuous-time singularly perturbed systems by their slow time version [1]-[2].

The slow time version of (28) can be obtained by multiplying the matrix A_c^f by $1/\epsilon$ and matrix S_c^f by $1/\epsilon^2$. Introducing a notation for the compatible partitions of these matrices, we have

$$A_c = \begin{bmatrix} A_{11} & A_{12} \\ \frac{A_{21}}{\epsilon} & \frac{A_{22}}{\epsilon} \end{bmatrix}, \quad S_c = \begin{bmatrix} S_{11} & \frac{S_{12}}{\epsilon} \\ \frac{S_{12}^T}{\epsilon} & \frac{S_{22}}{\epsilon^2} \end{bmatrix} \tag{29}$$

By doing this, the required solution P from (23), obtained now from (24), will be multiplied by ϵ, that is

$$\epsilon P = P_c = \begin{bmatrix} P_1 & \epsilon P_2 \\ \epsilon P_2^T & \epsilon P_3 \end{bmatrix} \tag{30}$$

Going from the fast time version to the slow time version does not change the matrix Q_c. It is partitioned as

$$Q_c = \begin{bmatrix} Q_{11} & Q_{12} \\ Q_{12}^T & Q_{22} \end{bmatrix} = Q_c^f \tag{31}$$

It is important to notice that partitions defined in (27)-(29) have to be performed by a computer only, in the process of calculations, and there is no need for the corresponding analytical expressions.

The solution of (24) can be found in terms of the reduced-order problems by imposing standard stabilizability-detectability assumptions on the slow and fast subsystems. The efficient recursive reduced-order algorithm for solving (24) is obtained in [24]. It will be briefly summarized here taking into account the specific features of the problem under study.

First of all, we derive expressions for B_c and R_c so that the analogy between the discrete quantities (A, B, Q, R) and continuous ones (A_c, B_c, Q_c, R_c) is completed. By definition

$$S_c^f = B_c^f R_c^{-1} B_c^{f^T} \tag{32}$$

From (25) we have

$$S_c^f = 2(I + A)^{-1} S_d D^{-1}(I + A^T)(I + A^T)^{-1} \tag{33}$$

Since

$$S_d D^{-1}(I + A^T) = S_d \left[(I + A^T)^{-1} D \right]^{-1} =$$

$$= S_d \left[I + (I + A^T)^{-1} Q(I + A)^{-1} S_d \right]^{-1} =$$

$$= B \left[R + B^T(I + A^T)^{-1} Q(I + A)^{-1} B \right]^{-1} B^T \tag{34}$$

(the last step in this expression is justified in [23], we get

$$S_c^f = 2(I+A)^{-1}B\left[R+B^T(I+A^T)^{-1}Q(I+A)^{-1}B\right]^{-1}$$
$$\times B^T(I+A^T)^{-1} \tag{35}$$

Comparing (32) and (35) we conclude

$$B_c^f = (I+A)^{-1}B = \begin{bmatrix} \epsilon B_1^f \\ B_2^f \end{bmatrix} \tag{36}$$

and

$$R_c = 0.5\left[R+B^T(I+A^T)^{-1}Q(I+A)^{-1}B\right] \tag{37}$$

Note that R_c is positive definite. The slow time version of (36) is

$$B_c = \frac{1}{\epsilon}B_c^f = \begin{bmatrix} B_1^f \\ \frac{B_2^f}{\epsilon} \end{bmatrix} \tag{38}$$

The $O(\epsilon)$ approximation of (24) subject to (27)-(29) and (36)-(37) can be obtained from the following reduced-order algebraic equations

$$0 = \bar{P}_1\underline{A} + \underline{A}^T\bar{P}_1 + \underline{Q} - \bar{P}_1\underline{S}\bar{P}_1, \quad \underline{S} = B_0R_0^{-1}B_0^T$$
$$0 = \bar{P}_3A_{22} + A_{22}^T\bar{P}_3 + Q_{22} - \bar{P}_3S_{22}\bar{P}_3 \tag{39}$$
$$\bar{P}_2 = \bar{P}_1Z_1 - Z_2$$

where newly defined matrices can be obtained easily using results from [24].

The unique positive semidefinite stabilizing solution of (39) exists under the following assumption.

Assumption 3 The triples $\left(\underline{A}, B_0, \sqrt{\underline{Q}}\right)$ and $\left(A_{22}, B_2, \sqrt{Q_{22}}\right)$ are stabilizable-detectable.

Defining the approximation errors as

$$P_i = \bar{P}_i + \epsilon E_i, \quad i = 1, 2, 3. \tag{40}$$

the recursive reduced-order algorithm, with the rate of convergence of $O(\epsilon)$, can be derived as follows.

Algorithm 2:

$$E_1^{(j+1)}D_1 + D_1^T E_1^{(j+1)} = D^T H_1^{(j)^T} + H_1^{(j)}D + D^T H_3^{(j)}D + \epsilon H_2^{(j)}$$
$$E_2^{(j+1)}D_3 + E_1^{(j+1)}D_{21} + D_{22}^T E_3^{(j+1)} = H_1^{(j,j+1)} \tag{41}$$
$$E_3^{(j+1)}D_3 + D_3^T E_3^{(j+1)} = H_3^{(j)}$$

with $j = 0, 1, 2, \ldots$, and $E_1^{(0)} = 0$, $E_2^{(0)} = 0$, $E_3^{(0)} = 0$, where newly defined matrices are given by

$$D_1 = A_{11} - S_{11}\bar{P}_1 - S_{12}\bar{P}_2^T - D_{21}D_3^{-1}D_{22} = D_{11} - D_{21}D_3^{-1}D_{22}$$
$$D_3 = A_{22} - S_{22}\bar{P}_3, \quad D = D_3^{-1}D_{22} \qquad (42)$$
$$D_{21} = A_{12} - S_{12}\bar{P}_3, \quad D_{22} = A_{21} - S_{12}^T\bar{P}_1 - S_{22}\bar{P}_2^T$$

$$H_1^{(j,j+1)} = A_{21}^T P_2^{(j)} - P_1^{(j+1)}S_{11}P_2^{(j)} - P_2^{(j)}S_{12}^T P_2^{(j)}$$
$$\qquad - \epsilon\left(E_1^{(j+1)}S_{12}E_3^{(j+1)} + E_2^{(j)}S_{22}E_2^{(j)} \right)$$

$$H_2^{(j)} = E_1^{(j)}S_{11}E_1^{(j)} + E_1^{(j)}S_{12}E_2^{(j)^T} + E_2^{(j)}S_{12}^T E_2^{(j)} + E_2^{(j)}S_{22}E_2^{(j)^T} \qquad (43)$$
$$H_3^{(j)} = -P_2^{(j)^T}A_{12} - A_{12}^T P_2^{(j)} + \epsilon\left(P_2^{(j)^T}S_{11}P_2^{(j)} + E_3^{(j)}S_{22}E_3^{(j)} \right)$$
$$\qquad + P_2^{(j)^T}S_{12}P_3^{(j)} + P_3^{(j)}S_{12}^T P_2^{(j)}$$

It is important to point out that D_1 and D_3 are stable matrices [24].

The rate of convergence of (41) is $O(\epsilon)$, that is

$$\left\| P_i - P_i^{(j)} \right\| = O(\epsilon^j), \qquad i = 1, 2, 3; \quad j = 0, 1, 2, \ldots . \qquad (44)$$

where

$$P_i^{(j)} = \bar{P}_i + \epsilon E_i^{(j)} \qquad i = 1, 2, 3; \quad j = 0, 1, 2, \ldots . \qquad (45)$$

In summary, the proposed algorithm for the reduced-order solution of the singularly perturbed discrete algebraic Riccati equation has the following form:

1) Transform (20) into (24) by using the bilinear transformation defined in (25).

2) Solve (24) by using the recursive reduced-order parallel algorithm defined by (39)-(45).

C. APPROXIMATE LINEAR REGULATOR PROBLEM FOR DISCRETE SYSTEMS

The positive semidefinite stabilizing solution of the algebraic discrete Riccati equation (20), produces the answer to the optimal linear-quadratic steady state control problem. Namely, a quadratic criterion

$$J = \frac{1}{2}\sum_{k=0}^{\infty}\left(x^T(k)Qx(k) + u^T(k)Ru(k) \right) \qquad (46)$$

is minimized along trajectories of a linear dynamic system

$$x(k+1) = Ax(k) + Bu(k) \tag{47}$$

by using the control input of the form

$$u(k) = -\left(R + B^T P B\right)^{-1} B^T P A x(k) \tag{48}$$

where P is obtained from (20), [25]. This problem has been studied in the context of singular perturbations in [5], where the fast time version has been adopted, so that (46) is multiplied by a small perturbation parameter, that is,

$$J_f = \epsilon J \tag{49}$$

It is proved in [4] that the near-optimal control given by

$$u^{(j)}(k) = -\left(R + B^T P^{(j)} B\right)^{-1} B^T P^{(j)} A x(k) = -F^{(j)} x(k) \tag{50}$$

where $P^{(j)}$ satisfies

$$P^{(j)} - P^{opt} = O\left(\epsilon^j\right) \tag{51}$$

is near-optimal in the sense

$$J_f^{(j)} - J_f^{opt} = O\left(\epsilon^{2j}\right) \tag{52}$$

The approximate performance $J^{(j)}$ can be obtained from the discrete algebraic Lyapunov equation

$$K^{(j)} = \left(A - BF^{(j)}\right)^T K^{(j)} \left(A - BF^{(j)}\right) + Q + F^{(j)^T} R F^{(j)} \tag{53}$$

so that

$$J^{(j)} = \frac{1}{2} x^T(0) K^{(j)} x(0) \tag{54}$$

In the previous section we have developed a very efficient technique for generating $P^{(j)}$ by using the recursive reduced-order schemes (39)-(45), such that each iteration improves the accuracy by an order of magnitude. Thus, the proposed algorithm and the theoretical results obtained in [4]-[5] and given in (50)-(52) comprise an efficient method for solving the linear-quadratic control problem of singularly perturbed discrete systems.

The efficiency of this method is demonstrated on a real world example in the next section.

1. CASE STUDY: DISCRETE MODEL OF AN F-8 AIRCRAFT

A linearized model of an F-8 aircraft is considered in [26]. By a proper scaling this model was presented in the singularly perturbed continuous-time form (fast time version) in [6], with the system matrix

$$\begin{bmatrix} -0.015 & -0.0805 & -0.0011666 & 0 \\ 0 & 0 & 0 & 0.03333 \\ -2.28 & 0 & -0.84 & 1 \\ 0.6 & 0 & -4.8 & -0.49 \end{bmatrix}$$

and the control matrix

$$\begin{bmatrix} -0.0000916 & 0.0007416 \\ 0 & 0 \\ -0.11 & 0 \\ -8.7 & 0 \end{bmatrix}$$

Small elements in the first two rows indicate two slow variables in contrary to relatively big elements in the third and forth rows corresponding to fast variables. The small perturbation parameter ϵ is chosen as $\epsilon = 1/30$. This model is discretized in [6] by using the sampling period $T = 1$, leading to

$$A = \begin{bmatrix} 0.98475 & -0.079903 & 0.0009054 & -0.0010765 \\ 0.041588 & 0.99899 & -0.035855 & 0.012684 \\ -0.54662 & 0.044916 & -0.32991 & 0.19318 \\ 2.6624 & -0.10045 & -0.92455 & -0.26325 \end{bmatrix}$$

$$B = \begin{bmatrix} 0.0037112 & 0.00073610 \\ -0.087051 & 0.0000093411 \\ -1.19844 & -0.00041378 \\ -3.1927 & 0.00092535 \end{bmatrix}$$

The linear-quadratic control problem is solved for weighting matrices $R = I_2$, $Q = 10^{-2}I_4$ and the initial condition $x(0) = [1, 0, 0.008, 0]^T$.

The eigenvalues of the matrix A_4 are $-0.297 \pm j0.442$, so that Assumption 2 is satisfied. Simulation results for the reduced-order solution for the approximate values of the criterion are presented in Table II.

Table II: Near optimality of the approximate criterion

j	$J_{apr}^{(j)} - J_{opt}$
0	0.208 x 10^{-2}
1	0.885 x 10^{-5}
2	0.155 x 10^{-7}
3	0.534 x 10^{-10}

III. PARALLEL REDUCED-ORDER CONTROLLER FOR STOCHASTIC SYSTEMS

The continuous-time LQG problem of singularly perturbed systems is solved in [27] by using the power-series expansion approach, and later on in [24] by using the fixed point theory. In this section, we solve the discrete-time LQG problem of singularly perturbed system by using the results obtained in Section II.

This section presents the approach to the decomposition and approximation of the linear-quadratic Gaussian control problem of singularly perturbed discrete systems by treating the decomposition and approximation tasks separately from each other. The decoupling transformation of [28] is used for the exact block diagonalization of the global Kalman filter. The approximate feedback control law is then obtained by approximating the coefficients of the optimal regulator and the optimal local filters with the accuracy of $O(\epsilon^N)$. The resulting feedback control law is shown to be a near-optimal solution of the LQG by studying the corresponding closed-loop system as a system driven by white noise. It is shown that the order of approximation of the optimal system trajectories is $O(\epsilon^{N+1/2})$ in the case of slow variables and $O(\epsilon^N)$ in the case of fast variables. All required coefficients of desired accuracy are easily obtained by using the recursive reduced-order fixed point type numerical techniques developed in Section II. Obtained numerical algorithms converge to the required optimal coefficients with the rate of convergence of $O(\epsilon)$. In addition, only low-order subsystems are involved in the algebraic computations and no analyticity requirements are imposed on the system coefficients — which is the standard assumption in the power-series expansion method. As a consequence of these, under very mild conditions (coefficients are bounded functions of a small perturbation parameter), in addition to the standard stabilizability-detectability subsystem assumptions, we have achieved the reduction in both off-line and on-line computational requirements.

The results presented in this section are mostly based on the doctoral dissertation [29] and the recent research papers of [7]-[8].

Consider the discrete linear singularly perturbed stochastic system represented in the fast time scale by

$$
\begin{aligned}
x_1(n+1) &= (I_{n_1} + \epsilon A_{11})x_1(n) + \epsilon A_{12}x_2(n) \\
&\quad + \epsilon B_1 u(n) + \epsilon G_1 w(n) \\
x_2(n+1) &= A_{21}x_1(n) + A_{22}x_2(n) + B_2 u(n) + G_2 w(n)
\end{aligned}
\tag{55}
$$

$$
y(n) = C_1 x_1(n) + C_2 x_2(n) + v(n)
$$

with the performance criterion

$$
J = \frac{1}{2}E\left\{\sum_{n=0}^{\infty}\left[z^T(n)z(n) + u^T(n)Ru(n)\right]\right\}, \quad R > 0
\tag{56}
$$

where $x_i \in \Re^{n_i}$, $i = 1, 2$, comprise slow and fast state vectors, respectively, $u \in \Re^m$ is the control input, $y \in \Re^l$ is the observed output, $w \in \Re^r$ and $v \in \Re^l$ are independent zero-mean stationary Gaussian mutually uncorrelated white noise processes with intensities $W > 0$ and $V > 0$, respectively, and $z \in \Re^s$ is the controlled output given by

$$
z(n) = D_1 x_1(n) + D_2 x_2(n)
\tag{57}
$$

All matrices are bounded functions of a small positive parameter ϵ [3] having appropriate dimensions.

The optimal control law is given by [30]

$$
u(n) = -F\hat{x}(n)
\tag{58}
$$

with

$$
\hat{x}(n+1) = A\hat{x}(n) + Bu(n) + K[y(n) - C\hat{x}(n)]
\tag{59}
$$

where

$$
A = \begin{bmatrix} I_{n_1} + \epsilon A_{11} & \epsilon A_{12} \\ A_{21} & A_{22} \end{bmatrix}, \quad B = \begin{bmatrix} \epsilon B_1 \\ B_2 \end{bmatrix}, \quad C = [\,C_1 \quad C_2\,]
$$

$$
K = \begin{bmatrix} \epsilon K_1 \\ K_2 \end{bmatrix}, \quad F = [\,F_1 \quad F_2\,]
\tag{60}
$$

The regulator gain F and filter gain K are obtained from

$$
F = (R + B^T P B)^{-1} B^T P A
\tag{61}
$$

$$K = AQC^T \left(V + CQC^T \right)^{-1} \tag{62}$$

where P and Q are positive semidefinite stabilizing solutions of the discrete-time algebraic regulator and filter Riccati equations, respectively given by

$$P = D^T D + A^T P A - A^T P B \left(R + B^T P B \right)^{-1} B^T P A \tag{63}$$

$$Q = AQA^T - AQC^T \left(V + CQC^T \right)^{-1} CQA^T + GWG^T \tag{64}$$

where

$$D = [\, D_1 \quad D_2 \,], \quad G = \begin{bmatrix} \epsilon G_1 \\ G_2 \end{bmatrix} \tag{65}$$

Due to the singularly perturbed structure of the problem matrices the required solutions P and Q in the fast time scale version have the forms

$$P = \begin{bmatrix} P_{11}/\epsilon & P_{12} \\ P_{12}^T & P_{22} \end{bmatrix}, \quad Q = \begin{bmatrix} \epsilon Q_{11} & \epsilon Q_{12} \\ \epsilon Q_{12}^T & Q_{22} \end{bmatrix} \tag{66}$$

In order to obtain required solutions of (63)-(64) in terms of the reduced-order problems and overcome the complicated partitioned form of the discrete-time algebraic Riccati equation, we have used the method developed in Section II (based on a bilinear transformation), to transform the discrete-time algebraic Riccati equations (63)-(64) into continuous-time algebraic Riccati equations of the forms

$$A_R^T \mathbf{P} + \mathbf{P} A_R - \mathbf{P} S_R \mathbf{P} + D_R^T D_R = 0, \quad S_R = B_R R_R^{-1} B_R^T \tag{67}$$

$$A_F \mathbf{Q} + \mathbf{Q} A_F^T - \mathbf{Q} S_F \mathbf{Q} + G_F W_F G_F^T = 0, \quad S_F = C_F^T V_F^{-1} C_F \tag{68}$$

such that the solutions of (63)-(64) are equal to the solutions of (67)-(68), that is

$$P = \mathbf{P}, \quad Q = \mathbf{Q} \tag{69}$$

where

$$\begin{aligned}
A_R &= I - 2 \left(\triangle_R^{-1} \right)^T \\
B_R R_R^{-1} B_R^T &= 2(I + A)^{-1} B R^{-1} B^T \triangle_R^{-1} \\
D_R^T D_R &= 2 \triangle_R^{-1} D^T D (I + A)^{-1} \\
\triangle_R &= \left(I + A^T \right) + D^T D (I + A)^{-1} B R^{-1} B^T
\end{aligned} \tag{70}$$

and

$$A_F = I - 2\left(\triangle_F^{-1}\right)$$
$$C_F^T V_F^{-1} C_F = 2\left(I + A^T\right)^{-1} C^T V^{-1} C \triangle_F^{-1}$$
$$G_F W_F G_F^T = 2\triangle_F^{-1} G W G^T \left(I + A^T\right)^{-1} \tag{71}$$
$$\triangle_F = (I + A) + G W G^T \left(I + A^T\right)^{-1} C^T V^{-1} C$$

It is shown in Section II that Eqs. (67)-(68) preserve the structure of singularly perturbed systems. These equations can be solved in terms of the reduced-order problems very efficiently by using the recursive method developed in Section II, which converges with the rate of convergence of $O(\epsilon)$ under the following assumption.

Assumption 4 The matrix A_{22} has no eigenvalues located at -1.

Under this assumption the matrices $\triangle_R$ and $\triangle_F$ are invertible.

Solutions of (67) and (68) are found in terms of the reduced-order problems by imposing standard stabilizability-detectability assumptions on subsystems.

Getting approximate solutions for P and Q in terms of reduced-order problems will produce savings in off-line computations. However, in the case of stochastic systems, where an additional dynamical system — filter — has to be built, one is particularly interested in the reduction of on-line computations. In this section, the savings of on-line computation will be achieved by using a decoupling transformation introduced in [28]. The Kalman filter is viewed as a system driven by the innovation process [27]. However, one might study the filter form when it is driven by both measurements and control, that is

$$\hat{x}_1(n+1) = (I_{n_1} + \epsilon A_{11} - \epsilon B_1 F_1)\hat{x}_1(n)$$
$$+\epsilon(A_{12} - B_1 F_2)\hat{x}_2(n) + \epsilon K_1 v(n) \tag{72}$$
$$\hat{x}_2(n+1) = (A_{21} - B_2 F_1)\hat{x}_1(n) + (A_{22} - B_2 F_2)\hat{x}_2(n) + K_2 v(n)$$

with the innovation process

$$v(n) = y(n) - C_1 \hat{x}_1(n) - C_2 \hat{x}_2(n) \tag{73}$$

The nonsingular state transformation of [28] will block diagonalize (72). The transformation is given by

$$\begin{bmatrix} \hat{\eta}_1(n) \\ \hat{\eta}_2(n) \end{bmatrix} = \begin{bmatrix} I_{n_1} - \epsilon H L & -\epsilon H \\ L & I_{n_2} \end{bmatrix} \begin{bmatrix} \hat{x}_1(n) \\ \hat{x}_2(n) \end{bmatrix} = \mathbf{T}_1 \begin{bmatrix} \hat{x}_1(n) \\ \hat{x}_2(n) \end{bmatrix} \tag{74}$$

with

$$\mathbf{T}_1^{-1} = \begin{bmatrix} I_{n_1} & \epsilon H \\ -L & I_{n_2} - \epsilon L H \end{bmatrix} \tag{75}$$

where matrices L and H satisfy equations

$$\epsilon L a_{11} + (I - a_{22})L + a_{21} - \epsilon L a_{12} L = 0 \tag{76}$$

$$H(I - a_{22} - \epsilon L a_{12}) + \epsilon(a_{11} - a_{12}L)H + a_{12} = 0 \tag{77}$$

with

$$\begin{aligned}
a_{11} &= A_{11} - B_1 F_1, & a_{12} &= A_{12} - B_1 F_2 \\
a_{21} &= A_{21} - B_2 F_1, & a_{22} &= A_{22} - B_2 F_2
\end{aligned} \tag{78}$$

The optimal feedback control, expressed in the new coordinates, has the form

$$u(n) = -f_1 \hat{\eta}_1(n) - f_2 \hat{\eta}_2(n) \tag{79}$$

with

$$\begin{aligned}
\hat{\eta}_1(n+1) &= \alpha_1 \hat{\eta}_1(n) + \epsilon \beta_1 v(n) \\
\hat{\eta}_2(n+1) &= \alpha_2 \hat{\eta}_2(n) + \beta_2 v(n)
\end{aligned} \tag{80}$$

where

$$\begin{aligned}
f_1 &= F_1 - F_2 L, & f_2 &= F_2 + \epsilon(F_1 - F_2 L)H \\
\alpha_1 &= I_{n_1} + \epsilon(a_{11} - a_{12}L), & \alpha_2 &= a_{22} + \epsilon L a_{12} \\
\beta_1 &= K_1 - H(K_2 + \epsilon L K_1), & \beta_2 &= K_2 + \epsilon L K_1
\end{aligned} \tag{81}$$

The innovation process $v(n)$ is now given by

$$v(n) = y(n) - d_1 \hat{\eta}_1(n) - d_2 \hat{\eta}_2(n) \tag{82}$$

where

$$d_1 = C_1 - \epsilon C_2 L, \quad d_2 = C_2 + \epsilon(C_1 - C_2 L)H \tag{83}$$

Near-optimum control law is defined by perturbing coefficients F_i, K_i, $i, j = 1, 2, L$ and H by $O(\epsilon^k)$, $k = 1, 2, ...$, in other words by using k-th approximations for these coefficients, where k stands for the required order of accuracy, that is

$$u^{(k)}(n) = -f_1^{(k)} \hat{\eta}_1^{(k)}(n) - f_2^{(k)} \hat{\eta}_2^{(k)}(n) \tag{84}$$

with

$$\begin{aligned}
\hat{\eta}_1^{(k)}(n+1) &= \alpha_1^{(k)} \hat{\eta}_1^{(k)}(n) + \epsilon \beta_1^{(k)} v^{(k)}(n) \\
\hat{\eta}_2^{(k)}(n+1) &= \alpha_2^{(k)} \hat{\eta}_2^{(k)}(n) + \beta_2^{(k)} v^{(k)}(n)
\end{aligned} \tag{85}$$

where

$$v^{(k)}(n) = y(n) - d_1^{(k)} \hat{\eta}_1^{(k)}(n) - d_2^{(k)} \hat{\eta}_2^{(k)}(n) \tag{86}$$

and

$$f_i^{(k)} = f_i + O(\epsilon^k), \quad d_i^{(k)} = d_i + O(\epsilon^k)$$
$$\alpha_i^{(k)} = \alpha_i + O(\epsilon^k), \quad \beta_i^{(k)} = \beta_i + O(\epsilon^k) \tag{87}$$
$$i = 1, 2$$

The approximate values of $J^{(k)}$ are obtained from the following expression

$$J^{(k)} = \frac{1}{2} E \left\{ \sum_{n=0}^{\infty} \left[x^{(k)^T}(n) D^T D x^{(k)}(n) + u^{(k)^T}(n) R u^{(k)}(n) \right] \right\}$$
$$= \frac{1}{2} tr \left\{ D^T D q_{11}^{(k)} + f^{(k)^T} R f^{(k)} q_{22}^{(k)} \right\} \tag{88}$$

where

$$q_{11}^{(k)} = Var \left\{ \left(x_1^{(k)} \ x_2^{(k)} \right)^T \right\}, \quad q_{22}^{(k)} = Var \left\{ \left(\hat{\eta}_1^{(k)} \ \hat{\eta}_2^{(k)} \right)^T \right\}$$
$$f^{(k)} = \left[f_1^{(k)} \ f_2^{(k)} \right] \tag{89}$$

Quantities $q_{11}^{(k)}$ and $q_{22}^{(k)}$ can be obtained by studying the variance equation of the following system driven by white noise

$$\begin{bmatrix} x^{(k)}(n+1) \\ \hat{\eta}^{(k)}(n+1) \end{bmatrix} = \begin{bmatrix} A & -Bf^{(k)} \\ \beta^{(k)}C & \alpha^{(k)} - \beta^{(k)}d^{(k)} \end{bmatrix} \begin{bmatrix} x^{(k)}(n) \\ \hat{\eta}^{(k)}(n) \end{bmatrix}$$
$$+ \begin{bmatrix} G & 0 \\ 0 & \beta^{(k)} \end{bmatrix} \begin{bmatrix} w(n) \\ v(n) \end{bmatrix} \tag{90}$$

where

$$\alpha^{(k)} = \begin{bmatrix} \alpha_1^{(k)} & 0 \\ 0 & \alpha_2^{(k)} \end{bmatrix}, \quad \beta^{(k)} = \begin{bmatrix} \epsilon \beta_1^{(k)} \\ \beta_2^{(k)} \end{bmatrix}, \quad d^{(k)} = \begin{bmatrix} d_1^{(k)} & d_2^{(k)} \end{bmatrix} \tag{91}$$

Equation (90) can be represented in a composite form

$$\Gamma^{(k)}(n+1) = \Lambda^{(k)} \Gamma^{(k)}(n) + \Pi^{(k)} w(n) \tag{92}$$

with obvious definitions for $\Lambda^{(k)}$, $\Pi^{(k)}$, $\Gamma^{(k)}(n)$, and $w(n)$. The variance of $\Gamma^{(k)}(n)$ at steady state denoted by $q^{(k)}$, is given by the discrete algebraic Lyapunov equation [30]

$$q^{(k)}(n+1) = \Lambda^{(k)} q^{(k)} \Lambda^{(k)^T} + \Pi^{(k)} \overline{W} \Pi^{(k)^T}, \quad \overline{W} = diag(W, V) \tag{93}$$

with $q^{(k)}$ partitioned as

$$q^{(k)} = \begin{bmatrix} q_{11}^{(k)} & q_{12}^{(k)} \\ q_{12}^{(k)^T} & q_{22}^{(k)} \end{bmatrix} \qquad (94)$$

On the other hand, the optimal value of J has the very well-known form, [30]

$$J^{opt} = \frac{1}{2} tr \left[D^T D \mathbf{Q} + \mathbf{P} K \left(C \mathbf{Q} C^T + V \right) K^T \right] \qquad (95)$$

where $\mathbf{P}$, $\mathbf{Q}$, F, and K are obtained from (61)-(64).

The near-optimality of the proposed approximate control law (84) is established in the following theorem.

Theorem 3 *Let x_1 and x_2 be optimal trajectories and J be the optimal value of the performance criterion. Let $x_1^{(k)}$, $x_2^{(k)}$, and $J^{(k)}$ be corresponding quantities under the approximate control law $u^{(k)}$ given by (84). Under the condition stated in Assumption 4 and the stabilizability-detectability subsystem assumptions, the following hold*

$$J^{opt} - J^{(k)} = O\left(\epsilon^k\right)$$
$$Var\left\{x_1 - x_1^{(k)}\right\} = O\left(\epsilon^{2k+1}\right) \qquad (96)$$
$$Var\left\{x_2 - x_2^{(k)}\right\} = O\left(\epsilon^{2k}\right) \qquad k = 0, 1, 2, \dots.$$

The proof of this theorem is rather lengthy and is omitted. It follows the ideas of Theorems 1 and 2 from [27]. In addition, due to the discrete nature of the problem, the proof of Theorem 3 utilizes the bilinear transformation from [31] which transforms the discrete Lyapunov equation (93) into the continuous one and compares it with the corresponding equation under the optimal control law. More about the proof can be found in [29].

A. CASE STUDY: DISCRETE STEAM POWER SYSTEM

A real world physical example, a fifth-order discrete model of a steam power system [32] demonstrates the efficiency of the proposed method. The problem matrices A and B are given by

$$A = \begin{bmatrix} 0.9150 & 0.0510 & 0.0380 & 0.015 & 0.038 \\ -0.030 & 0.889 & -0.0005 & 0.046 & 0.111 \\ -0.006 & 0.468 & 0.247 & 0.014 & 0.048 \\ -0.715 & -0.022 & -0.0211 & 0.240 & -0.024 \\ -0.148 & -0.003 & -0.004 & 0.090 & 0.026 \end{bmatrix}$$

$$B^T = \begin{bmatrix} 0.0098 & 0.122 & 0.036 & 0.562 & 0.115 \end{bmatrix}$$

Remaining matrices are chosen as

$$C = \begin{bmatrix} 1 & 1 & 0 & 0 & 0 \\ 0 & 0 & 1 & 1 & 1 \end{bmatrix}, \quad D^T D = diag\{5 \quad 5 \quad 5 \quad 5 \quad 5\}, \quad R = 1$$

It is assumed that $G = B$ and that white noise intensity matrices are given by

$$W = 5, \quad V_1 = 5, \quad V_2 = 5$$

It is shown [32] that this model possesses the singularly perturbed property with $n_1 = 2$, $n_2 = 3$, and $\epsilon = 0.264$.

The simulation results are presented in the following table.

Table III: Approximate values for the criterion

k	$J^{(k)}$	$J^{(k)} - J$
0	13.4918	0.229×10^{-1}
1	13.4825	0.136×10^{-1}
2	13.4700	0.110×10^{-2}
3	13.4695	0.600×10^{-3}
4	13.4690	1.000×10^{-4}
5	13.4689	$< 10^{-4}$
optimal	13.4689	

It can be seen from this table that the approximate solution has quite rapid convergence to the optimal solution. This table justifies the result of Theorem 3, that $J^{(k)} - J^{opt} = O(\epsilon)$. Notice that $(0.246)^6 = 3 \times 10^{-4}$.

IV. NEW METHODS FOR OPTIMAL CONTROL AND FILTERING

In this section, we introduce a completely new approach pretty much different than all other methods used so far in the theory of singular perturbations. It is well known that the main goal in the control theory of singular perturbations is to achieve the problem decomposition into slow and fast time scales. Our approach is based on a closed-loop decomposition technique which guarantees complete decomposition of the optimal filters and regulators and distribution of all required off-line and on-line computations.

In the regulation problem (optimal linear-quadratic control problem), we show how to decompose exactly the ill-defined discrete-time singularly perturbed algebraic Riccati equation into two reduced-order pure-slow and pure-fast well-defined continuous-time algebraic Riccati equations. Note that the reduced-order continuous-time algebraic Riccati equations are nonsymmetric, but their $O(\epsilon)$ approximations are symmetric ones. We show that the Newton method is very efficient for their solution since the initial guesses close an $O(\epsilon)$ to the exact solutions can be easily obtained from the results already available in [5].

In the filtering problem, in addition of using duality between filter and regulator to solve the discrete-time ill-defined filter algebraic Riccati equation in terms of the reduced-order pure-slow and pure-fast well-defined continuous-time algebraic Riccati equations, we have obtained completely independent pure-slow and pure-fast Kalman filters both driven by the system measurements and the system optimal control input. In the literature of linear stochastic singularly perturbed systems, it is possible to find exactly decomposed slow and fast Kalman filters ([27] for continuous-time systems, and [8] for discrete-time systems), but those filters are driven by the innovation process so that the additional communication channels have to be formed in order to construct the innovation process.

In the last part of this section we use the separation principle to solve the linear-quadratic Gaussian control problem of singularly perturbed discrete systems. Corresponding block diagram, with all required matrix coefficients, which represents clearly the proposed method, is given in Figure 1.

A. LINEAR-QUADRATIC CONTROL PROBLEM

In this section, we present a new approach in the study of the linear-quadratic control problem of singularly perturbed discrete systems. By applying the new algorithm developed, the discrete algebraic Riccati equation of singularly perturbed systems is completely and exactly decomposed into two reduced-order continuous-time algebraic Riccati equations. This decomposition allows us to design the linear controllers for slow and fast subsystems completely independently of each other and thus, to achieve the complete and exact separation for the linear-quadratic regulator problem.

Consider the singularly perturbed linear time-invariant discrete system described by [4]-[6]

$$x_1(k+1) = (I_{n_1} + \epsilon A_1)x_1(k) + \epsilon A_2 x_2(k) + \epsilon B_1 u(k), \quad x_1(0) = x_{10}$$

$$x_2(k+1) = A_3 x_1(k) + A_4 x_2(k) + B_2 u(k), \quad x_2(0) = x_{20} \qquad (97)$$

with slow variables $x_1 \in R^{n_1}$, fast state variables $x_2 \in R^{n_2}$, and control inputs $u \in R^m$, where ϵ is a small positive parameter. The performance criterion of the corresponding linear-quadratic control problem is represented by

$$J = \frac{1}{2} \sum_{k=0}^{\infty} \left[x(k)^T Q x(k) + u(k)^T R u(k) \right] \tag{98}$$

where

$$x(k) = \begin{bmatrix} x_1(k) \\ x_2(k) \end{bmatrix}, \quad Q = \begin{bmatrix} Q_1 & Q_2 \\ Q_2^T & Q_3 \end{bmatrix} \geq 0, \quad R > 0 \tag{99}$$

It is well known that the solution of the above optimal regulation problem is given by

$$\begin{aligned} u(k) &= -R^{-1} B^T \lambda(k+1) \\ &= -(R + B^T P B)^{-1} B^T P A x(k) \end{aligned} \tag{100}$$

where $\lambda(k)$ is a costate variable and P is the positive semi-definite stabilizing solution of the discrete algebraic Riccati equation [25]-[33] given by

$$\begin{aligned} P &= Q + A^T P [I + SP]^{-1} A \\ &= Q + A^T P A - A^T P B \left[R + B^T P B \right]^{-1} B^T P A \end{aligned} \tag{101}$$

The Hamiltonian form of (97) and (98) can be written as the forward recursion [33]

$$\begin{bmatrix} x(k+1) \\ \lambda(k+1) \end{bmatrix} = \mathbf{H} \begin{bmatrix} x(k) \\ \lambda(k) \end{bmatrix} \tag{102}$$

with

$$\mathbf{H} = \begin{bmatrix} A + BR^{-1}B^T A^{-T} Q & -BR^{-1}B^T A^{-T} \\ -A^{-T} Q & A^{-T} \end{bmatrix} \tag{103}$$

where $\mathbf{H}$ is the symplectic matrix which has the property that the eigenvalues of $\mathbf{H}$ can be grouped into two disjoint subsets Γ_1 and Γ_2, such that for every $\lambda_c \in \Gamma_1$ there exists $\lambda_d \in \Gamma_2$, which satisfies $\lambda_c \times \lambda_d = 1$, and we can choose either Γ_1 or Γ_2 to contain only the stable eigenvalues [34].

For the singularly perturbed discrete systems, corresponding matrices in (100)-(103) are given by

$$A = \begin{bmatrix} I_{n_1} + \epsilon A_1 & \epsilon A_2 \\ A_3 & A_4 \end{bmatrix}, \quad B = \begin{bmatrix} \epsilon B_1 \\ B_2 \end{bmatrix}, \quad S = BR^{-1}B^T = \begin{bmatrix} \epsilon^2 S_1 & \epsilon Z \\ \epsilon Z^T & S_2 \end{bmatrix}$$
$$S_1 = B_1 R^{-1} B_1^T, \quad S_2 = B_2 R^{-1} B_2^T, \quad Z = B_1 R^{-1} B_2^T \tag{104}$$

In this section, it is shown that the optimal LQ control problem of singularly perturbed discrete systems can be solved in terms of the completely independent slow and fast continuous-time algebraic Riccati equations.

The optimal open-loop control problem is a two-point boundary value problem with the associated state-costate equations forming the Hamiltonian matrix. For singularly perturbed discrete systems, after modifying some variables, the Hamiltonian matrix retains the singularly perturbed form by interchanging some state and costate variables so that it can be block diagonalized via the nonsingular transformation introduced in [28] (see also, [35]). In the following, we show how to get the solution of the discrete-time algebraic Riccati equation of singularly perturbed systems exactly from the solutions of two reduced-order continuous-time, slow and fast, algebraic Riccati equations.

Partitioning vector $\lambda(k)$ such that $\lambda(k) = \begin{bmatrix} \lambda_1^T(k) & \lambda_2^T(k) \end{bmatrix}^T$ with $\lambda_1(k) \in R^{n_1}$ and $\lambda_2(k) \in R^{n_2}$, we get

$$\begin{bmatrix} x_1(k+1) \\ x_2(k+1) \\ \lambda_1(k+1) \\ \lambda_2(k+1) \end{bmatrix} = \mathbf{H} \begin{bmatrix} x_1(k) \\ x_2(k) \\ \lambda_1(k) \\ \lambda_2(k) \end{bmatrix} \tag{105}$$

It has been shown in [3] that the Hamiltonian matrix (103) has the following form

$$\mathbf{H} = \begin{bmatrix} I_{n_1} + \epsilon \overline{A_1} & \epsilon \overline{A_2} & \epsilon^2 \overline{S_1} & \epsilon \overline{S_2} \\ \overline{A_3} & \overline{A_4} & \epsilon \overline{S_3} & \overline{S_4} \\ \overline{Q_1} & \overline{Q_2} & I_{n_1} + \epsilon \overline{A_{11}^T} & \overline{A_{21}^T} \\ \overline{Q_3} & \overline{Q_4} & \epsilon \overline{A_{12}^T} & \overline{A_{22}^T} \end{bmatrix} \tag{106}$$

Note that in the following there is no need for the analytical expressions for matrices with "bar". Those matrices have to be formed by the computer in the process of calculations, which can be done easily.

Interchanging second and third rows in (106) and setting $\begin{bmatrix} p_1(k) & p_2(k) \end{bmatrix}^T = \begin{bmatrix} \epsilon \lambda_1(k) & \lambda_2(k) \end{bmatrix}^T$ in (105) yield

$$\begin{bmatrix} x_1(k+1) \\ p_1(k+1) \\ x_2(k+1) \\ p_2(k+1) \end{bmatrix} = \begin{bmatrix} I_{n_1} + \epsilon \overline{A_1} & \epsilon \overline{S_1} & \epsilon \overline{A_2} & \epsilon \overline{S_2} \\ \epsilon \overline{Q_1} & I_{n_1} + \epsilon \overline{A_{11}^T} & \epsilon \overline{Q_2} & \epsilon \overline{A_{21}^T} \\ \overline{A_3} & \overline{S_3} & \overline{A_4} & \overline{S_4} \\ \overline{Q_3} & \overline{A_{12}^T} & \overline{Q_4} & \overline{A_{22}^T} \end{bmatrix} \begin{bmatrix} x_1(k) \\ p_1(k) \\ x_2(k) \\ p_2(k) \end{bmatrix}$$

$$= \begin{bmatrix} I + \epsilon T_1 & \epsilon T_2 \\ T_3 & T_4 \end{bmatrix} \begin{bmatrix} x_1(k) \\ p_1(k) \\ x_2(k) \\ p_2(k) \end{bmatrix} \tag{107}$$

where

$$T_1 = \begin{bmatrix} \overline{A_1} & \overline{S_1} \\ \overline{Q_1} & \overline{A_{11}^T} \end{bmatrix}, \quad T_2 = \begin{bmatrix} \overline{A_2} & \overline{S_2} \\ \overline{Q_2} & \overline{A_{21}^T} \end{bmatrix}$$

$$T_3 = \begin{bmatrix} \overline{A_3} & \overline{S_3} \\ \overline{Q_3} & \overline{A_{12}^T} \end{bmatrix}, \quad T_4 = \begin{bmatrix} \overline{A_4} & \overline{S_4} \\ \overline{Q_4} & \overline{A_{22}^T} \end{bmatrix} \tag{108}$$

Introducing the notation

$$U(k) = \begin{bmatrix} x_1(k) \\ p_1(k) \end{bmatrix}, \quad V(k) = \begin{bmatrix} x_2(k) \\ p_2(k) \end{bmatrix} \tag{109}$$

we have the singularly perturbed discrete system under new notation

$$\begin{aligned} U(k+1) &= (I + \epsilon T_1)U(k) + \epsilon T_2 V(k) \\ V(k+1) &= T_3 U(k) + T_4 V(k) \end{aligned} \tag{110}$$

Applying Chang's transformation [28] defined by

$$\mathbf{T_1} = \begin{bmatrix} I - \epsilon HL & -\epsilon H \\ L & I \end{bmatrix}, \quad \mathbf{T_1^{-1}} = \begin{bmatrix} I & \epsilon H \\ -L & I - \epsilon LH \end{bmatrix}$$
$$\begin{bmatrix} \eta(k) \\ \xi(k) \end{bmatrix} = \mathbf{T_1} \begin{bmatrix} U(k) \\ V(k) \end{bmatrix} \tag{111}$$

to (110) produces two completely decoupled subsystems

$$\begin{bmatrix} \eta_1(k+1) \\ \eta_2(k+1) \end{bmatrix} = \eta(k+1) = (I + \epsilon T_1 - \epsilon T_2 L)\eta(k) \tag{112}$$

$$\begin{bmatrix} \xi_1(k+1) \\ \xi_2(k+1) \end{bmatrix} = \xi(k+1) = (T_4 + \epsilon LT_2)\xi(k) \tag{113}$$

where L and H satisfy

$$H + T_2 - HT_4 + \epsilon(T_1 - T_2 L)H - \epsilon HLT_2 = 0 \tag{114}$$

$$-L + T_4 L - T_3 - \epsilon L(T_1 - T_2 L) = 0 \tag{115}$$

The unique solutions of (114) and (115) exist under condition that $(T_4 - I)$ is a nonsingular matrix. The algebraic equations (114) and (115) can be solved by using the Newton method [36], which converges quadratically, that is, with $O(\epsilon^{2^i})$ rate of convergence, in the neighborhood of the sought solution. The initial guess required in the Newton recursive scheme is easily obtained with the accuracy of $O(\epsilon)$, by setting $\epsilon = 0$ in that equation, that is

$$L^{(0)} = (T_4 - I)^{-1} T_3 = L + O(\epsilon) \tag{116}$$

Thus, the Newton sequence will be $O(\epsilon^2)$, $O(\epsilon^4)$, $O(\epsilon^8)$, ..., $O(\epsilon^{2^i})$ close to the exact solution, respectively, in each iteration.

The Newton-type algorithm can be constructed by setting $L^{(i+1)} = L^{(i)} + \Delta L^{(i)}$ and neglecting $O(\Delta L)^2$ terms. This will produce a Lyapunov-type equation of the form

$$D_1^{(i)} L^{(i+1)} + L^{(i+1)} D_2^{(i)} = Q^{(i)} \tag{117}$$

where

$$D_1^{(i)} = T_4 - I + \epsilon L^{(i)} T_2, \quad D_2^{(i)} = -\epsilon\left(T_1 - T_2 L^{(i)}\right)$$
$$Q^{(i)} = T_3 + \epsilon L^{(i)} T_2 L^{(i)}, \quad i = 0, 1, 2, \ldots \tag{118}$$

with the initial condition given by (116).

Having found the solution of (114) up to the desired degree of accuracy, one can get the solution of (115) by solving directly the algebraic Lyapunov equation of the form

$$H^{(i)} D_1^{(i)} + D_2^{(i)} H^{(i)} = T_2 \tag{119}$$

which implies $H^{(i)} = H + O\left(\epsilon^{2^i}\right)$.

The rearrangement and modification of variables in (107) is done by using the permutation matrix E_1 of the form

$$\begin{bmatrix} x_1(k) \\ p_1(k) \\ x_2(k) \\ p_2(k) \end{bmatrix} = \begin{bmatrix} I_{n_1} & 0 & 0 & 0 \\ 0 & 0 & \epsilon I_{n_1} & 0 \\ 0 & I_{n_2} & 0 & 0 \\ 0 & 0 & 0 & I_{n_2} \end{bmatrix} \begin{bmatrix} x_1(k) \\ x_2(k) \\ \lambda_1(k) \\ \lambda_2(k) \end{bmatrix} = E_1 \begin{bmatrix} x(k) \\ \lambda(k) \end{bmatrix} \tag{120}$$

From (109), (111)-(113), and (120), we obtain the relationship between the original coordinates and the new ones

$$\begin{bmatrix} \eta_1(k) \\ \xi_1(k) \\ \eta_2(k) \\ \xi_2(k) \end{bmatrix} = E_2^T T_1 E_1 \begin{bmatrix} x(k) \\ \lambda(k) \end{bmatrix} = \Pi \begin{bmatrix} x(k) \\ \lambda(k) \end{bmatrix} = \begin{bmatrix} \Pi_1 & \Pi_2 \\ \Pi_3 & \Pi_4 \end{bmatrix} \begin{bmatrix} x(k) \\ \lambda(k) \end{bmatrix} \tag{121}$$

where E_2 is a permutation matrix of the form

$$E_2 = \begin{bmatrix} I_{n_1} & 0 & 0 & 0 \\ 0 & 0 & I_{n_1} & 0 \\ 0 & I_{n_2} & 0 & 0 \\ 0 & 0 & 0 & I_{n_2} \end{bmatrix} \tag{122}$$

Since $\lambda(k) = Px(k)$, where P satisfies the discrete algebraic Riccati equation (101), it follows from (121) that

$$\begin{bmatrix} \eta_1(k) \\ \xi_1(k) \end{bmatrix} = (\Pi_1 + \Pi_2 P)x(k), \qquad \begin{bmatrix} \eta_2(k) \\ \xi_2(k) \end{bmatrix} = (\Pi_3 + \Pi_4 P)x(k) \qquad (123)$$

In the original coordinates, the required optimal solution has a closed-loop nature. We have the same characteristic for the new systems (112) and (113), that is,

$$\begin{bmatrix} \eta_2(k) \\ \xi_2(k) \end{bmatrix} = \begin{bmatrix} P_1 & 0 \\ 0 & P_2 \end{bmatrix} \begin{bmatrix} \eta_1(k) \\ \xi_1(k) \end{bmatrix} \qquad (124)$$

Then (123) and (124) yield

$$\begin{bmatrix} P_1 & 0 \\ 0 & P_2 \end{bmatrix} = (\Pi_3 + \Pi_4 P)(\Pi_1 + \Pi_2 P)^{-1} \qquad (125)$$

It can be easily shown from (121) that the inversion defined in (125) exists. Following the same logic, we can find P reversely by introducing

$$E_1^{-1}T_1^{-1}E_2 = \Omega = \begin{bmatrix} \Omega_1 & \Omega_2 \\ \Omega_3 & \Omega_4 \end{bmatrix} \qquad (126)$$

and it yields

$$P = \left(\Omega_3 + \Omega_4 \begin{bmatrix} P_1 & 0 \\ 0 & P_2 \end{bmatrix} \right) \left(\Omega_1 + \Omega_2 \begin{bmatrix} P_1 & 0 \\ 0 & P_2 \end{bmatrix} \right)^{-1} \qquad (127)$$

It is easy to show that required matrix in (127) is invertible. Partitioning (112) and (113) as

$$\begin{bmatrix} \eta_1(k+1) \\ \eta_2(k+1) \end{bmatrix} = \begin{bmatrix} a_1 & a_2 \\ a_3 & a_4 \end{bmatrix} \begin{bmatrix} \eta_1(k) \\ \eta_2(k) \end{bmatrix} = (I + \epsilon T_1 - \epsilon T_2 L) \begin{bmatrix} \eta_1(k) \\ \eta_2(k) \end{bmatrix} \qquad (128)$$

$$\begin{bmatrix} \xi_1(k+1) \\ \xi_2(k+1) \end{bmatrix} = \begin{bmatrix} b_1 & b_2 \\ b_3 & b_4 \end{bmatrix} \begin{bmatrix} \xi_1(k) \\ \xi_2(k) \end{bmatrix} = (T_4 + \epsilon L T_2) \begin{bmatrix} \xi_1(k) \\ \xi_2(k) \end{bmatrix} \qquad (129)$$

and using (129) yield to two reduced-order nonsymmetric algebraic Riccati equations

$$P_1 a_1 - a_4 P_1 - a_3 + P_1 a_2 P_1 = 0 \qquad (130)$$

$$P_2 b_1 - b_4 P_2 - b_3 + P_2 b_2 P_2 = 0 \qquad (131)$$

where

$$
\begin{bmatrix} a_1 & a_2 \\ a_3 & a_4 \end{bmatrix} = \begin{bmatrix} I_{n_1} + \epsilon\overline{A_1} - \epsilon\overline{A_2}L_1 - \epsilon\overline{S_2}L_3 & \epsilon\overline{S_1} - \epsilon\overline{A_2}L_2 - \epsilon\overline{S_2}L_4 \\ \epsilon\overline{Q_1} - \epsilon\overline{Q_2}L_1 - \epsilon\overline{A_{21}^T}L_3 & I_{n_1} + \epsilon\overline{A_{11}^T} - \epsilon\overline{A_2}L_2 - \epsilon\overline{S_2}L_4 \end{bmatrix}
$$
(132)

$$
\begin{bmatrix} b_1 & b_2 \\ b_3 & b_4 \end{bmatrix} = \begin{bmatrix} \overline{A_4} + \epsilon L_1\overline{A_2} + \epsilon L_2\overline{Q_2} & \overline{S_4} + \epsilon L_1\overline{S_2} + \epsilon L_2\overline{A_{21}^T} \\ \overline{Q_4} + \epsilon L_3\overline{A_2} + \epsilon L_4\overline{Q_2} & \overline{A_{22}^T} + \epsilon L_3\overline{S_2} + \epsilon L_4\overline{A_{21}^T} \end{bmatrix}
$$
(133)

with

$$
L = \begin{bmatrix} L_1 & L_2 \\ L_3 & L_4 \end{bmatrix}
$$
(134)

It is very interesting that the algebraic Riccati equation of singularly perturbed discrete-time control systems is completely and exactly decomposed into two reduced-order nonsymmetric continuous-time algebraic Riccati equations (130)-(131).

The pure-slow algebraic Riccati equation (130) is nonsymmetric and it is given by

$$
P_1\left(I_{n_1} + \epsilon\overline{A_1} - \epsilon\overline{A_2}L_1 - \epsilon\overline{S_2}L_3\right) - \left(I_{n_1} + \epsilon\overline{A_{11}^T} - \epsilon\overline{A_2}L_2 - \epsilon\overline{S_2}L_4\right)P_1
$$
$$
-\epsilon\left(\overline{Q_1} - \overline{Q_2}L_1 - \overline{A_{21}^T}L_3\right) + \epsilon P_1\left(\overline{S_1} - \overline{A_2}L_2 - \overline{S_2}L_4\right)P_1 = 0
$$
(135)

Simplifying (135) by cancelling P_1 in the first and the second terms, and dividing both sides by ϵ, we get the following

$$
P_1\left(\overline{A_1} - \overline{A_2}L_1 - \overline{S_2}L_3\right) - \left(\overline{A_{11}^T} - \overline{A_2}L_2 - \overline{S_2}L_4\right)P_1
$$
$$
- \left(\overline{Q_1} - \overline{Q_2}L_1 - \overline{A_{21}^T}L_3\right) + P_1\left(\overline{S_1} - \overline{A_2}L_2 - \overline{S_2}L_4\right)P_1 = 0
$$
(136)

The pure-fast algebraic Riccati equation (131) is also nonsymmetric

$$
P_2\left(\overline{A_4} + \epsilon\left(L_1\overline{A_2} + L_2\overline{Q_2}\right)\right) - \left(\overline{A_{22}^T} + \epsilon\left(L_3\overline{S_2} + L_4\overline{A_{21}^T}\right)\right)P_2
$$
$$
- \left(\overline{Q_4} + \epsilon\left(L_3\overline{A_2} + L_4\overline{Q_2}\right)\right) + P_2\left(\overline{S_4} + \epsilon\left(L_1\overline{S_2} + L_2\overline{A_{21}^T}\right)\right)P_2 = 0
$$
(137)

It can be shown that $O(\epsilon)$ perturbations of (136) and (137) lead to the symmetric reduced-order slow and fast algebraic Riccati equations obtained in [5]. The solutions of these equations, [5], can be used as very good initial guesses for the Newton method for solving the obtained nonsymmetric Riccati equations (136) and (137).

The Newton algorithm for (136) is given by

$$P_1^{(i+1)}\left(a_1 + a_2 P_1^{(i)}\right) - \left(a_4 - P_1^{(i)} a_2\right) P_1^{(i+1)} = a_3 + P_1^{(i)} a_2 P_1^{(i)}$$
$$i = 0, 1, 2, \dots \tag{138}$$

The Newton algorithm for (137) is similarly obtained as

$$P_2^{(i+1)}\left(b_1 + b_2 P_2^{(i)}\right) - \left(b_4 - P_2^{(i)} b_2\right) P_2^{(i+1)} = b_3 + P_2^{(i)} b_2 P_2^{(i)}$$
$$i = 0, 1, 2, \dots \tag{139}$$

Note that the nonsymmetric algebraic Riccati equations have been studied in [37]. An efficient algorithm for solving the general nonsymmetric algebraic Riccati equation is derived in [38].

The proposed method is very suitable for parallel computations since it allows complete parallelism. In addition, due to complete and exact decomposition of the discrete algebraic Riccati equation, the optimal control at steady state can be performed independently and in parallel in both slow and fast time scales. Pure-slow and pure-fast subsystems in the new coordinates are, respectively, given by

$$\eta_1(k + 1) = (a_1 + a_2 P_1)\eta_1(k) \tag{140}$$

$$\xi_1(k + 1) = (b_1 + b_2 P_2)\xi_1(k) \tag{141}$$

In summary, the optimal strategy and the optimal performance value are obtained by using the following algorithm.

Algorithm 3:

1) Solve Chang decoupling equations (114)-(115).
2) Find coefficients a_i, b_i, $i = 1, 2, 3, 4$ by using (128)-(129).
3) Solve the reduced-order exact pure-slow and pure-fast algebraic Riccati equations (130)-(131) which leads to P_1 and P_2.
4) Find the global solution of Riccati equation in terms of P_1 and P_2 by using (127).
5) Find the optimal regulator gain from (100) and the optimal performance criterion as $J_{opt} = 0.5x^T(t_0) P x(t_0)$.

1. CASE STUDY: DISCRETE MODEL OF AN F-8 AIRCRAFT

In order to demonstrate the efficiency of the proposed method a linearized model of an F-8 aircraft, presented in Section II, C, 1, is considered. The problem matrices are given in Section II, C, 1.

The optimal global solution of the discrete algebraic Riccati equation is obtained as

$$P_{exact} = \begin{bmatrix} 1.8460 & -0.0562 & -0.0111 & -0.0106 \\ -0.0562 & 2.0479 & -0.0594 & 0.0123 \\ -0.0111 & -0.0594 & 0.0228 & 0.0008 \\ -0.0106 & 0.0123 & 0.0008 & 0.0115 \end{bmatrix}$$

Solutions of the pure-slow and pure-fast algebraic Riccati equations obtained from (138) and (139) are

$$P_1 = \begin{bmatrix} 0.0608 & -0.0018 \\ -0.0007 & 0.0684 \end{bmatrix}, \quad P_2 = \begin{bmatrix} 0.0211 & 0.0012 \\ 0.0012 & 0.0114 \end{bmatrix}$$

By using the formula of (127), the obtained solution for P is found to be identical to P_{exact} and the error between the solution of the proposed method and the exact one which is obtained by using the classical global method for solving algebraic Riccati equation is given by

$$P_{exact} - P = O(10^{-15})$$

Assuming the initial conditions as $x^T(t_0) = \begin{bmatrix} 1 & 1 & 1 & 1 \end{bmatrix}$ the optimal performance value is $J_{opt} = 0.5x^T(0)Px(0) = 1.83995$.

B. NEW FILTERING METHOD FOR SINGULARLY PERTURBED LINEAR DISCRETE SYSTEMS

The continuous-time filtering problem of linear singularly perturbed systems has been studied by several researchers — in [27] by using the power series expansion approach, and later in [24] by using the fixed-point theory. The fixed-point approach to the discrete-time filtering of singularly perturbed systems has been presented in [8]. In this section, we will solve the filtering problem of linear discrete-time singularly perturbed system. The new method is based on the exact decomposition of the global singularly perturbed algebraic Riccati equation into pure-slow and pure-fast local algebraic Riccati equations. The optimal filter gain will be completely determined in terms of the exact pure-slow and exact pure-fast reduced-order continuous-time algebraic Riccati equations, based on the duality property between the optimal filter and regulator. Even more, we have obtained the exact expressions for the optimal pure-slow and pure-fast local filters both driven by the system measurements. This is an important advantage over the results of [8], [24], [27] where the local filters are driven by the innovation process so that the additional communication channels have to be used in order to construct the innovation process.

Consider the linear discrete-time invariant singularly perturbed stochastic system

$$
\begin{aligned}
x_1(k+1) &= (I_{n_1} + \epsilon A_1)x_1(k+1) + \epsilon A_2 x_2(k) + \epsilon G_1 w_1\,(k) \\
x_2(k+1) &= A_3 x_1(k) + A_4 x_2(k) + G_2 w_1(k) \\
x_1(0) &= x_{10}, \quad x_2(0) = x_{20}
\end{aligned}
\tag{142}
$$

with the corresponding measurements

$$
y(k) = C_1 x_1(k) + C_2 x_2(k) + w_2(k)
\tag{143}
$$

where $x_1 \in \mathbf{R}^{n_1}$ and $x_2 \in \mathbf{R}^{n_2}$ are state vectors, $w_1 \in \mathbf{R}^r$ and $w_2 \in \mathbf{R}^l$ are zero-mean stationary, white Gaussian noise stochastic processes with intensities $W_1 > 0$ and $W_2 > 0$, respectively, and $y \in \mathbf{R}^l$ are the system measurements. In the following A_i, G_j, C_j, $i = 1, 2, 3, 4$, $j = 1, 2$ are constant matrices.

The optimal Kalman filter, driven by the innovation process, is given by, [30]

$$
\hat{x}(k+1) = A\hat{x}(k) + K[y(k) - C\hat{x}(k)]
\tag{144}
$$

where

$$
A = \begin{bmatrix} I_{n_1} + \epsilon A_1 & \epsilon A_2 \\ A_3 & A_4 \end{bmatrix}, \quad C = [C_1 \ C_2], \quad K = \begin{bmatrix} \epsilon K_1 \\ K_2 \end{bmatrix}
\tag{145}
$$

The filter gain K is obtained from

$$
K = A P_F C^T (W_2 + C P_F C^T)^{-1}
\tag{146}
$$

where P_F is the positive semidefinite stabilizing solution of the discrete-time filter algebraic Riccati equation given by

$$
P_F = A P_F A^T - A P_F C^T (W_2 + C P_F C^T)^{-1} C P_F A^T + G W_1 G^T
\tag{147}
$$

where

$$
G = \begin{bmatrix} \epsilon G_1 \\ G_2 \end{bmatrix}
\tag{148}
$$

Due to the singularly perturbed structure of the problem matrices the required solution P_F in the fast time scale version has the form

$$
P_F = \begin{bmatrix} \epsilon P_{F1} & \epsilon P_{F2} \\ \epsilon P_{F2}^T & P_{F3} \end{bmatrix}
\tag{149}
$$

Partitioning the discrete-time filter Riccati equation given by (147), in the sense of singularly perturbation methodology, will produce a lot of terms and make the corresponding problem numerically inefficient, even though the problem order-reduction is achieved.

Using the decomposition procedure proposed in Section IV, A and the duality property between the optimal filter and regulator, we propose a new decomposition scheme such that the slow and fast filters of the singularly perturbed discrete systems are completely decoupled and both of them are driven by the system measurements. The new method is based on the exact pure-slow pure-fast decomposition technique, which is proposed in the previous section for solving the regulator algebraic Riccati equation of singularly perturbed discrete systems.

The results of interest which can be deduced from Section IV, A are summarized in the form of the following lemma.

Lemma 1 *Consider the optimal <u>closed-loop</u> linear discrete system*

$$
\begin{aligned}
x_1(k+1) &= (I + \epsilon A_1 - \epsilon B_1 F_1)x_1(k) + \epsilon(A_2 - B_1 F_2)x_2(k) \\
x_2(k+1) &= (A_3 - B_2 F_1)x_1(k) + (A_4 - B_2 F_2)x_2(k)
\end{aligned}
\tag{150}
$$

then there exists a nonsingular transformation **T**

$$
\begin{bmatrix} \xi_s(k) \\ \xi_f(k) \end{bmatrix} = \mathbf{T} \begin{bmatrix} x_1(k) \\ x_2(k) \end{bmatrix}
\tag{151}
$$

such that

$$
\begin{aligned}
\xi_s(k+1) &= (a_1 + a_2 P_{Rs})\xi_s(k) \\
\xi_f(k+1) &= (b_1 + b_2 P_{Rf})\xi_f(k)
\end{aligned}
\tag{152}
$$

where P_{Rs} and P_{Rf} are the unique solutions of the exact pure-slow and pure-fast completely decoupled algebraic Riccati equations presented in the previous section, that is

$$
\begin{aligned}
P_{Rs}a_1 - a_4 P_{Rs} - a_3 + P_{Rs}a_2 P_{Rs} &= 0 \\
P_{Rf}b_1 - b_4 P_{Rf} - b_3 + P_{Rf}b_2 P_{Rf} &= 0
\end{aligned}
\tag{153}
$$

Matrices a_i, b_i, $i = 1, 2, 3, 4$, can be found from (128)-(129). The nonsingular transformation **T** *is given by*

$$
\mathbf{T} = (\Pi_1 + \Pi_2 P_R)
\tag{154}
$$

Even more, the global solution P_R can be obtained from the reduced-order exact pure-slow and pure-fast algebraic Riccati equations, that is

$$
P_R = \left(\Omega_3 + \Omega_4 \begin{bmatrix} P_{Rs} & 0 \\ 0 & P_{Rf} \end{bmatrix} \right) \left(\Omega_1 + \Omega_2 \begin{bmatrix} P_{Rs} & 0 \\ 0 & P_{Rf} \end{bmatrix} \right)^{-1}
\tag{155}
$$

Known matrices Ω_i, $i = 1, 2, 3, 4$ and Π_1, Π_2 are given in terms of the solutions of the Chang decoupling equations in (114)-(115).

The desired slow-fast decomposition of the Kalman filter (144) will be obtained by using a duality between the optimal filter and regulator, and the same decomposition method developed in Section IV, B. Consider the optimal closed-loop Kalman filter (144) driven by the system measurements, that is

$$
\begin{aligned}
\hat{x}_1(k+1) &= (I + \epsilon A_1 - \epsilon K_1 C_1)\hat{x}_1(k) + \epsilon(A_2 - K_1 C_2)\hat{x}_2(k) + \epsilon K_1 y(k) \\
\hat{x}_2(k+1) &= (A_3 - K_2 C_1)\hat{x}_1(k) + (A_4 - K_2 C_2)\hat{x}_2(k) + K_2 y(k)
\end{aligned}
\tag{156}
$$

By using (142) and duality between the optimal filter and regulator, that is

$$
\begin{aligned}
A \to A^T, \quad Q &\to GW_1 G^T, \quad B \to C^T \\
BR^{-1}B^T &\to C^T W_2^{-1} C
\end{aligned}
\tag{157}
$$

the filter "state-costate equation" can be defined as

$$
\begin{bmatrix} x(k+1) \\ \lambda(k+1) \end{bmatrix} = \overline{\mathbf{H}} \begin{bmatrix} x(k) \\ \lambda(k) \end{bmatrix}
\tag{158}
$$

where

$$
\overline{\mathbf{H}} = \begin{bmatrix} A^T + C^T W_2^{-1} C A^{-1} GW_1 G^T & -C^T W_2^{-1} C A^{-1} \\ -A^{-1} GW_1 G^T & A^{-1} \end{bmatrix}
\tag{159}
$$

Partitioning $\lambda(k)$ as $\lambda(k) = \begin{bmatrix} \lambda_1^T(k) & \lambda_2^T(k) \end{bmatrix}^T$ with $\lambda_1(k) \in R^{n_1}$ and $\lambda_2(k) \in R^{n_2}$, (159) can be rewritten as following (see Appendix A)

$$
\begin{bmatrix} x_1(k+1) \\ x_2(k+1) \\ \lambda_1(k+1) \\ \lambda_2(k+1) \end{bmatrix} = \begin{bmatrix} I_{n_1} + \epsilon \overline{A_1^T} & \overline{A_3^T} & \overline{S_1} & \overline{S_2} \\ \epsilon \overline{A_2^T} & \overline{A_4^T} & \overline{S_3} & \overline{S_4} \\ \epsilon^2 \overline{Q_1} & \epsilon \overline{Q_2} & I_{n_1} + \epsilon \overline{A_{11}} & \epsilon \overline{A_{12}} \\ \epsilon \overline{Q_3} & \overline{Q_4} & \overline{A_{21}} & \overline{A_{22}} \end{bmatrix} \begin{bmatrix} x_1(k) \\ x_2(k) \\ \lambda_1(k) \\ \lambda_2(k) \end{bmatrix}
\tag{160}
$$

Interchanging the second and third rows in (160) and introducing $x(k) = \begin{bmatrix} \epsilon x_1^T(k) & x_2^T(k) \end{bmatrix}^T$ yield

$$
\begin{aligned}
\begin{bmatrix} \epsilon x_1(k+1) \\ \lambda_1(k+1) \\ x_2(k+1) \\ \lambda_2(k+1) \end{bmatrix} &= \begin{bmatrix} I_{n_1} + \epsilon \overline{A_1^T} & \epsilon \overline{S_1} & \epsilon \overline{A_3^T} & \epsilon \overline{S_2} \\ \epsilon \overline{Q_1} & I_{n_1} + \epsilon \overline{A_{11}} & \epsilon \overline{Q_2} & \epsilon \overline{A_{12}} \\ \overline{A_2^T} & \overline{S_3} & \overline{A_4^T} & \overline{S_4} \\ \overline{Q_3} & \overline{A_{21}} & \overline{Q_4} & \overline{A_{22}} \end{bmatrix} \begin{bmatrix} \epsilon x_1(k) \\ \lambda_1(k) \\ x_2(k) \\ \lambda_2(k) \end{bmatrix} \\
&= \begin{bmatrix} I + \epsilon T_1 & \epsilon T_2 \\ T_3 & T_4 \end{bmatrix} \begin{bmatrix} \epsilon x_1(k) \\ \lambda_1(k) \\ x_2(k) \\ \lambda_2(k) \end{bmatrix}
\end{aligned}
\tag{161}
$$

where

$$T_1 = \begin{bmatrix} \overline{A_1^T} & \overline{S_1} \\ \overline{Q_1} & \overline{A_{11}} \end{bmatrix}, \quad T_2 = \begin{bmatrix} \overline{A_3^T} & \overline{S_2} \\ \overline{Q_2} & \overline{A_{12}} \end{bmatrix}$$

$$T_3 = \begin{bmatrix} \overline{A_2^T} & \overline{S_3} \\ \overline{Q_3} & \overline{A_{21}} \end{bmatrix}, \quad T_4 = \begin{bmatrix} \overline{A_4^T} & \overline{S_4} \\ \overline{Q_4} & \overline{A_{22}} \end{bmatrix} \qquad (162)$$

These matrices comprise the system matrix of a standard singularly perturbed discrete system, namely

$$\begin{bmatrix} I + \epsilon T_1 & \epsilon T_2 \\ T_3 & T_4 \end{bmatrix}$$

so that the slow-fast decomposition can be achieved by applying the Chang transformation to (161), which yields two completely decoupled subsystems

$$\begin{bmatrix} \eta_1(k+1) \\ \eta_2(k+1) \end{bmatrix} = \begin{bmatrix} a_{1F} & a_{2F} \\ a_{3F} & a_{4F} \end{bmatrix} \begin{bmatrix} \eta_1(k) \\ \eta_2(k) \end{bmatrix} = [I + \epsilon(T_1 - T_2 L)] \begin{bmatrix} \eta_1(k) \\ \eta_2(k) \end{bmatrix} \quad (163)$$

$$\begin{bmatrix} \xi_1(k+1) \\ \xi_2(k+1) \end{bmatrix} = \begin{bmatrix} b_{1F} & b_{2F} \\ b_{3F} & b_{4F} \end{bmatrix} \begin{bmatrix} \xi_1(k) \\ \xi_2(k) \end{bmatrix} = [T_4 + \epsilon L T_2] \begin{bmatrix} \xi_1(k) \\ \xi_2(k) \end{bmatrix} \qquad (164)$$

Note that the decoupling transformation has the form of (111) with H and L matrices obtained from (114)-(115) with T_i's taken from (162). By duality and Lemma 1 the following reduced-order nonsymmetric algebraic Riccati equations exist

$$P_s a_{1F} - a_{4F} P_s - a_{3F} + P_s a_{2F} P_s = 0 \qquad (165)$$

$$P_f b_{1F} - b_{4F} P_f - b_{3F} + P_f b_{2F} P_f = 0 \qquad (166)$$

By using the permutation matrices

$$\begin{bmatrix} x_1(k) \\ \lambda_1(k) \\ x_2(k) \\ \lambda_2(k) \end{bmatrix} = E_1 \begin{bmatrix} x_1(k) \\ x_2(k) \\ \lambda_1(k) \\ \lambda_2(k) \end{bmatrix} \qquad (167)$$

with (note that E_1 is different than the corresponding one from Section IV, A)

$$E_1 = \begin{bmatrix} \epsilon I_{n1} & 0 & 0 & 0 \\ 0 & 0 & I_{n1} & 0 \\ 0 & I_{n_2} & 0 & 0 \\ 0 & 0 & 0 & I_{n2} \end{bmatrix}, \quad E_2 = \begin{bmatrix} I_{n1} & 0 & 0 & 0 \\ 0 & 0 & I_{n1} & 0 \\ 0 & I_{n2} & 0 & 0 \\ 0 & 0 & 0 & I_{n2} \end{bmatrix} \qquad (168)$$

we can define

$$\Pi_F = \begin{bmatrix} \Pi_{1F} & \Pi_{2F} \\ \Pi_{3F} & \Pi_{4F} \end{bmatrix} = E_2^T \begin{bmatrix} I - \epsilon H L & -\epsilon H \\ L & I \end{bmatrix} E_1 \qquad (169)$$

Then, the desired transformation is given by

$$T_2 = (\Pi_{1F} + \Pi_{2F} P_F) \qquad (170)$$

The transformation T_2 applied to the filter variables (156) as

$$\begin{bmatrix} \hat{\eta}_s \\ \hat{\eta}_f \end{bmatrix} = T_2^{-T} \begin{bmatrix} \hat{x}_1 \\ \hat{x}_2 \end{bmatrix} \qquad (171)$$

produces

$$\begin{bmatrix} \hat{\eta}_s(k+1) \\ \hat{\eta}_f(k+1) \end{bmatrix} = T_2^{-T} \begin{bmatrix} I + \epsilon A_1 - \epsilon K_1 C_1 & \epsilon(A_2 - K_1 C_2) \\ A_3 - K_2 C_1 & A_4 - K_2 C_2 \end{bmatrix} T_2^T \begin{bmatrix} \hat{\eta}_s(k) \\ \hat{\eta}_f(k) \end{bmatrix}$$
$$+ T_2^{-T} \begin{bmatrix} \epsilon K_1 \\ K_2 \end{bmatrix} y(k)$$

$$(172)$$

such that the complete closed-loop decomposition is achieved, that is

$$\hat{\eta}_s(k+1) = (a_{1F} + a_{2F} P_s)^T \hat{\eta}_s(k) + K_s y(k)$$
$$\hat{\eta}_f(k+1) = (b_{1F} + b_{2F} P_f)^T \hat{\eta}_f(k) + K_f y(k) \qquad (173)$$

where

$$\begin{bmatrix} K_s \\ K_f \end{bmatrix} = T_2^{-T} \begin{bmatrix} \epsilon K_1 \\ K_2 \end{bmatrix} \qquad (174)$$

It is important to point out that the matrix P_F in (170) can be obtained in terms of P_s and P_f instead of P_{Rs} and P_{Rf} by using (155), and Ω_{1F}, Ω_{2F}, Ω_{3F}, Ω_{4F} obtained from

$$\Omega_F = \begin{bmatrix} \Omega_{1F} & \Omega_{2F} \\ \Omega_{3F} & \Omega_{4F} \end{bmatrix} = E_1^{-1} \begin{bmatrix} I & \epsilon H \\ -L & I - \epsilon L H \end{bmatrix} E_2 \qquad (175)$$

A lemma dual to Lemma 1 can be now formulated as follows.

Lemma 2 *Given the* closed-loop *optimal Kalman filter (156) of a linear discrete singularly perturbed system. Then there exists a nonsingular transformation matrix (170), which completely decouples (156) into pure-slow and pure-fast local filters (173) both driven by the system measurements. Even more, the decoupling transformation (170) and the filter coefficients given in (173) can be* *obtained in terms of the exact pure-slow and pure-fast reduced-order completely decoupled Riccati equations (165) and (166).*

It should be noted that the new filtering method allows complete decomposition and parallelism between pure-slow and pure-fast filters. The complete solution to our problem can be summarized by the following algorithm.

Algorithm 4:

1)   Find T_1, T_2, T_3, and T_4 from (161).
2)   Calculate L and H from (114)–(115).
3)   Find a_{iF}, b_{iF}, for $i = 1, 2, 3, 4$ from (163)-(164).
4)   Solve for P_s and P_f from (165) and (166).
5)   Find $\mathbf{T_2}$ from (170) with P_F obtained from (155).
6)   Calculate K_s and K_f from (174).
7)   Find the pure-slow and pure-fast filter system matrices by using (173).

C. LINEAR-QUADRATIC GAUSSIAN OPTIMAL CONTROL PROBLEM

This section presents a new approach in the study of the LQG control problem of singularly perturbed discrete systems when the performance index is defined on an infinite-time period. The discrete-time LQG problem of a singularly perturbed system has been studied for the full state feedback in [8] and for the output feedback in [21]. We solve the LQG problem by using the results obtained in Sections IV, A and B. That is, the discrete algebraic Riccati equation, which is the main equation in the optimal control problem of the singularly perturbed discrete system, is completely and exactly decomposed into two reduced-order continuous-time algebraic Riccati equations.

Consider the singularly perturbed discrete linear stochastic system represented in the fast time scale by [8]

$$
\begin{aligned}
x_1(k+1) &= (I_{n_1} + \epsilon A_1)x_1(k) + \epsilon A_2 x_2(k) + \epsilon B_1 u(k) + \epsilon G_1 w_1(k) \\
x_2(k+1) &= A_3 x_1(k) + A_4 x_2(k) + B_2 u(k) + G_2 w_1(k) \\
y(k) &= C_1 x_1(k) + C_2 x_2(k) + w_2(k)
\end{aligned}
\tag{176}
$$

with the performance criterion

$$
J = \frac{1}{2}E\left\{ \sum_{k=0}^{\infty} \left[z^T(k)z(k) + u^T(k)Ru(k) \right] \right\}, \quad R > 0
\tag{177}
$$

where $x_i \in R^{n_i}$, $i = 1, 2$, comprise slow and fast state vectors respectively. $u \in R^m$ is the control input, $y \in R^l$ is the observed output, $w_1 \in R^r$ and

$w_2 \in R^l$ are independent zero-mean stationary Gaussian mutually uncorrelated white noise processes with intensities $W_1 > 0$ and $W_2 > 0$, respectively, and $z \in R^s$ is the controlled output given by

$$z(k) = D_1 x_1(k) + D_2 x_2(k) \tag{178}$$

All matrices are of appropriate dimensions and assumed to be constant.

The optimal control law of the system (176) with performance criterion (177) is given by [30]

$$u(k) = -F\hat{x}(k) \tag{179}$$

with the time-invariant filter

$$\hat{x}(k+1) = A\hat{x}(k) + Bu(k) + K[y(k) - C\hat{x}(k)] \tag{180}$$

where

$$A = \begin{bmatrix} I_{n_1} + \epsilon A_1 & \epsilon A_2 \\ A_3 & A_4 \end{bmatrix}, \quad B = \begin{bmatrix} \epsilon B_1 \\ B_2 \end{bmatrix}, \quad C = [C_1 \ C_2], \quad K = \begin{bmatrix} \epsilon K_1 \\ K_2 \end{bmatrix} \tag{181}$$

The regulator gain F and filter gain K are obtained from

$$F = (R + B^T P_R B)^{-1} B^T P_R A \tag{182}$$

$$K = A P_F C^T (W_2 + C P_F C^T)^{-1} \tag{183}$$

where P_R and P_F are positive semidefinite stabilizing solutions of the discrete-time algebraic regulator and filter Riccati equations [25], respectively given by

$$P_R = D^T D + A^T P_R A - A^T P_R B (R + B^T P_R B)^{-1} B^T P_R A \tag{184}$$

$$P_F = A P_F A^T - A P_F C^T (W_2 + C P_F C^T)^{-1} C P_F A^T + G W_1 G^T \tag{185}$$

where

$$D = [D_1 \ D_2], \quad G = \begin{bmatrix} \epsilon G_1 \\ G_2 \end{bmatrix} \tag{186}$$

The required solutions P_R and P_F in the fast time scale version have the forms

$$P_R = \begin{bmatrix} P_{R1}/\epsilon & P_{R2} \\ P_{R2}^T & P_{R3} \end{bmatrix}, \quad P_F = \begin{bmatrix} \epsilon P_{F1} & \epsilon P_{F2} \\ \epsilon P_{F2}^T & P_{F3} \end{bmatrix} \tag{187}$$

In obtaining the required solutions of (184) and (185) in terms of the reduced-order problems, in [7] a bilinear transformation technique introduced in [19] has used. The discrete-time algebraic Riccati equation is transformed into the continuous-time algebraic Riccati equation. In the method presented in this section, the exact decomposition method of the discrete algebraic regulator and filter Riccati equations produces two sets of two reduced-order nonsymmetric algebraic Riccati equations, that is, for the regulator

$$P_1 a_1 - a_4 P_1 - a_3 + P_1 a_2 P_1 = 0 \qquad (188)$$

$$P_2 b_1 - b_4 P_2 - b_3 + P_2 b_2 P_2 = 0 \qquad (189)$$

and for the filter

$$P_s a_{1F} - a_{4F} P_s - a_{3F} + P_s a_{2F} P_s = 0 \qquad (190)$$

$$P_f b_{1F} - b_{4F} P_f - b_{3F} + P_f b_{2F} P_f = 0 \qquad (191)$$

where the unknown coefficients are obtained from previous sections. The Newton algorithm can be used efficiently in solving the reduced-order nonsymmetric Riccati equations (188)-(191).

It has shown in Section IV, B that the optimal global Kalman filter, based on the exact decomposition technique, is decomposed into pure-slow and pure-fast local optimal filters both driven by the system measurements. As a result, the coefficients of the optimal pure-slow filter are functions of the solution of the pure-slow Riccati equation only and those of the pure-fast filter are functions of the solution of the pure-fast Riccati equation only. Thus, these two filters can be implemented independently in the different time scales (slow and fast). The pure-slow and pure-fast filters are, respectively, given by

$$\hat{\eta}_s(k+1) = (a_{1F} + a_{2F} P_s)^T \hat{\eta}_s(k) + K_s y(k) + B_s u(k)$$
$$\hat{\eta}_f(k+1) = (b_{1F} + b_{2F} P_f)^T \hat{\eta}_f(k) + K_f y(k) + B_f u(k) \qquad (192)$$

where

$$\begin{bmatrix} B_s \\ B_f \end{bmatrix} = T_2^{-T} B = (\Pi_{1F} + \Pi_{2F} P_F)^{-T} B \qquad (193)$$

It should be noted that the filtering method proposed for singularly perturbed linear discrete systems allows complete decomposition and parallelism between pure-slow and pure-fast filters.

The optimal control in the new coordinates has been obtained as

$$u(k) = -F\hat{x}(k) = -F T_2^T \begin{bmatrix} \hat{\eta}_s(k) \\ \hat{\eta}_f(k) \end{bmatrix} = -[F_s \quad F_f] \begin{bmatrix} \hat{\eta}_s(k) \\ \hat{\eta}_f(k) \end{bmatrix} \qquad (194)$$

where F_s and F_f are obtained from

$$[F_s \quad F_f] = F T_2^T = (R + B^T P_R B)^{-1} B^T P_R A (\Pi_{1F} + \Pi_{2F} P_F)^T \quad (195)$$

The optimal value of J is given by the very well-known form [30]

$$J_{opt} = \frac{1}{2} tr [D^T D P_F + P_R K (C P_F C^T + W_2) K^T] \quad (196)$$

where F, K, P_R, and P_F are obtained from (182)-(185).

The proposed scheme is given in Figure 1.

Figure 1: New method for filtering and control
of singularly perturbed linear discrete systems

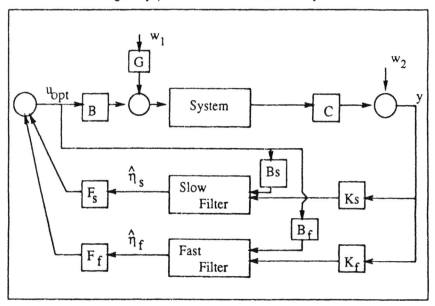

1. CASE STUDY: A STEAM POWER SYSTEM

In order to demonstrate the efficiency of the proposed method, we consider a real world control system — a fifth-order discrete model of a steam power system [32]. The system matrices are given by

$$A = \begin{bmatrix} 0.9150 & 0.0510 & 0.0380 & 0.0150 & 0.0380 \\ -0.0300 & 0.8890 & -0.0005 & 0.0460 & 0.1110 \\ -0.0060 & 0.4680 & 0.2470 & 0.0140 & 0.0480 \\ -0.7150 & -0.0220 & -0.0211 & 0.2400 & -0.0240 \\ -0.1480 & -0.0030 & -0.0040 & 0.0900 & 0.0260 \end{bmatrix}$$

$$B^T = [0.0098 \quad 0.1220 \quad 0.0360 \quad 0.5620 \quad 0.1150]$$

and other matrices are chosen as

$$C = \begin{bmatrix} 1 & 1 & 0 & 0 & 0 \\ 0 & 0 & 1 & 1 & 1 \end{bmatrix}, \quad D^T D = diag\{5 \quad 5 \quad 5 \quad 5 \quad 5\}, \quad R = I$$

It is assumed that $G = B$ and that the white noise processes are independent and have intensities

$$W_1 = 5, \quad W_2 = diag\{5 \quad 5\}$$

It is shown [32] that this model possesses the singularly perturbed property with $n_1 = 2$, $n_2 = 3$, and $\epsilon = 0.264$.

The obtained solutions for the LQG control problem are summarized as following. The completely decoupled filters driven by measurements y are given as

$$\hat{\eta}_s(k+1) = \begin{bmatrix} 0.8804 & 0.0428 \\ -0.0481 & 0.7824 \end{bmatrix} \hat{\eta}_s(k) + \begin{bmatrix} 0.1045 & 0.0643 \\ 0.1717 & 0.2780 \end{bmatrix} y(k)$$
$$+ \begin{bmatrix} 0.0629 \\ 0.3650 \end{bmatrix} u(k)$$

$$\hat{\eta}_f(k+1) = \begin{bmatrix} 0.2606 & -0.0112 & -0.0158 \\ -0.0533 & 0.1822 & -0.0585 \\ -0.0224 & 0.0662 & 0.0069 \end{bmatrix} \hat{\eta}_f(k)$$
$$+ \begin{bmatrix} -0.0044 & -0.0163 \\ 0.0164 & 0.0741 \\ 0.0067 & 0.0296 \end{bmatrix} y(k) + \begin{bmatrix} -0.0458 \\ 0.5590 \\ 0.1157 \end{bmatrix} u(k)$$

The feedback control in the new coordinates is given

$$u(k) = [0.1407 \quad -0.3068] \hat{\eta}_s(k) - [0.1918 \quad 0.3705 \quad 0.1019] \hat{\eta}_f(k)$$

The difference of the performance criterion between the optimal value, J_{opt}, and the one of the proposed method, J, is given by

$$J_{opt} = 6.73495$$
$$J - J_{opt} = 0.7727 \times 10^{-13}$$

It should be noted that the results represented here in solving (via the Newton method) recursively both of L and H matrices in (114)-(115), and reduced-order nonsymmetric algebraic Riccati equations for filter and regulator (188)-(191), are obtained by using the number of iterations of $i = 5$, respectively.

APPENDIX A

In Eq. (159) is given that

$$\overline{\mathbf{H}} = \begin{bmatrix} A^T + C^T W_2^{-1} C A^{-1} G W_1 G^T & -C^T W_2^{-1} C A^{-1} \\ -A^{-1} G W_1 G^T & A^{-1} \end{bmatrix} \quad \text{(A.1)}$$

From the structure of A^{-1} it is easy to see that the matrix A^T has the form

$$A^T = \begin{bmatrix} I_{n_1} + O(\epsilon) & O(1) \\ O(\epsilon) & O(1) \end{bmatrix} \quad \text{(A.2)}$$

so that

$$\begin{aligned} C^T W_2^{-1} C A^{-1} G W_1 G^T \\ = \begin{bmatrix} O(1) & O(1) \\ O(1) & O(1) \end{bmatrix} \begin{bmatrix} I_{n_1} + O(\epsilon) & O(\epsilon) \\ O(1) & O(1) \end{bmatrix} \begin{bmatrix} O(\epsilon^2) & O(\epsilon) \\ O(\epsilon) & O(1) \end{bmatrix} \\ = \begin{bmatrix} O(1) & O(1) \\ O(1) & O(1) \end{bmatrix} \begin{bmatrix} O(\epsilon^2) & O(\epsilon) \\ O(\epsilon) & O(1) \end{bmatrix} = \begin{bmatrix} O(\epsilon) & O(1) \\ O(\epsilon) & O(1) \end{bmatrix} \end{aligned}$$

$$A^T + C^T W_2^{-1} C A^{-1} G W_1 G^T = \begin{bmatrix} I_{n_1} + O(\epsilon) & O(1) \\ O(\epsilon) & O(1) \end{bmatrix}$$

$$C^T W_2^{-1} C A^{-1} = \begin{bmatrix} O(1) & O(1) \\ O(1) & O(1) \end{bmatrix} \begin{bmatrix} I_{n_1} + O(\epsilon) & O(\epsilon) \\ O(1) & O(1) \end{bmatrix} = \begin{bmatrix} O(1) & O(1) \\ O(1) & O(1) \end{bmatrix}$$

$$A^{-1} G W_1 G^T = \begin{bmatrix} I_{n_1} + O(\epsilon) & O(\epsilon) \\ O(1) & O(1) \end{bmatrix} \begin{bmatrix} O(\epsilon^2) & O(\epsilon) \\ O(\epsilon) & O(1) \end{bmatrix} = \begin{bmatrix} O(\epsilon^2) & O(\epsilon) \\ O(\epsilon) & O(1) \end{bmatrix}$$

$$\overline{\mathbf{H}} = \begin{bmatrix} I_{n_1} + \epsilon \overline{A_1^T} & \overline{A_3^T} & \overline{S_1} & \overline{S_2} \\ \epsilon \overline{A_2^T} & \overline{A_4^T} & \overline{S_3} & \overline{S_4} \\ \epsilon^2 \overline{Q_1} & \epsilon \overline{Q_2} & I_{n_1} + \epsilon \overline{A_{11}} & \overline{A_{12}} \\ \epsilon \overline{Q_3} & \overline{Q_4} & \overline{A_{21}} & \overline{A_{22}} \end{bmatrix} \quad \text{(A.3)}$$

Note that it is easy to obtain overlined matrices in the process of programming and it is of no interest to obtain corresponding analytical expressions.

V. REFERENCES

1. P. Kokotovic, H. Khalil, and J. O'Reilly, *Singular Perturbation Methods in Control: Analysis and Design*, Academic Press, Orlando, (1986).
2. P. Kokotovic and H. Khalil, *Singular Perturbations in Systems and Control*, IEEE Press, New York, 1986.
3. Z. Gajic and X. Shen, *Parallel Algorithms for Optimal Control of Large Scale Linear Systems*, Springer Verlag, London, (1993).
4. B. Litkouhi and H. Khalil, "Multirate and composite control of two-time-scale discrete systems," *IEEE Trans. Automatic Control* **AC-30**, pp. 645–651, (1985).
5. B. Litkouhi and H. Khalil, "Infinite-time regulators for singularly perturbed difference equations," *Int. J. Control* **39**, pp. 587–598, (1984).
6. B. Litkouhi, *Sampled-Data Control of Systems with Slow and Fast Modes*, Ph.D. Dissertation, Michigan State University, (1983).
7. Z. Gajic and X. Shen, "Study of the discrete singularly perturbed linear-quadratic control problem by a bilinear transformation," *Automatica* **27**, pp. 1025–1028, (1991).
8. Z. Gajic and X. Shen, "Parallel reduced-order controllers for stochastic linear singularly perturbed discrete systems," *IEEE Trans. Automatic Control* **AC-35**, pp. 87–90, (1991).
9. V. Butuzov and A. Vasileva, "Differential and difference equation systems with small parameter for the case in which the unperturbed (singular) system is in the spectrum," *J. Differential Equations* **6**, pp. 499–510, (1971).
10. F. Hoppensteadt and W. Miranker, "Multitime methods for systems of difference equations," *Studies Appl. Math.* **56**, pp. 273–289, (1977).
11. G. Blankenship, "Singularly perturbed difference equations in optimal control problems," *IEEE Trans. Automatic Control* **AC-26**, pp. 911–917, (1981).
12. M. Mahmoud, "Stabilization of discrete systems with multiple–time scales," *IEEE Trans. Automatic Control* **AC-31**, pp. 159–162, (1986).
13. H. Oloomi and W. Sawan, "The observer-based controller design of discrete-time singularly perturbed systems," *IEEE Trans. Automatic Control* **AC-32**, pp. 246–248, (1987).
14. K. Khorasani and M. Azimi-Sadjadi, "Feedback control of two-time scale block implemented discrete-time systems," *IEEE Trans. Automatic Control,* **AC-32**, pp. 69–73, (1987).
15. R. Phillips, "Reduced order modeling and control of two-time scale discrete control systems, *Int. J. Control* **31**, pp. 761–780, (1980).
16. D. Naidu and A. Rao, *Singular Perturbation Analysis of Discrete Control Systems*, Springer Verlag, Berlin, (1985).

17. H. Othman, N. Khraishi, and M. Maumoud, "Discrete regulators with time scale separation," *IEEE Trans. Automatic Control* **AC-30**, pp. 293–297, (1985).
18. M. Mahmoud, Y. Chen, and M. Singh, "Discrete two–time–scale systems," *Int. J. Systems Science* **17**, pp. 1187–1207, (1986).
19. R. Kondo and K. Furuta, "On the bilinear transformation of Riccati equations," *IEEE Trans. Automatic Control*, AC-31, pp. 50–54, (1986).
20. X. Shen, Z. Gajic, and D. Petkovski, "Parallel reduced–order algorithms for Lyapunov equations of large scale linear systems," *Proc. IMACS Symp. MCTS*, Lille, France, pp. 697–702, (1991).
21. M. Qureshi, X. Shen, and Z. Gajic, "Output feedback control of discrete linear singularly perturbed stochastic systems," *Int. J. Control* **55**, pp. 361–371, (1992).
22. G. Stewart, *Introduction to Matrix Computations*, New York, (1973).
23. Y. Bar-Ness and A. Halbersberg, "Solution of the singular discrete regulator problem using eigenvector methods," *Int. J. Control* **30**, pp. 615–625, (1980).
24. Z. Gajic, "Numerical fixed–point solution of linear quadratic guassian control problem for singularly perturbed systems," *Int. J. Control* **43**, pp. 373–387, (1986).
25. P. Dorato and A. Levis, "Optimal linear regulator: the discrete time case," *IEEE Trans. Automatic Control* **AC-16**, pp. 613–620, (1971).
26. J. Elliott, "NASA's advanced control law program for the F-8 digital fly-by-wire aircraft," *IEEE Trans. Automatic Control*, **AC-22**, pp. 753–757, (1977).
27. H. Khalil and Z. Gajic, "Near-optimum regulators for stochastic linear singularly perturbed systems," *IEEE Trans. Automatic Control*, **AC-29**, pp. 531–541, (1984).
28. K. Chang, "Singular Perturbations of a general boundary value problem," *SIAM J. Math. Anal.* **3**, pp. 502–526, 1972.
29. X. Shen, *Near-Optimum Reduced-Order Stochastic Control of Linear Discrete and Continuous Systems with Small Parameters,* Ph. D. Dissertation, Rutgers University, (1990).
30. H. Kwakernaak and R. Sivan, *Linear Optimal Control Systems*, Wiley Interscience, New York, (1972).
31. H. Power, "Equivalence of Lyapunov matrix equations for continuous and discrete systems," *Electronic Letters* **3**, 83, (1967).
32. M. Mahmoud, "Order reduction and control of discrete systems," *Proc. IEE*, Part D **129**, pp. 129–135, (1982).
33. F. Lewis, *Optimal Control*, Wiley, New York, (1986).
34. M. Salgado, R. Middleton, and G. Goodwin, "Connection between continuous and discrete Riccati equation with applications to Kalman filtering," *Proc. IEE*, Part D **135**, pp. 28–34, (1988).

35. M. Qureshi and Z. Gajic, "A new version of the Chang transformation," *IEEE Trans. Automatic Control* **AC-37**, pp. 800–801, (1992).

36. T. Grodt and Z. Gajic, "The recursive reduced order numerical solution of the singularly perturbed matrix differential Riccati equation," *IEEE Trans. Automatic Control* **AC-33**, pp. 751–754, (1988).

37. J. Medanic, "Geometric properties and invariant manifolds of the Riccati equation," *IEEE Trans. Automatic Control* **AC-27**, pp. 670–677, (1982).

38. B. Avramovic, P. Kokotovic, J. Winkelman, and J. Chow, "Area decomposition for electromechanical models of power systems," *Automatica* **16**, pp. 637–648, (1980).

Modeling Techniques and Control Architectures for Machining Intelligence

Allan D. Spence*
Yusuf Altintas

Department of Mechanical Engineering
The University of British Columbia
Vancouver, BC, CANADA

I. INTRODUCTION

To successfully participate in a worldwide market, manufacturing companies must rapidly adopt new technologies as they emerge. Failure to do so will leave the enterprise disadvantaged in comparison to its more responsive competitors. Companies in the machining or material removal processes industry, which produce industrial machinery parts, aircraft structural components and engines, and die casting and plastic injection molds, are an important part of the manufacturing sector, and must keep pace with technological advances [1].

Traditionally, machining has relied on skilled tradespeople. Since batch sizes were large, a trial and error approach to setting up the machine tools could be afforded. Cam operated mechanisms reproduced the same part time after time. After manually inspecting the parts the attending machine operator would, if necessary, readjust the machine tool until acceptable production was achieved. Today's customer, however, demands a customized product, which cannot be produced by the large batch manufacturing methods of the past. In addition, a low price, short delivery time, and high quality are needed to remain competitive.

*now with McMaster University, Hamilton, ON, CANADA

Current industry and research emphasis is therefore directed towards rapidly producing an acceptable part the first time, with maximum automation and minimum labor. To achieve these requirements, a sophisticated, automatic machining process realization is necessary. Such an implementation is referred to as an intelligent machining center, and has the following characteristics:

- Sensors are used to measure the important outputs (cutting force, torque, etc.) of the process (drilling, turning, milling, etc.) which is being performed.

- A part and process model is used to determine whether the process is progressing within specified norms. The machining center can decide to invoke a reaction strategy to correct a deviation from the norms.

- Actuators are used to adjust the process inputs to maintain the process within the specified norms.

- Memory and learning are used to automatically adjust the machining center's reaction strategies and correct its norms in response to feedback from longer term measurements such as cutting tool wear and final part shape. That is, the machine tool is self-improving.

All of these characteristic components are required for intelligent machining to be successful.

This chapter describes geometric and process models for machining processes, and discusses their implementation to provide useful knowledge to the machine tool controller. In the next section, the current status of machining automation is reviewed. Subsequent sections deal with geometric and process models, and the integration of the models with factory floor controllers to create an intelligent machining center. Continuing research areas are suggested in Section V, followed by a closing summary and references.

II. MACHINING PROCESS AUTOMATION

A. CURRENT STATUS

Existing machine tool automation is generally limited to open loop Computer Numerical Control (CNC), in which preprogrammed axes movements are sent to the machine tool controller as an instruction batch (Fig. 1). Coordinated motion between axes of the machine tool is accomplished by interpolation [2]. Control compensation to limit deviation of the actual path from the programmed path (contouring error) is automatically performed, but remains an area of active research interest [3].

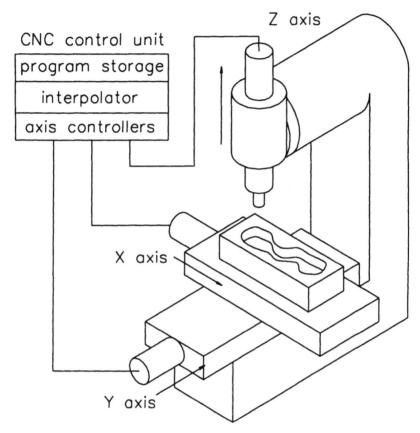

Fig. 1. Computer Numerical Control (CNC). Part contours are cut by executing preprogrammed motions which are coordinated by a central interpolator.

More recently, increased attention has been directed to monitoring and controlling the actual machining process, with the goal of reducing the need for operator attention, improving the Material Removal Rate (MRR), or decreasing the overall process cost. A brief example from each of these focus areas is given in the following sections.

1.   REDUCING THE NEED FOR OPERATOR ATTENTION

Due to the high cost of labor, efforts to safely reduce the need for operator attention at a machine tool remain a research priority. Two examples are the detection of tool breakage and excessive wear:

a.   Tool Breakage.   Suppose that the cutting tool edge breaks during a machining operation. It is important for an untended machine tool to monitor for this event, and, should it occur, to quickly halt the process. Otherwise, the now dull tool will continue advancing into the part, causing further damage to the part, tool holder, and machine. Cutting force monitoring [4] and acoustic emission techniques [5] have been proposed to detect this type of breakage.

b.   Tool Wear.   Heat from the cutting process, mechanical rubbing, and diffusion will wear down the edge of the tool. As this progresses, cutting forces will increase and the dimensional accuracy of the machined part will deteriorate. Representative research papers on detecting wear have been published for cutting forces [6], part diameter change measurement in turning [7], and vision measurement of the tool (out of cut) [8].

2.   IMPROVING THE MATERIAL REMOVAL RATE

The Material Removal Rate (MRR) achieved by offline Computer-Aided Machining (CAM) software rarely is the maximum since it is limited by the programmer's skill and incomplete knowledge of the actual cutting forces, tool deflection, chatter limit, etc. that will be experienced during the process. Automatic methods of providing fine tuning of the cutting conditions to maintain safe and stable cutting have therefore been developed. Some approaches for chatter avoidance and controlling cutting forces are discussed below:

a.   Chatter Avoidance.   Chatter is a severe vibration which occurs between the cutting tool and part. To avoid poor surface finish or tool breakage, it must rapidly be detected and the cutting conditions modified to prevent its recurrence. In high-speed milling, Smith and Delio [9] detect chatter by using an acoustic microphone. The dominant chatter frequency is recognized and the spindle speed then adjusted to a more stable choice. Assuming that a stable spindle speed is found, the feed rate is readjusted to maintain the originally specified feed per revolution. This is a valuable approach to automatically avoiding chatter in cases where the knowledge of the machine tool stiffness, etc. is limited. An extension to the approach would be to record stable spindle speeds vs. the radial immersion and depth of cut so that learning can occur. In the event that stable cutting cannot be achieved by spindle speed adjustment alone, the ability to reprogram a new cutter immersion would increase autonomy. Teltz and Elbestawi [10] implement this strategy for turning.

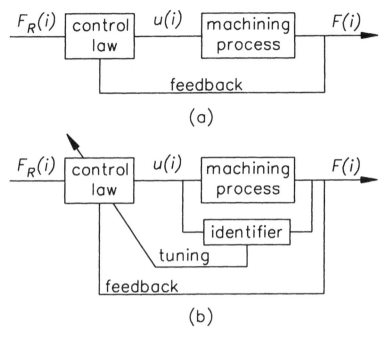

Fig. 2. (a) Feedback force control example. At each control interval i, the cutting force $F(i)$ is compared to its specified set point $F_R(i)$. The difference is used to compute a new feed rate $u(i + 1)$ which will bring $F(i + 1)$ closer to $F_R(i + 1)$. (b) Self-tuning control example. In addition to feedback control, the machining process transfer function is identified. This information is used to tune the controller so that a specified closed loop transfer function is maintained.

b.   Cutting Force Control.   Excessive cutting force will result in unacceptable part or tool deflection. In extreme cases either a thin part or slender cutting tool will break. To maintain a constant cutting force, adaptive control has frequently been the focus of manufacturing laboratory research. The most common approach is to use Adaptive Control with Constraints (ACC), with the measured cutting force $F(i)$ being fed back to adjust the machine tool feed rate $u(i)$ so that a reference cutting force level $F_R(i)$ is maintained (Fig. 2(a)).

Because the machining process transfer function varies depending on the intersection geometry between the cutting tool and the part, self-tuning regulators (Fig. 2(b)) have also been investigated for application to machining. In a self-tuning regulator implementation, the machining process transfer function is identified using a method such as Recursive Least Squares (RLS).

The control law is then modified to maintain a specified closed loop performance [11]. This type of implementation, however, has the practical difficulty that the controller has no advance warning of upcoming rapid changes in the machining process. When part geometry leads to a sudden increase in the cutter immersion, a large amplitude spike occurs in the cutting force, which will break the tool [12]. This issue is further discussed in Section IV, E.

In addition to the concerns listed above, other process limitations, such as available machine tool power and torque, must be respected. From a part geometry perspective, thermal errors are important. Additionally, collisions between the cutting tool and part during non-cutting motions must be avoided or at least detected. These issues are beyond the scope of the chapter.

3. DECREASING THE OVERALL PROCESS COST

Beyond in cut considerations, minimizing the overall cost per part requires that items such as the cutting tool cost, wear rate, and replacement time be evaluated. This is referred to as Adaptive Control with Optimization (ACO). For example, the unit cost accounting for tool wear and replacement may be written as [2, Ch. 8]

$$C = \frac{c_1 + (c_1 t_1 + c_2 \beta) W' / W_0}{MRR} \qquad (1)$$

where C is the unit volume cost; c_1 is the machine tool and operator cost per unit time; t_1 is the tool replacement time; c_2 is the tool replacement cost; W' is the tool wear rate per unit time; W_0 is the level of tool wear at which replacement is required and MRR is the Material Removal Rate. The parameter $0 \leq \beta \leq 1$ allows weighting to emphasize minimal unit volume cost ($\beta = 1$) or maximum production rate ($\beta = 0$). An attempt to implement ACO was made under the supervision of the United States Air Force [13].

B.   REVIEW

At present, many machine tools operate with CNC motion control only. Process sensors or adaptive control systems are not in widespread use. The reviewed machining process monitoring and control implementations are deficient in three key areas:

1. The control systems have been developed for independent use, although a concurrent implementation is required for industrial application. For example, a chatter avoidance procedure would likely

cause erratic behaviour when used in conjunction with a cutting force adaptive controller.

2. Knowledge of the workpiece and part shape to be machined is not exploited. This limits, for example, the diagnostic capabilities during tooth breakage detection, which might be mistaken for an entry or exit from cut transient. In the case that tooth breakage is confirmed, a good recovery strategy would be to move the tool along the normal away from the part surface to a safe replacement location. Without both workpiece and part knowledge, these strategic motions cannot be made.

3. There is no potential for learning in existing systems. Adaptive controllers have only a short-term memory, and no real knowledge of the underlying machining process. Recording of cutting conditions for comparison with tool wear rates is not performed. Without a longer term memory, the self-improvement or intelligence potential of the machine tool controller cannot be developed [14, Fig. 2-1].

The common element to overcoming the above deficiencies is to provide the machine tool controller with knowledge about the workpiece and part geometry, and the machining process to be used. Models are recommended as a well structured means of conveying this knowledge.

III.   MACHINING MODELS

Machining knowledge can be represented by two types of models:

1. geometric models of the initial workpiece shape, the desired final part shape, and the geometry of the cutting tools to be used

2. process models of the machining operation, which allow the relevant variables (cutting forces, torques, etc.) to be predicted and reconciled with online sensor observations and measured part shapes

Knowledge of both geometry and the machining process forms the skill of an experienced tradesworker [15, Ch. 7]. The challenge for implementing machining intelligence is to capture this knowledge in a model form which can be accessed by the machine tool controller. Methods for accomplishing this, for each of the model types, are reviewed in the following sections.

A.   GEOMETRIC MODELS

1.   SURVEY

A machine tool controller needs knowledge about both the initial workpiece shape, and the final part shape to be achieved. These descriptions are best represented by a solid model, for which three modeling approaches are prevalent [16]:

a. Constructive Solid Geometry (CSG). CSG is the most intuitive solid modeling approach. It describes space using primitive shapes such as blocks and cylinders, which are combined using Boolean set operations to build up a complex part. It has the advantage of always retaining an implicit, exact description of the part. Evaluation of the overall shape can be deferred until a calculation requires it.

b. Boundary Representation (B-rep). A B-rep solid model uses a directed graph linking surfaces, edges and vertices to represent the part. A surface normal defines the interior of the solid. This approach provides an explicit definition, and is useful when surface intersection calculations are required. A sculptured surface is represented by methods such as Non-Uniform Rational B-splines (NURBS) [17].

c. Spatial Enumeration (Octrees). This approach recursively subdivides a volume into cube regions, called voxels, until the space is exactly described or a specified tolerance is reached. An eight child hierarchical data structure, in which each voxel node is marked as occupied, vacant, or further subdivided, is used as the representation data structure. Spatial enumeration does not provide an exact mathematical description for non-prismatic part shapes. It is, however, well suited for applications where objects are measured using a regular grid.

2. REVIEW

Because it is exact, the CSG representation is likely to remain important in solid modeling. B-reps are frequently used is commercially available Computer-Aided Design (CAD) programs. Dual CSG/B-rep modelers offer the advantages of both schemes. Spatial decomposition is less exact, but is potentially the most convenient representation for localized calculations. In Section IV implementation of a geometric modeling method for machining simulation, and for use in an intelligent machine tool controller, is further discussed.

B. MACHINING PROCESS MODELS

In machining of metals, a sharp tool is moved at a velocity v relative to the part being cut (Fig. 3). Near the cutting edge, material of an uncut thickness h $(= u \sin \phi$ for milling [18]) is sheared, and flows along the tool

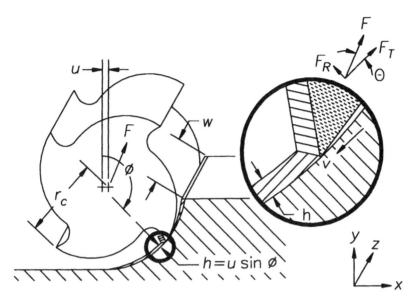

Fig. 3. Machining process overview. An end milling operation is depicted. Legend: r_c-cutter radius; u-feed rate per tooth; w-chip width (depth of cut); h-uncut chip thickness; v-cutting velocity; F-resultant cutting force; F_T-tangential cutting force component; F_R-radial cutting force component; x-parallel to feed rate direction; y-normal to feed rate direction; z-cutter axis direction; Θ-angle between F and F_T.

face, producing chips. The chip width is denoted by w. The resultant cutting force F is a function of conditions such as the part material, cutting velocity, and tool shape [19]. Heat generated during cutting is substantial, and is a primary contributor to wear of the tool edge. For common machining operations, the cutting velocity is adjusted by changing the rotational speed of either the tool (e.g. milling) or part (e.g. turning). The uncut chip thickness is adjusted by changing the feed rate per tooth u, and the chip width is adjusted by changing the depth of cut.

1. SURVEY

Prediction of the cutting force by analysis is complex due to the difficulty in predicting exact model coefficients, tool stiffness, and the machine tool stiffness and inertia. For this reason, mechanistic cutting force models are predominant. A review of milling process simulation models is presented in Smith and Tlusty [20]. The classifications used herein are based on this reference, and describe an end milling process. Other machining processes are modeled in a similar way.

a.   Rigid Force Model.   A widely used mechanistic model separates the cutting force into a component tangential to the cutting velocity, F_T, and a radial component F_R [21,22,23] (Fig. 3). The tangential force component is calculated using the expression

$$F_T = K_T \, wh \tag{2}$$

where K_T is a model coefficient obtained by experimental calibration. The corresponding radial force expression is

$$F_R = K_R \, F_T \tag{3}$$

where, again, K_R is an experimentally calibrated model coefficient. For a multiple toothed helical cutter, differential depth of cut slices dz are taken and the individual cutting force elements added vectorially.

The cutting torque T is determined from the relationship

$$T = r_c \sum_{j=1}^{J} F_{T,j} \tag{4}$$

where J is the total number of teeth. The resultant force F is the vector sum of all force components from all teeth. This force is frequently the controlled variable of an ACC control loop intended to prevent end mill shank breakage. Alternatively, the normal force component F_y can be controlled to adhere to a part dimensional error constraint.

b.   Deflection Feedback Model.   Since an end mill is flexible, the cutting force will deflect it back as it advances into the workpiece. To model this effect, Eq. (2) is changed to

$$F_T = K_T \, w(h - kF) \tag{5}$$

where k is the cutter compliance.

c.   Dynamic Models.   Dynamic milling process models also account for the cutting force variations caused by chip thickness waviness, and the machine tool stiffness and inertia. These models require high computational load time domain simulations, but are required for accurate chatter prediction.

2. REVIEW

The rigid force model is adequate for monitoring applications and feed rate scheduling. Inclusion of tool deflection provides a good first order milling process transfer function for ACC implementation. Chatter simulation requires use of dynamic models, which are beyond the scope of the chapter.

In the next section, the rigid force model is implemented with a CSG solid modeler to predict cutting forces and to automatically schedule feed rates which maintain specified cutting forces and limit part dimensional errors. An ACC force control system, based on the deflection feedback model, is described, and solid modeler data is used to avoid cutting force transients which otherwise would break a slender end mill.

IV. MODELING AND CONTROL INTEGRATION

In this section, an implementation of geometric and process modeling with online monitoring and control is described. The example application is rough end milling of aluminum parts with several geometric features. Cutting forces and tool deflections are calculated to schedule open loop feed rates, and to assist an ACC maximum resultant cutting force controller.

A. GEOMETRIC MODEL IMPLEMENTATION

To implement the milling process model, the uncut chip geometry must first be determined. For rough end milling, the initial part shape generally is a rectangular block or a cylinder. A flat end tool is used, and material is removed by a series of 2 1/2 dimensional terracing motions (Fig. 4). The chip width w is the axial immersion of the cutter in the part, and the instantaneous uncut chip thickness is $h = u \sin \phi$. End mills frequently have a helix angle ψ to reduce cutting force pulsations.

1. MODEL UPDATING

The CSG representation for this situation begins with a "primitive" block or cylinder which corresponds to the initial part shape. For linear interpolation, the material removed by each cutter motion is represented by a "swept volume" M of height w, and the plane area described by offsetting the motion line segment L in all directions by the cutter radius r_c (Fig. 5) [24]. A similar representation is used for circular interpolation motions.

After each cutter motion is completed, the CSG part model representation is updated by performing a regularized Boolean subtraction $-*$ [25]

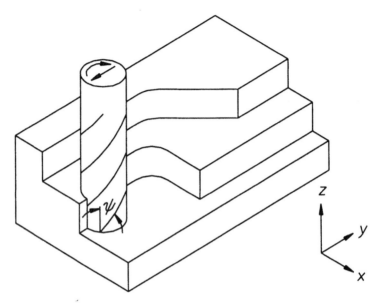

Fig. 4. Terracing motions. Rough end milling is frequently performed by a series of 2 1/2 D terracing motions. At each constant z level, the machine tool x and y axes move in coordinated motion to produce that level's contour. End mills frequently have a helix angle ψ to reduce the amplitude of cutting force pulsations.

of the swept volume from the previous representation. For example, Fig. 6 shows the initial part shape P after completion of two cutter motions with swept volumes M_1 and M_2. The CSG expression for this shape is $(P-*M_1)-*M_2$. For common operations such as pocketing, repeated part model updating is avoided by representing the swept volumes corresponding to a series of cutter motions by a single CSG subtree S_1 (Fig. 7). One update of the part model is then made at the completion of the cutter motion series.

In an integrated CAD/CAM modeling system, a programming facility to automatically plan the cutter paths for design features such as pockets is available. In this case, the subtree of swept volumes S_1 is replaced by the equivalent feature design subtree, which usually will be comprised of fewer primitives. A second alternative is to define an "extruded volume". This volume has height w and area bounded by the plane projection of the design feature. This approach requires the solid modeler to support a general extruded volume primitive, but greatly reduces the number of primitives needed in the part model CSG expression.

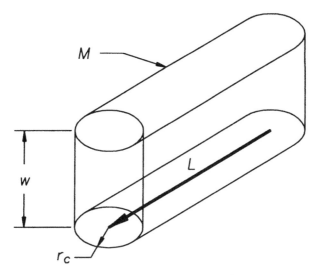

Fig. 5. Cutter motion swept volumes. For 2 1/2 D rough end milling, the material removed by a cutter motion is represented by a swept volume M of height w, and the plane area described by offsetting the motion line segment L by the cutter radius r_c.

2.   IMMERSION GEOMETRY SOLUTION

For 2 1/2 D applications the chip width is a constant w. The instantaneous uncut chip thickness, however, varies with the feed rate per tooth u and the rotational angle ϕ according to the expression $h = u \sin \phi$. This expression is valid only during the angular immersion intervals $\phi \in [\phi_{st,i}, \phi_{ex,i})$ within which the cutting edge is actually engaged in the part material. Otherwise $h = 0$. The important problem to be solved by the solid modeler is the determination of these angular immersion intervals.

For a flat end mill, it is convenient to model the tool as a 180 degree arc with radius equal to the tool radius r_c. This arc advances along the cutter motion path in increments of u, the feed rate per tooth. At each path increment the angular immersion intervals $[\phi_{st,i}, \phi_{ex,i})$ are determined by evaluating the local portion of the part CSG model. This procedure is briefly described below. For more complete details refer to [26,27].

First, the CSG part model is reduced to its locally relevant portion by using computational geometry techniques such as S-bounds [28], Active Zones [29], and isoboxes [30]. This "filtering" step typically eliminates much of the part model from consideration, and hence accelerates subsequent calculations. Design features and CSG subtrees are also ex-

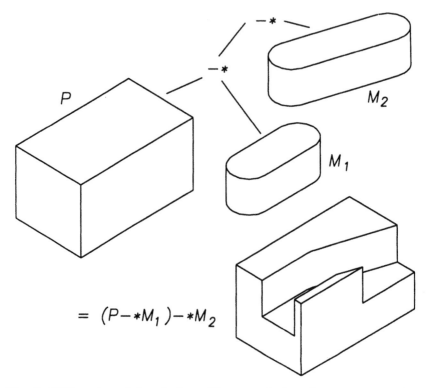

Fig. 6. CSG part model updating. After completion of two cutter motions with swept volumes M_1 and M_2, the resulting part shape expression is $(P-*M_1)-*M_2$.

ploited. To illustrate, Fig. 8 shows an initial part P after features S_1, S_2 and S_3 have been machined. The updated part model expression is therefore $((P-*S_1)-*S_2)-*S_3$. Since the cutter path swept volumes for feature S_4 are disjoint from the first three features, the part model representation can be simplified to P for local cutter arc calculations. Furthermore, since S_4 is entirely within horizontal extents of P, P itself can be replaced by the half space H which lies below the top surface of P.

The 180 degree cutter arc is then intersected with each primitive in the reduced part model, and the angular immersion intervals within which the arc is IN/OUT of cut with each primitive are recorded. The immersion intervals from each primitive are then recursively merged using the rules listed in Table I (Fig. 9). This process continues until the root of the CSG part model data structure tree is reached. The speed at which these calculations are performed depends on the success of the CSG primitive filtering

$S_1 =$

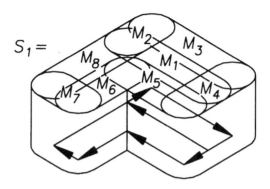

Fig. 7. Swept volume subtrees. Swept volumes M_i for a series of cutter motions, such as in pocketing, are grouped into a subtree S_1. In this example $S_1 = M_1 \cup_* M_2 \cup_* \ldots \cup_* M_8$, where $\cup_*$ is the regularized Boolean union operator. A single update $P \leftarrow P-_*S_1$ is then applied to the part model.

Table I. Angular immersion interval merging rules. The angular immersion interval merging rules for the regularized Boolean union $\cup_*$, intersection $\cap_*$, and difference $-_*$ operators are listed below. For an illustration, see Fig. 9.

A	B	$A \cup_* B$	$A \cap_* B$	$A -_* B$
IN	IN	IN	IN	OUT
IN	OUT	IN	OUT	IN
OUT	IN	IN	OUT	OUT
OUT	OUT	OUT	OUT	OUT

step. More sophistication is required to filter effectively, but the result is a substantial reduction in the time to perform the cutter arc intersection and recursive immersion interval merging calculations.

B.   PROCESS MODEL IMPLEMENTATION

After the cutter-part intersection geometry has been obtained, the milling process model can be implemented. For the purpose of obtaining monitoring and feed rate scheduling data in 2 1/2 dimensional rough machining applications, the mechanistic rigid force model (Eqs. (2) and (3) in Section III, B) is adequate.

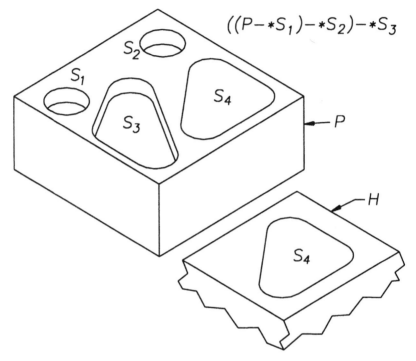

Fig. 8. Locally relevant CSG part models. This example illustrates an initial part P after features S_1, S_2 and S_3 have been machined. The updated part model expression is therefore $((P-*S_1)-*S_2)-*S_3$. Since the cutter path swept volumes for feature S_4 are disjoint from the first three features, the part model expression can be simplified to P for local cutter arc calculations. Furthermore, since S_4 is entirely within the horizontal extents of P, P itself can be replaced by the half space H which lies below the top surface of P.

1.   CALIBRATION

Calibration of the model coefficients K_T and K_R is accomplished by taking average cutting force measurements at several feed rates and angular immersion interval geometries. Specifically

$$\overline{F}_x = \frac{1}{2\pi} \int_0^{2\pi} F_x(\phi)\, d\phi; \quad \overline{F}_y = \frac{1}{2\pi} \int_0^{2\pi} F_y(\phi)\, d\phi \qquad (6)$$

Since the cutting forces are periodic and any end mill helix angle has no influence [31], it is sufficient to consider a single, zero helix cutting edge at

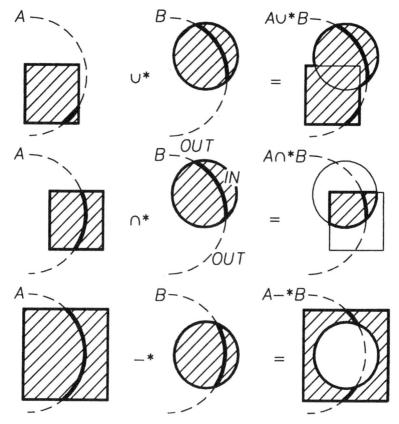

Fig. 9. Angular immersion interval merging. The immersion intervals from each primitive are recursively merged using the rules listed in Table I. Examples are shown for each of the regularized Boolean union ∪∗, intersection ∩∗ and difference −∗ operators. The broken lines indicate OUT intervals. Wider, solid lines represent IN intervals. The process continues until the root of the CSG part model data structure is reached.

rotation angle ϕ, and write

$$\overline{F}_x = \sum_i \frac{J}{2\pi} \int_{\phi_{st,i}}^{\phi_{ex,i}} F_{x,0}(\phi)\,d\phi \qquad (7)$$

where J is the number of (equally spaced) teeth on the end mill and i indexes the angular immersion intervals $[\phi_{st,i}, \phi_{ex,i})$. The contribution by the single tooth is denoted by $F_{x,0}(\phi)$.

From Fig. 3, the uncut chip thickness is $h = u \sin \phi$ and hence

$$
\begin{aligned}
F_{x,0}(\phi) &= -F_{T,0}(\phi) \cos \phi - F_{R,0}(\phi) \sin \phi \\
&= -F_{T,0}(\phi)[\cos \phi + K_R \sin \phi] \\
&= -K_T wu \sin \phi [\cos \phi + K_R \sin \phi] \\
&= -K_T wu[\sin \phi \cos \phi + K_R \sin^2 \phi] \\
&= -0.5 K_T wu[\sin 2\phi + K_R(1 - \cos 2\phi)]
\end{aligned} \tag{8}
$$

Similarly

$$
F_{y,0}(\phi) = 0.5 K_T wu[(1 - \cos 2\phi) - K_R \sin 2\phi] \tag{9}
$$

and hence

$$
\overline{F}_x = K_T u(\mathcal{R} + \mathcal{S} K_R); \quad \overline{F}_y = K_T u(-\mathcal{S} + \mathcal{R} K_R) \tag{10}
$$

where the geometric parameters $\mathcal{R}$ and $\mathcal{S}$ are defined as

$$
\mathcal{R} = \frac{wJ}{8\pi} \sum_i [\cos 2\phi]_{\phi_{st,i}}^{\phi_{ex,i}}; \quad \mathcal{S} = \frac{wJ}{8\pi} \sum_i [\sin 2\phi - 2\phi]_{\phi_{st,i}}^{\phi_{ex,i}} \tag{11}
$$

Simultaneously solving Eqs. (10) yields

$$
K_T = \frac{\overline{F}_x \mathcal{R} - \overline{F}_y \mathcal{S}}{u(\mathcal{R}^2 + \mathcal{S}^2)}; \quad K_R = \frac{\overline{F}_y \mathcal{R} + \overline{F}_x \mathcal{S}}{\overline{F}_x \mathcal{R} - \overline{F}_y \mathcal{S}} \tag{12}
$$

This solution for K_T and K_R is specific to the corresponding average chip thickness $\overline{h}$ at that point along the cutter path, defined as

$$
\begin{aligned}
\overline{h} &= \frac{1}{2\pi} \int_0^{2\pi} h(\phi)\, d\phi \\
&= \frac{1}{2\pi} \int_0^{2\pi} u \sin \phi\, d\phi \\
&= \frac{\sum_i \int_{\phi_{st,i}}^{\phi_{ex,i}} u \sin \phi\, d\phi}{\sum_i (\phi_{ex,i} - \phi_{st,i})} \\
&= -u \frac{\sum_i (\cos \phi_{ex,i} - \cos \phi_{st,i})}{\sum_i (\phi_{ex,i} - \phi_{st,i})}
\end{aligned} \tag{13}
$$

Calibration can now be performed by taking several average force measurements $\overline{F}_x$ and $\overline{F}_y$, solving Eqs. (12) for K_T and K_R, calculating the average chip thickness $\overline{h}$ from Eq. (13), and fitting the data to the model coefficient expressions

$$
\begin{aligned}
\log K_T(\overline{h}) &= \log M_T - P_T \log \overline{h} \\
\log K_R(\overline{h}) &= \log M_R - P_R \log \overline{h}
\end{aligned} \tag{14}
$$

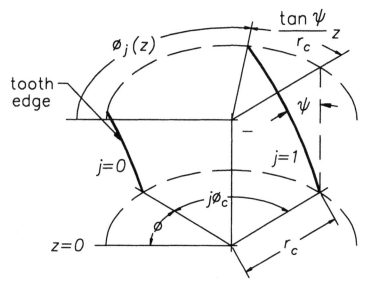

Fig. 10. End mill helix angles. End mills frequently have a helix angle ψ. In such a case at level $z = 0$ tooth j is at immersion angle $\phi_j(0) = \phi + j\phi_c$, where $\phi_c = 2\pi/J$ is the angular spacing between successive teeth. At an arbitrary level z the immersion angle is $\phi_j(z) = \phi + j\phi_c - k_\psi z$ where $k_\psi = \tan\psi/r_c$ and r_c is the cutter radius.

in the least squares sense. The calibrated model can now be used to predict machining process quantities such as the average torque, power, etc.

2. INSTANTANEOUS FORCE CALCULATION

Calculation of the instantaneous cutting forces and tool deflection requires consideration of the end mill helix angle ψ so that the in cut portion of each tooth can be determined (Fig. 10). In such a case at level $z = 0$ tooth j is at immersion angle $\phi_j(0) = \phi + j\phi_c$, where $\phi_j = 2\pi/J$ is the angular spacing between successive teeth and J is the total number of cutter teeth. At an arbitrary level z the immersion angle is $\phi_j(z) = \phi_j(0) - k_\psi z = \phi + j\phi_c - k_\psi z$ where $k_\psi = \tan\phi/r_c$ and r_c is the cutter radius.

For 2 1/2 dimensional angular immersion intervals $[\phi_{st,i}, \phi_{ex,i})$, the in cut portion of each tooth is determined by imagining that the part wall has been "unrolled" to form flat faces bounded by vertical lines $\phi_j = \phi_{st,i}$ and $\phi_j = \phi_{ex,i}$, and horizontal lines $z = 0$ and $z = w$. Depending on the instantaneous tooth rotation angle $\phi_j(0)$ and helix angle ψ, the tooth intersects each face in one of five cases (Fig. 11). The corresponding axial limits $z_{j,1}$ and $z_{j,2}$ for each case are:

case 0 $z_{j,1} = 0$; $z_{j,2} = w$

case 1 $z_{j,1} = 0$; $z_{j,2} = (1/k_\psi)[\phi + j\phi_c - \phi_{st,i}]$

case 2 $z_{j,1} = (1/k_\psi)[\phi + j\phi_c - \phi_{ex,i}]$; $z_{j,2} = w$

case 3 $z_{j,1} = (1/k_\psi)[\phi + j\phi_c - \phi_{ex,i}]$; $z_{j,2} = (1/k_\psi)[\phi + j\phi_c - \phi_{st,i}]$

case 4 the flute is not engaged with this face

Note that when the axial depth of cut is very large the full set of in cut flute sections is found by stepping ϕ_j in multiples of 2π. Critical angles occur during the rotation period each time a tooth j changes its intersection case with an immersion interval part face.

To calculate the instantaneous x and y forces, consider an element at rotation angle ϕ and height z. The elemental tangential $dF_{T,j}(\phi, z)$ and radial $dF_{T,j}(\phi, z)$ forces on tooth edge j are expressed as

$$dF_{T,j}(\phi, z) = K_T u \sin\phi_j(z); \quad dF_{R,j}(\phi, z) = K_R dF_{T,j}(\phi, z) \quad (15)$$

where u is the feed rate per tooth. At any specific location along the cutter path being simulated the angular immersion interval set $[\phi_{st,i}, \phi_{ex,i})$, and the average chip thickness $\bar{h}$, are unchanging. The angle Θ between $dF_{T,j}$ and the resultant dF_j is therefore a fixed $\Theta = \arctan K_R$ (Fig. 3).

Resolving into x and y directions

$$\begin{aligned} dF_{x,j}(\phi, z) &= -dF_j(\phi, z)\cos(\phi_j(z) - \Theta) \\ dF_{y,j}(\phi, z) &= dF_j(\phi, z)\sin(\phi_j(z) - \Theta) \end{aligned} \quad (16)$$

where

$$\begin{aligned} dF_j(\phi, z) &= [dF_{T,j}^2(\phi, z) + dF_{R,j}^2(\phi, z)]^{1/2} \\ &= (1 + K_R^2)^{1/2} K_T u \sin\phi_j(z)\, dz \end{aligned} \quad (17)$$

Integrating

$$\begin{aligned} F_{x,j}(\phi) &= -0.5 u K_T (1 + K_R^2)^{1/2} \cdot \\ &\quad \int_{z_{j,1}}^{z_{j,2}} [\sin 2\phi_j(z)\cos\Theta + (1 - \cos 2\phi_j(z))\sin\Theta]\, dz \\ F_{y,j}(\phi) &= 0.5 u K_T (1 + K_R^2)^{1/2} \cdot \\ &\quad \int_{z_{j,1}}^{z_{j,2}} [(1 - \cos 2\phi_j(z))\cos\Theta - \sin 2\phi_j(z)\sin\Theta]\, dz \end{aligned} \quad (18)$$

For each angular immersion interval $[\phi_{st,i}, \phi_{ex,i})$, the values for the integration limits $z_{j,1}$ and $z_{j,2}$ are determined using the cases shown in Fig. 11, substituted into Eqs. (18), and the resulting expressions simplified. For

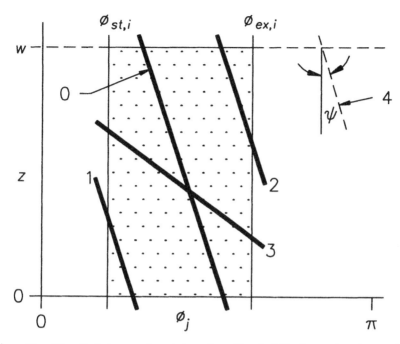

Fig. 11. Unrolled immersion intervals. For 2 1/2 dimensional angular immersion intervals $[\phi_{st,i}, \phi_{ex,i})$, the in cut portion of each tooth is determined by imagining that the part wall has been "unrolled" to form flat faces bounded by vertical lines $\phi_j = \phi_{st,i}$ and $\phi_j = \phi_{ex,i}$, and horizontal lines $z = 0$ and $z = w$. Depending on the instantaneous tooth rotation angle $\phi_j(0)$ and helix angle ψ, the tooth intersects each face in one of the cases 0 through 4.

computer implementation geometric summation coefficients $\mathcal{A}$–$\mathcal{E}$ are defined for each case. The total coefficient value is determined by summing the contributions from each combination (see below)

case 0

$$\mathcal{A} \leftarrow \mathcal{A} - w$$
$$\mathcal{B} \leftarrow \mathcal{B} + (1/2k_\psi)[\sin 2(j\phi_c - k_\psi w) - \sin 2j\phi_c]$$
$$\mathcal{D} \leftarrow \mathcal{D} + (1/2k_\psi)[\cos 2(j\phi_c - k_\psi w) - \cos 2j\phi_c]$$

case 1

$$\mathcal{A} \leftarrow \mathcal{A} - (1/2k_\psi)\sin 2\phi_{st,i} - (1/k_\psi)[j\phi_c - \phi_{st,i}]$$
$$\mathcal{B} \leftarrow \mathcal{B} - (1/2k_\psi)\sin 2j\phi_c$$
$$\mathcal{C} \leftarrow \mathcal{C} + (1/2k_\psi)\cos 2\phi_{st,i}$$
$$\mathcal{D} \leftarrow \mathcal{D} - (1/2k_\psi)\cos 2j\phi_c$$

$$\mathcal{E} \quad \leftarrow \quad \mathcal{E} - 1/k_\psi$$

case 2

$$
\begin{aligned}
\mathcal{A} &\leftarrow \mathcal{A} - w + (1/2k_\psi)\sin 2\phi_{ex,i} + (1/k_\psi)[j\phi_c - \phi_{ex,i}] \\
\mathcal{B} &\leftarrow \mathcal{B} + (1/2k_\psi)\sin 2(j\phi_c - k_\psi w) \\
\mathcal{C} &\leftarrow \mathcal{C} - (1/2k_\psi)\cos 2\phi_{ex,i} \\
\mathcal{D} &\leftarrow \mathcal{D} + (1/2k_\psi)\cos 2(j\phi_c - k_\psi w) \\
\mathcal{E} &\leftarrow \mathcal{E} + 1/k_\psi
\end{aligned}
$$

case 3

$$
\begin{aligned}
\mathcal{A} &\leftarrow \mathcal{A} - (1/k_\psi)[\phi_{ex,i} - \phi_{st,i}] + (1/2k_\psi)[\sin 2\phi_{ex,i} - \sin 2\phi_{st,i}] \\
\mathcal{C} &\leftarrow \mathcal{C} - (1/2k_\psi)[\cos 2\phi_{ex,i} - \cos 2\phi_{st,i}]
\end{aligned}
$$

case 4

this case makes no contribution

and substituting the totals into

$$
\begin{aligned}
F_x(\phi) &= 0.5uK_T(1 + K_R^2)^{1/2}[(\mathcal{A} + \mathcal{E}\phi)\sin\Theta - \mathcal{C}\cos\Theta \\
&\quad -\mathcal{D}\cos(2\phi - \Theta) + \mathcal{B}\sin(2\phi - \Theta)] \\
F_y(\phi) &= 0.5uK_T(1 + K_R^2)^{1/2}[-(\mathcal{A} + \mathcal{E}\phi)\cos\Theta - \mathcal{C}\sin\Theta \\
&\quad +\mathcal{D}\sin(2\phi - \Theta) + \mathcal{B}\cos(2\phi - \Theta)]
\end{aligned}
\tag{19}
$$

In practice the zone combinations are few in number, an hence this method is much more efficient than numerical integration using a large number of force elements.

By substituting from Eq. (19), the resultant force $F(\phi) = [F_x^2(\phi) + F_y^2(\phi)]^{1/2}$ can be written as

$$
\begin{aligned}
F(\phi) &= 0.5uK_T(1 + K_R^2)^{1/2}[\mathcal{C}^2 + (\mathcal{A} + \mathcal{E}\phi)^2 + \mathcal{B}^2 + \mathcal{D}^2 \\
&\quad -2(\mathcal{A} + \mathcal{E}\phi)(\mathcal{D}\sin 2\phi + \mathcal{B}\cos 2\phi) \\
&\quad +2\mathcal{C}(\mathcal{D}\cos 2\phi - \mathcal{B}\sin 2\phi)]^{1/2}
\end{aligned}
\tag{20}
$$

Since in Eq. (20) K_T and K_R appear only in the multiplier at the beginning of the expression, the computationally intensive geometric calculations can be carried out independently from model calibration. This approach permits rapid online recalibration (learning).

3. SUMMARY

In this section, several expressions were derived to implement the mechanistic rigid force milling force model for 2 1/2 dimensional rough end milling. Wherever possible, the cutter-part intersection geometry was analyzed to avoid time consuming numerical integration. Use of these expressions for online monitoring, feed rate scheduling, and adaptive control applications is illustrated in the following sections.

C.   MONITORING

In the following sections a number of experimental cutting tests are presented to illustrate use of the solid modeler simulation method and its integration with online monitoring and control applications. The experiments were conducted using an in-house retrofitted vertical milling machine. Instantaneous cutting forces were measured using a piezoelectric dynamometer and charge amplifiers. Maximum and average forces per tooth period were obtained using analog circuitry and a synchronizing spindle mounted shaft encoder. An Intel 80286 based microcomputer data acquisition system recorded these force values as well as the instantaneous feed rate and spindle speed.

The part material was aluminum 7075-T6 alloy. The cutting tool for the monitoring and feed rate scheduling examples was a 25.4 mm diameter High-Speed Steel end mill with four flutes and a helix angle ψ of 30 degrees. The cutting depth w was 19.05 mm, and the selected spindle speed was 473 rpm. Preliminary calibration tests established the mechanistic model cutting parameters

$$
\begin{aligned}
\log K_T &= \log 0.317 - 0.424 \log \overline{h} \\
\log K_R &= \log 0.212 - 0.217 \log \overline{h}
\end{aligned}
\tag{21}
$$

with K_T measured in kN/mm^2 and $\overline{h}$ measured in mm.

1.   INSTANTANEOUS FORCES

The instantaneous x and y cutting forces for a constant angular immersion interval $[\phi_{st}, \phi_{ex}) = [\pi/2, \pi)$ (one half immersion down milling) are plotted in Fig. 12. The feed rate was a constant $u = 0.087$ mm/tooth. A good agreement in trend was obtained, although some amplitude variation caused by eccentricity (radial runout) in the rotating tool is apparent. An intelligent machining center controller would ideally measure these forces and, with the aid of the solid modeler predictions, be able to distinguish a broken tooth from normal cutting conditions.

D.   FEED RATE SCHEDULING

The feed rate scheduling experiments were conducted using the part shape shown in Fig. 13. Model predictions were made at 1 mm increments along the cutter path. Example cutting tests for both dimensional surface error and maximum resultant force are presented.

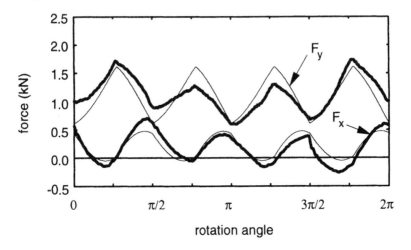

Fig. 12. Instantaneous forces. This plot shows the instantaneous x and y forces for one half immersion down milling using a 25.4 mm diameter, 30 degree helix angle High-Speed Steel (HSS) end mill. The cutting depth was $w = 19.05$ mm and the feed rate was $u = 0.087$ mm/tooth. The process model predictions are shown with the thinner lines and the experimental measurements are shown with the thicker lines.

1. DIMENSIONAL SURFACE ERROR

If the cutting force component normal to the cutter path is excessive, tool deflection will cause a significant dimensional surface error. This may result in gouging of the part during rough up milling (when the angular immersion interval $[\phi_{st}, \phi_{ex})$ satisfies $0 = \phi_{st} < \phi_{ex} < \pi$). In finish contour milling of airframe structures at a large cutting depth w, an unacceptable deviation of the finished surface from its nominal location will occur. A complete model of dimensional surface error considers part deflection as well, and models the tool as a cantilever with a distributed load along the z axis [32]. This level of detail is beyond the scope of the chapter.

A good first approximation is to assume a rigid part, and that the maximum cutter deflection occurs at the tool tip when the normal or y force is at an extreme value over one rotation. Applying this maximum force $F_{y,max}$ at a point midway along the cutting depth w, the deflection at the tip, and hence the maximum dimensional surface error, is

$$\delta_{y,max} = \frac{F_{y,max} A^2}{6EI}(A - 3L) \qquad (22)$$

where E is Young's modulus for the tool, L is the cantilever length of the

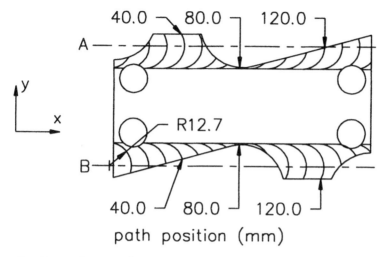

Fig. 13. Example part shape-feed rate scheduling. The part shape and cutter paths shown above were used for the feed rate scheduling examples. Again the cutting depth w was 19.05 mm.

tool extending from the collet, and $A = L - w/2$.

In the example presented here, the combined end mill, tool holder and spindle compliance was measured to be $k = 0.17$ mm/kN and the dimensional surface error predicted using the relationship

$$\delta_{y,max} = kF_{y,max} \qquad (23)$$

At a constant feed rate of 0.087 mm/tooth, the predicted and experimentaly measured dimensional surface error is plotted in Fig. 14(a).

Next, Eqs. (19) and (23) are numerically solved for the feed rate u using a threshold value of $\delta_{y,max} = 0.1$ mm. The feed rate schedule is then corrected to allow for geometric tolerances of the part, and the rise time of the table axis control system. The predicted and experimentally measured dimensional surface error for this feed rate schedule is shown in Fig. 14(b). Within the capabilities of the process model and the experimental equipment, a good adherence to the 0.1 mm threshold is obtained. In practice, a slightly conservative threshold would be used, but this threshold would be much higher than is typically employed by tool path planners having no access to a process modeling system. Further, the process modeling system automatically calculates a *varying* feed rate along the cutter path to continuously maximize the material removal rate.

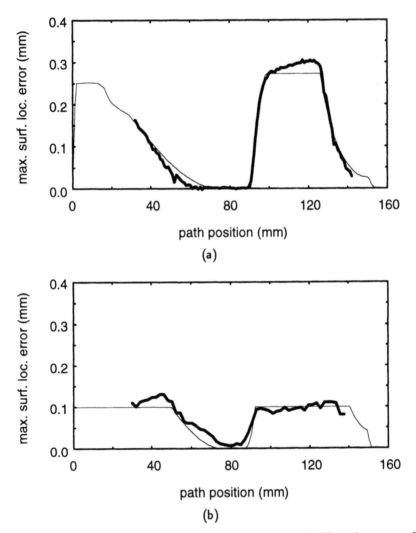

Fig. 14. Dimensional surface error. Cutter path B in Fig. 13 was used. The process model predictions are shown with the thinner lines and the experimental measurements are shown with the thicker lines. (a) Constant feed rate $u = 0.087$ mm/tooth. (b) Scheduled feed rate for a dimensional surface error threshold of $\delta_{y,max} = 0.100$ mm.

2.   MAXIMUM RESULTANT FORCE

An excessive resultant cutting force $F = (F_x^2 + F_y^2)^{1/2}$ must be avoided to prevent end mill shank breakage. This type of tool failure most frequently occurs when the cutting depth w is large and the cutter is small in diameter.

For this constraint, as noted in Section IV, B, the geometric portion of Eq. (20) can be solved independently, and the process model parameters K_T and K_R substituted just prior to actual machining. For a threshold value $F = 1.2$ kN, Fig. 15 shows the scheduled feed rate and experimentally measured maximum resultant force. The jitter in the force signal is again due to a small eccentricity of the cutter relative to the spindle axis (radial runout). The desired limiting of the resultant force was achieved.

In the next section, the use of adaptive control techniques as an alternative method of cutting force control is explored.

E.   ADAPTIVE CONTROL

An alternative and widely researched approach to controlling cutting forces in milling is to use feedback control systems. The basic approach is to measure the cutting force online and feed this measurement back into a controller. The controller compares it to a reference value, and adjusts the feed rate with the aim of regulating the next cutting force measurement to a value nearer the reference. An illustrative block diagram of the approach was shown in Fig. 2(a).

Because the machining process transfer function varies significantly with the cutting depth and angular immersion, self-tuning or adaptive controllers are preferred. By using a self-tuning controller, the closed loop gain is maintained near a level which provides rapid response without oscillation. Frequently a first order pole is included in the process to account for tool deflection. Higher order transfer functions are not warranted. For a review of adaptive control methods applied to machining, see [33,34].

Adaptive control is most useful during finish cutting passes, when no unanticipated sudden changes in cutter immersion are expected. In these cases it is difficult to calibrate a model for the cutting process, and hence adaptive control is attractive alternative. To expand on these assertions, an example of adaptive control for end milling is presented below.

1.   PROCESS MODEL

Because a linear milling process model is desireable as a transfer function, a slightly modified version of the deflection feedback model (Eq. (5)) is adopted. To implement a maximum resultant force controller the expres-

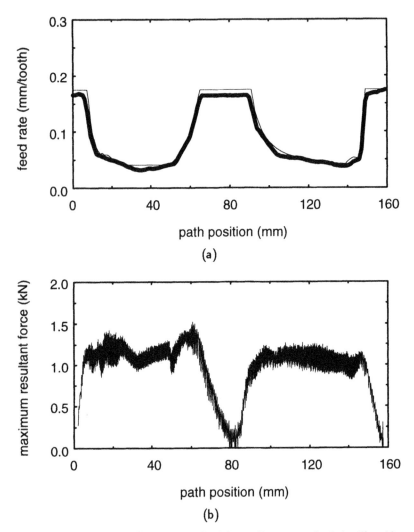

Fig. 15. Maximum resultant force limiting. Cutter path A in Fig. 13 was used. (a) Scheduled feed rate. The plot above shows the solid modeler feed rate schedule for a maximum resultant force threshold $F = 1.2$ kN (thinner line) The feed rate measured during the cutting test is also included (thicker line). (b) Measured maximum resultant force. The plot above shows the maximum resultant force achieved using the feed rate schedule from (a). The jitter in the force signal is caused by radial runout.

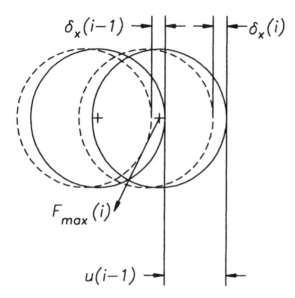

Fig. 16. Cutter compliance in adaptive control. Due to cutter compliance, the maximum resultant force $F_{max}(i)$ will deflect the tool by an amount $\delta_x(i)$ in the feed direction x to obey Eq. (25). The machining process model is therefore first order.

sion that is used is (derived from Yellowley and Kusiak [35])

$$F_{max}(i) = wK_s(h_{max}(i) + h^*) \qquad (24)$$

where i indexes the sampling intervals, K_s denotes the specific cutting pressure (similar to K_T in the rigid force model), $F_{max}(i)$ is the maximum resultant force over the sampling interval, $h_{max}(i)$ is the effective maximum chip thickness (see below), and h^* represents the frictional flank or edge forces on the tool.

First order dynamics enters the model because of cutter deflection which is caused by the machining forces [36,37] (Fig. 16). This is modeled using the expression

$$\delta_x(i) = k_x F_{max}(i) \qquad (25)$$

where k_x is the component of end mill compliance in the feed or x direction. The effective maximum chip thickness is therefore

$$h_{max}(i) = b[u(i-1) - \delta_x(i) + \delta_x(i-1)] \qquad (26)$$

where the parameter b varies depending on the angular immersion geometry.

Combining Eqs. (24), (25) and (26) yields

$$\frac{F_{max}(i)}{K_s\, wb} = u(i-1) - k_x\, F_{max}(i) + k_x\, F_{max}(i-1) + \frac{h^*}{b} \qquad (27)$$

Defining $\mu = k_x\, K_s\, wb$ and rearranging further yields

$$\left(\frac{1}{\mu}+1\right) F_{max}(i) - F_{max}(i-1) = \frac{1}{k_x}\left(u(i-1) + \frac{h^*}{b}\right) \qquad (28)$$

Introducing the backward shift operator q^{-1} [38] and defining

$$\alpha = \frac{-\mu}{1+\mu}; \quad \beta = \frac{\mu}{k_x\,(1+\mu)}; \quad \gamma = \frac{wh^*\,K_s}{1+\mu} \qquad (29)$$

provides the model form

$$(1+\alpha q^{-1})F_{max}(i) = \beta q^{-1}u(i) + \gamma \qquad (30)$$

2. CONTROLLER DESIGN

The planned cutting tests are to use a 25.4 mm diameter, zero helix, two tooth carbide insert end mill at a spindle speed of 615 rpm. This corresponds to a sampling period of 48.75 ms. The machine tool table drive servo zero order hold transfer function for this sampling period is

$$G_s(q^{-1}) = \frac{0.978(q^{-1} + 9.171 \times 10^{-4}q^{-2})}{1 - 2.079 \times 10^{-2}q^{-1}} \qquad (31)$$

The process model parameters were identified using the projection algorithm [39] which estimates $\hat{\theta}^T = [\hat{\alpha}\ \hat{\beta}\ \hat{\gamma}]$ from the recursive equation

$$\hat{\theta}^T(i) = \hat{\theta}^T(i-1) + \frac{0.15\varphi(i-1)}{\varphi^T(i-1)\varphi(i-1)}\left[F_{max}(i) - \varphi^T(i-1)\hat{\theta}(i-1)\right] \qquad (32)$$

where $\varphi^T(i-1) = [-F_{max}(i-1)\ u(i-1)\ 1]$ is the regressor vector. All initial estimates were set to zero except $\hat{\beta}$, which is set to a small positive value to avoid division by zero in the control law solution. The reference force is denoted by $F_{R,max}$.

After identifying the parameters, the steady state force component due to $\hat{\gamma}$ was eliminated by setting

$$F(i) = F_{max}(i) - \frac{\hat{\gamma}(i)}{1+\hat{\alpha}(i)}; \quad F_R(i) = F_{R,max}(i) - \frac{\hat{\gamma}(i)}{1+\hat{\alpha}(i)} \qquad (33)$$

in the controller implementation (Fig. 2(b)). This algebraically ensures correct reference force tracking. With this adjustment the machining process transfer function is

$$G_c(q^{-1}) = \frac{\hat{\beta}q^{-1}}{1 + \hat{\alpha}q^{-1}} \qquad (34)$$

Combining the table drive servo $G_s(q^{-1})$ and the machining process $G_c(q^{-1})$ yields

$$\frac{F}{u} = \frac{\hat{B}(q^{-1})}{\hat{A}(q^{-1})} = \frac{0.978(q^{-1} + 9.171 \times 10^{-4}q^{-2})}{1 - 2.079 \times 10^{-2}q^{-1}} \cdot \frac{\hat{\beta}q^{-1}}{1 + \hat{\alpha}q^{-1}} \qquad (35)$$

as the open loop transfer function. An indirect pole placement controller design was chosen with a closed loop damping ratio of 0.707 and a rise time of 195 ms (four sample periods). With this criteria the closed loop transfer function is

$$\frac{F}{F_R} = \frac{B_m(q^{-1})}{A_m(q^{-1})} = \frac{0.251q^{-2}}{1 - 1.162q^{-1} + 0.413q^{-2}} \qquad (36)$$

The control law solution $R(q^{-1})u(i) = T(q^{-1})F_R(i) - S(q^{-1})F(i)$ is obtained by solving for the polynomial coefficients in

$$\frac{\hat{B}(q^{-1})T(q^{-1})}{\hat{A}(q^{-1})R(q^{-1}) + \hat{B}(q^{-1})S(q^{-1})} = \frac{B_m(q^{-1})}{A_m(q^{-1})} \qquad (37)$$

The solution is

$$u(i) = -r_1 u(i-1) - r_2 u(i-2) + t_0 F_R(i) - s_0 F(i) - s_1 F(i-1) \qquad (38)$$

where $r_0 = -1.14121 - \hat{\alpha}$, $r_1 = -1.14029 + \hat{\alpha}$, $s_0 = (1/0.978\hat{\beta})[0.38927 + 1.14121\hat{\alpha} + \hat{\alpha}^2]$, $s_1 = (1/0.978\hat{\beta})[0.00208\hat{\alpha}(-1.14121 - \hat{\alpha})]$, and $t_0 = 0.251/0.978\hat{\beta}$.

3. UNASSISTED ADAPTIVE CONTROL

As an example of the difficulties experienced when a sudden increase in tool immersion is encountered, consider the cutter path shown in Fig. 17. Here the 25.4 mm diameter tool, in full radial immersion $[\phi_{st}, \phi_{ex}) = [0, \pi)$, is advanced through a part where the cutting depth w suddenly increases from 3 mm to 6 mm, and then down to 4 mm and to 2 mm. The experimentally measured maximum resultant force and feed rate are shown in Fig. 18. When the sudden increase in depth is encountered at path position 18 mm, a large amplitude cutting force "spike" from the reference level $F_{R,max} = 1.5$ kN to $F_{max} = 2.8$ kN occurs. To maintain a cost effective MRR in an industrial setting, the reference force would be set close to the

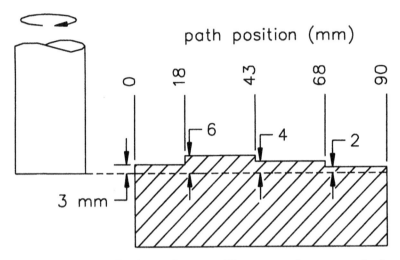

Fig. 17. Sudden depth of cut changes. The part and cutter path shown above are used to illustrate the force spike which will occur when an adaptive controller encounters a sudden increase in cutter immersion. The tool is in full radial immersion $[\phi_{st}, \phi_{ex}) = [0, \pi)$.

shank breakage threshold of the end mill. The sudden force spike would exceed this threshold, and break the end mill before the control system could react.

This is an inherent difficulty in applying feedback force control to machining processes. All control systems require time to recover after a disturbance is introduced, but, in milling, the machine tool structure is deliberately rigid to maintain dimensional accuracy. This leaves the tool itself as a weak component which will not withstand a sudden force increase.

Situations such as the one illustrated here perhaps are best handled by avoidance through careful tool path planning and feed rate scheduling. An application in mold machining, where avoidance of sudden immersion changes is difficult, and which invites employment of adaptive control with a form of solid modeler assistance, is discussed in the next section.

4. SOLID MODELER ASSISTANCE

Sculptured surface machining of dies and molds is an application where feedback control of the maximum resultant force is attractive. In this situation, ball and filleted end mills are used, and the surface shape is complex. Calculation of the cutter-part intersection geometry, and accurate process modeling, is difficult. These hindrances invite use of feedback and adaptive

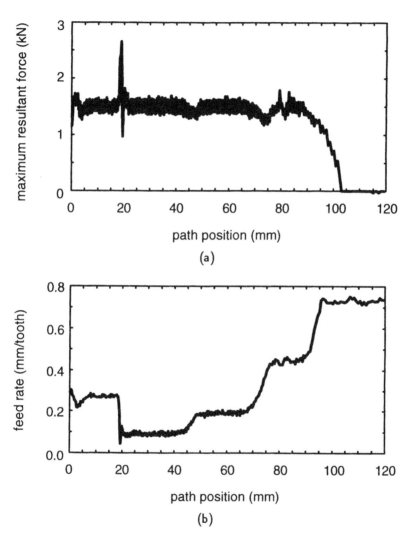

Fig. 18. Adaptive control force spike. The part shape is shown in Fig. 17. (a) When a sudden increase in cutting depth occurs at path position 18 mm, a maximum resultant force F_{max} "spike" above the reference level $F_{R,max} = 1.5$ kN occurs. In industrial practice this would break the end mill shank. (b) Feed rate per tooth u measurement.

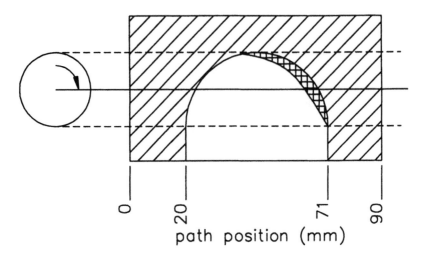

path position (mm)

Fig. 19. Surface interruptions. This part is used to demonstrate the difficulties that a feedback controller will encounter when passing over a surface interruption, such as a hole feature, and then encountering the sudden increase in immersion which follows. The cutter diameter is 25.4 mm and the cutting depth w is 4.5 mm. The cross-hatched area indicates the material added for solid modeler assistance (see Fig. 22).

control as alternatives.

If the cutter immersion varies only slowly then feedback control will likely function well. However, there may be interruptions on the surface, such as holes for plastic injection mold ejector pins, core pins, etc. (Fig. 19). In this event, the sudden increase in cutter immersion encountered after the hole is passed will cause a large amplitude cutting force spike, and break the tool (Fig. 20).

A promising solution to this problem is to use solid modeler assistance to avoid the hole feature force transient. The approach demonstrated here adds a disturbance force

$$\Delta F_{max}^{SMD} = \max \left(F_{max}^{SMD} - F_{max}, 0 \right) \tag{39}$$

to the adaptive controller block diagram near any geometric interruptions (Fig. 21). The force F_{max}^{SMD} corresponds to the maximum cutting force that would be expected if, instead of being removed, the solid modeler assistance region was still solid part material. By calculating the disturbance force in this way, the controller at all times remains responsive to the online force measurement F_{max}. Away from the interruption, the disturbance force is switched off.

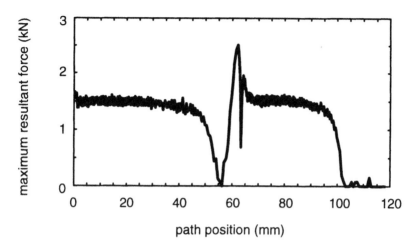

Fig. 20. Surface interruption force spike. The part shape is shown in Fig. 19. After passing over the hole feature surface interruption, the sudden increase in immersion near path position 65 mm causes a maximum resultant force spike which, in industrial practice, would break the milling cutter.

To simulate the controller response with this modification, pseudo-random binary signal (PRBS) tests were conducted to identify the model parameters. The values obtained were $k_x = 0.062$ mm/kN, $K_s = 1.411$ kN/mm^2, and $h^* = 0.074$ mm. A random disturbance force was introduced to the simulation by multiplying the maximum resultant force by zero mean, 0.05 variance Gaussian noise.

The controller simulation was carried out along the entire cutter path for the sample part using the added material indicated in Fig. 19. In this example the added material was determined by repeating the simulation until the controlled maximum resultant force no longer appeared to oscillate near path position 65 mm. This informal approach was taken to learn approximately the minimum added material required to achieve smooth control. For automatic calculation, a straightforward procedure is to simply imagine that the entire feature is still filled with solid part material. For larger features the tool paths should be adjusted to stop at the edge of the interruption to avoid useless machining of empty space.

Both the simulated and experimentally measured maximum resultant force, feed rate, and tuning parameter estimates are shown in Fig. 22 and Fig. 23. As anticipated, near path position 65 mm the maximum resultant force smoothly rises back to reference level $F_{R,max} = 1.5$ kN without oscilla-

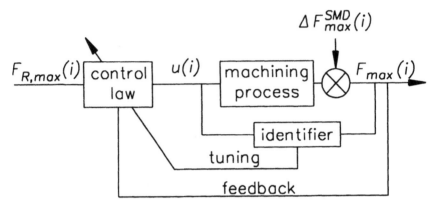

Fig. 21. Solid modeler assistance. The block diagram above includes a disturbance force injection ΔF_{max}^{SMD} where the solid modeler assistance is introduced. Compare to Fig. 2.

tion. In this example the tuning parameter estimates track with surprising agreement. The longer term stability of the estimates was not investigated due to the limited dimensions of the experimental equipment. Other trials suggest that the process gain $\hat{\beta}$ is most reliably identified, and that the process pole $\hat{\alpha}$ may drift. This drift occurs because of a lack of signal frequency richness during long periods with constant cutting conditions, and may lead to controller instability if not bounded.

F. SUMMARY

In this section examples demonstrating use of a CSG solid modeler to calculate cutter-part immersion geometry, implementation of a mechanistic milling force model, and adaptive cutting force control were presented. The strategy selections made intentionally exploited the relative ease with which geometric and process modeling, and control implementation, could be achieved.

For example, in 2 1/2 dimensional operations such as terracing, the cutter-part immersion can change suddenly as the tool reenters solid material after transiting an already machined region. Such a rapid change would wreak havoc for a feedback controller. The flat end mill shape and regular cutter path pattern, however, are simple to model geometrically. The constant cutting depth and easily identified angular immersion permit implementation of mechanistic process models. Updating of the part model is straightforward using the CSG representation.

In contrast, finish machining of sculptured die and mold surfaces is a

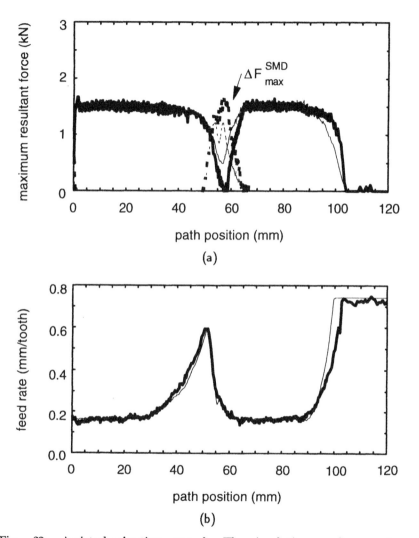

Fig. 22. Assisted adaptive control. The simulation results are shown with the thinner lines, and the experimental measurements are shown with the thicker lines. (a) maximum resultant force F_{max} and solid modeler assistance disturbance force ΔF_{max}^{SMD}. (b) feed rate per tooth u. See Fig. 23 for the tuning parameter estimates. .

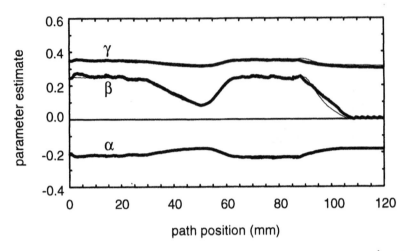

Fig. 23. Assisted adaptive control. Tuning parameter estimates $\hat{\alpha}$, $\hat{\beta}$ and $\hat{\gamma}$ corresponding to the maximum resultant force and feed rate results shown in Fig. 22.

process which is difficult to exactly model geometrically. The material removal rate is low, however, and sudden increases in cutter immersion do not occur. Feedback control of quantities such as the maximum resultant cutting force can therefore be implemented safely. In cases where interruptions in the cutter immersion occur, a heuristic form of control override is needed. Solid modelers can be used to identify these regions and locally predict the cutting forces to assist the online controller. Time consuming process model calculations are performed only near the interruption region, rather than over the entire, complex part surface.

The benefits of using adaptive control and higher order process models is questionable. It is unlikely that any controller could safely respond to rapid immersion changes using unaided force feedback. Some type of ancillary assistance is necessary, and should be used to smooth changes in the predicted cutting force so that dynamics beyond the process gain need not be modeled. In milling this is advantageous since it avoids the possibility of unacceptable surface finish and chatter which may occur if, to provide signal richness for identification, stable cutting conditions are modulated.

The examples considered are not exhaustive, but represent the diversity encountered in machining. Governing this diversity will require a machine tool controller with intelligence and integration beyond present industry standards. Continuing research areas in geometric and process modeling, and online controller implementation, are discussed in the next and final section of the chapter.

V. CONTINUING RESEARCH

This chapter has described a subset of the solid and process modeling, and monitoring and control applications in machining. Many areas require additional research before a tangible, intelligent machining controller can be realized. Part representation and process modeling–key areas in which more work is required–are briefly discussed below. A prototype machining intelligence architecture is described at the end of the section.

A. PART REPRESENTATION

The part and machining examples previously discussed in the chapter were in essence of a two dimensional nature. The 2 1/2 D terracing cutter paths were a collection of constant z contouring motions. Finish sculptured surface machining may also be considered as two dimensional, since a parametric representation of the form

$$p(u, v) = [x(u, v)\, y(u, v)\, z(u, v)]^T \tag{40}$$

where u and v are the parameters, is typically used [40].

The intermediate semi-finishing step, where the terraces are machined away to leave only a small final finishing allowance (Fig. 24), is more difficult to represent. Bridging between solid and surface modeling is recognized as a requirement to address such situations, but remains an open research problem [41].

B. PROCESS MODELING

Exact modeling of the milling process remains elusive because of the uncertainty in predicting fundamental parameters (rake face friction, shear angle, etc.). For this reason, mechanistic models, which must be calibrated for each tool and part material combination, are widely used. Work towards determining new tool model parameters from orthogonal cutting tests has begun [42], but is early in development. Additional research into developing models for ball end mills [43] and more complex cutter shapes is needed. Modeling of thin part deflection under machining forces [44] is also required. All of these models are needed to further understanding of the machining process.

For tool path planning and control assistance simulations, less detailed models will suffice. Adaptive force control of sculptured surface machining, for example, requires only recognition of areas where part surface interruptions and sudden force increases will occur. The feedback capabilities of

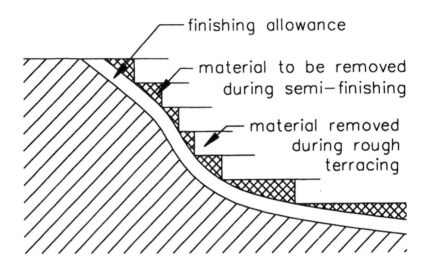

Fig. 24. Mold semi-finishing representation difficulties. Shown above is a mold cross-section labelled to indicate (1) material removed during rough terracing; (2) material to be removed during semi-finishing; (3) the finishing allowance. Both (1) and (3) are in essence 2 D geometric modeling problems. A difficult 3 D representation, however, is required to handle (2).

the controller will fine tune the process. By accepting more approximate models, there is potential to separate the geometric portion of the process model from the cutting mechanics portion. If this can be achieved with adequate model accuracy, then much of the computationally intensive simulation work can be completed and stored prior to actual part machining. This is necessary if online, time-critical cooperation between the machine tool controller and solid modeler is to be achieved.

C.   MACHINING INTELLIGENCE ARCHITECTURE

In order to accomodate control system improvements, future generation machine tool controllers must have a modular or open-architecture design [45]. Only in this way will new developments be rapidly accepted by conservative industrial users, who object to the high capital cost of equipment replacement.

Control loops, which in current research implementations are very strict [46], will become more heuristic so that they can be integrated with other components and modules within the machine tool controller. This is not possible with exact model feedback loops.

When required, any machine tool controller module will be able to directly communicate with any other module. Solid modeler based CAD/CAM assistance will be available on demand to adjust tool path plans, provide detailed part geometry information, etc.

Additional recording of monitoring data will permit longer term trends to be recognized and added to the machining process knowledge base. This data will be organized into a more convenient form for presentation to a human operator at either a local or remote control console. The machine tool modules will be able to consult the knowledge base for assistance in identifying and responding to machining process exceptions.

A Hierarchical Open Archicture Multiprocessor (HOAM) CNC system which allows modular integration of machining process monitoring and control functions is being developed by Altintas et al [47,48]. The hardware and software architecture of the current system is shown in Fig. 25.

The system is based on an ISA bus with an Intel 80486 processor on the master board. The motion control tasks, such as position and velocity control of machine the tool drives, interpolation, and sensory function interface software modules, run on a TMS320C30 based Digital Signal Processing (DSP) board residing in one of the ISA bus slots. Each machining process control and monitoring task has a dedicated single board computer module on the ISA bus. Chatter detection and suppression, tool condition monitoring, and adaptive control functions send cutting condition modification requests to the control module via the 80486 master. For example, the adaptive control module may request modification of the machining feed rate at each tooth or spindle period. A micro kernel running on the TMS320C30 board schedules each incoming machining process monitoring or control request in a job queue, which is updated in one millisecond time intervals according to request priority. Events such as collision and tool breakage have the highest priority. A feed rate change command coming from the adaptive control module has a lower priority. The CNC executive is designed in such a way that new process control functions can be introduced in a modular fashion without replacing the CNC executive software.

The existing system allows simultaneous execution of the regular NC program, and also supports chatter detection and suppression by in-process modification of the NC tool path, adaptive force control, and tool breakage detection. Presently the system is being ported to a VME bus, which is more robust for production machine tool use. Additional development would add a machining process knowledge base, which would learn from long term sensor data observations, and provide reference strategies in response to controller requests.

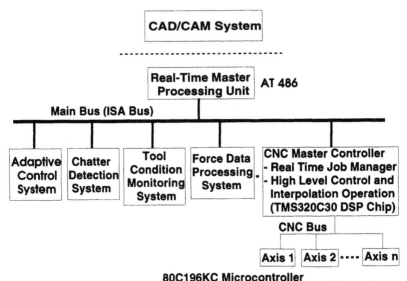

80C196KC Microcontroller

The global hardware architecture of the HOAM-CNC system

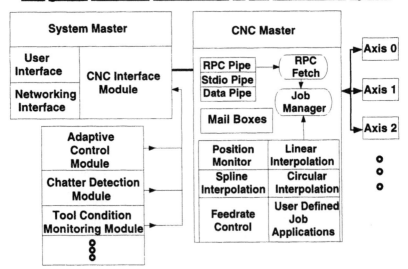

The software architecture of the HOAM-CNC system

Fig. 25. Open architecture CNC system

VI. SUMMARY

With special emphasis on end milling, this chapter has reviewed the machining process and the constraints hindering more rapid material removal rates.

Solid modelers used for part representation and tool path planning were described, with particular emphasis on the use of the Constructive Solid Geometry (CSG) method to obtain the cutter-part intersection geometry needed for 2 1/2 dimensional rough machining process simulation. It was noted that integration of solid modelers with sculptured surface design software is needed to fully handle die and mold machining. Realization of this integration remains an open research issue.

Using the milling process modeler, the instantaneous feed and normal forces were simulated. Feed rates were scheduled to avoid excessive part dimensional surface error and maximum resultant force, which can break the tool shank. Experimental results demonstrated the accuracy of the mechanistic process model used, and its ability to provide useful feed rate scheduling.

Feedback force control was also discussed. For rough machining, the sudden large cutting immersion increases and resulting force transients would break the cutting tool before the controller could respond. Feedback control is of questionable value under these conditions. Sculptured surface finish machining, however, is an application where online force control has potential. In this case sudden immersion changes should not occur, and hence adequate force regulation is possible. When part surface interruptions exist, solid modeler assistance can provide a local force simulation to maintain smooth reference force tracking.

A machining intelligence architecture was proposed which allows modular integration of monitoring and control components. A link to a solid modeler based CAD/CAM system was included to allow online query of detailed part geometry, force control assistance, etc. A machining process knowledge base should also be included.

A key to realizing successful implementation of the machining intelligence system lies in selection and development of suitable data representation schemes. Better and more integrated data representation methods are needed for solid and surface part modeling, and tool path and process data communication to the machine tool controller. An open issue is development of a method for storing machining process knowledge so that it can be effectively utilized during process planning, and online monitoring and control. Resolution of these issues is the cornerstone to achieving advanced machining process modeling, automation and control.

VII. REFERENCES

1. A.G. Ulsoy and Y. Koren, "Control of Machining Processes", *Trans. ASME, J. Dyn. Syst., Meas., Control*, **115**, pp. 301–308 (1993).

2. Y. Koren, *Computer Control of Manufacturing Systems*, McGraw-Hill, New York, N.Y. (1983).

3. Y. Koren, C.C. Lo and M. Shpitalni, "CNC Interpolators: Algorithms and Analysis", *Manufacturing Science and Engineering Symposia*, ASME, **PED-64**, pp. 83–92 (1993).

4. Y. Altintas, I. Yellowley and J. Tlusty, "The Detection of Tool Breakage in Milling Operations", *Trans. ASME, J. Eng. Ind.*, **110**, pp. 271–277 (1988).

5. M.S. Lan and D.A. Dornfield, "In-Process Tool Fracture Detection", *Trans. ASME, J. Eng. Mater. Technol.*, **106**, pp. 111–118 (1984).

6. Y. Koren et al., "Methods for Tool Wear Estimation from Force Measurements under Varying Cutting Conditions", *Symposium on Control Issues in Manufacturing Processes*, ASME, **DSC-18**, pp. 45–53 (1989).

7. J.I. El Gomayel and K.D. Bregger, "On-Line Tool Wear Sensing for Turning Operations", *Trans. ASME, J. Eng. Ind.*, **108**, pp. 44–47 (1986).

8. L.K. Daneshmand and H.A. Pak, "Performance Monitoring of a Computer Numerically Controlled (CNC) Lathe Using Pattern Recognition Techniques", *Third International Conference on Robot Vision and Sensory Controls (ROVISEC3)*, Cambridge, MA (1983).

9. S. Smith and T. Delio, "Sensor-Based Control for Chatter-Free Milling by Spindle Speed Selection", *Symposium on Control Issues in Manufacturing Processes*, ASME, **DSC-18**, pp. 107–114 (1989).

10. R. Teltz and M.A. Elbestawi, "Hierarchical, Knowledge-Based Control in Turning", *Trans. ASME, J. Dyn. Syst., Meas., Control*, **115**, pp. 122–132 (1993).

11. K.J. Åström and B. Wittenmark, *Adaptive Control*, Addison-Wesley, Reading, MA (1989).

12. A. Spence and Y. Altintas, "CAD Assisted Adaptive Control for Milling", *Trans. ASME, J. Dyn. Syst., Meas., Control*, **113**, pp. 444–450 (1991).

13. R. Centner, "Final Report on Development of Adaptive Control Technique for Numerically Controlled Milling Machine", *Tech. Documentary Report ML-TDR-64-279*, USAF, (1964).

14. National Research Council, *The Competitive Edge: Research Priorities for U.S. Manufacturing*, National Academy Press, Washington, DC (1991).

15. P.K. Wright and D.A. Bourne, *Manufacturing Intelligence*, Addison-Wesley, Reading, MA (1988).

16. C.M. Hoffman, *Solid and Geometric Modeling*, Morgan Kaufmann, Palo Alto, CA (1989).

17. L. Piegl and W. Tiller, "Curve and Surface Constructions using Rational B-splines", *Computer-Aided Design*, **19**(9), pp. 485–498 (1987).

18. M.E. Martellotti, "An Analysis of the Milling Process", *Trans. ASME*, **63**, pp. 677–700 (1941).

19. M.C. Shaw, *Metal Cutting Principles*, Oxford University Press, Oxford, UK (1984).

20. S. Smith and J. Tlusty, "An Overview of Modeling and Simulation of the Milling Process", *Trans. ASME, J. Eng. Ind.*, **113**, pp. 169–175 (1991).

21. F. Koenigsberger and A.J.P. Sabberwal, "An Investigation into the Cutting Force Pulsations During Milling Operations", *Int. J. Machine Tool Design and Research*, **1**, pp. 15–33 (1961).

22. J. Tlusty and P. MacNeil, "Dynamics of Cutting Forces in End Milling", *CIRP Ann.*, **24**, pp. 21–25 (1975).

23. W.A. Kline, R.E. DeVor and J.R. Lindberg, "The Prediction of Cutting Forces in End Milling with Application to Cornering Cuts", *Int. J. Machine Tool Design and Research*, **22**(1), pp. 7–22 (1982).

24. A.A.G. Requicha and H.B. Voelcker, "Boolean Operations in Solid Modeling: Boundary Evaluation and Merging Algorithms", *Proc. IEEE*, **73**(1), pp. 30–44 (1985).

25. R.B. Tilove, "Set Membership Classification: A Unified Approach to Geometric Intersection Problems", *IEEE Trans. Comput.*, **C-29**(10), pp. 874–883 (1980).

26. A.D. Spence, "Solid Modeller Based Milling Process Simulation", *Ph.D. Thesis*, The University of British Columbia, Vancouver, BC, CANADA, 1992.

27. A. Spence, Y. Altintas and D. Kirkpatrick, "Direct Calculation of Machining Parameters from a Solid Model", *Computers in Industry*, **14**, pp. 271–280 (1990).

28. S. Cameron, "Efficient Intersection Tests for Objects Defined Constructively", *Int. J. Robotics Research*, **8**(1), pp. 3–25 (1989).

29. J.R. Rossignac and H.B. Voelcker, "Active Zones in CSG for Accelerating Boundary Evaluation, Redundancy Elimination, Interference Detection and Shading Algorithms", *ACM Trans. Graphics*, **8**(1), pp. 51–87 (1989).

30. K. Mehlhorn, *Data Structures and Algorithms: Multi-dimensional Searching and Computational Geometry*, v. 3, Springer-Verlag, New York, N.Y. (1984).

31. I. Yellowley, "Observations of the Mean Values of Forces, Torque and Specific Power in the Peripheral Milling Process", *Int. J. Machine Tool Design and Research*, **25**(4), pp. 337–346 (1985).

32. W.A. Kline, R.E. DeVor and I.A. Shareef, "The Prediction of Surface Accuracy in End Milling", *Trans. ASME, J. Eng. Ind.*, **104**, pp. 272–278 (1982).

33. A.G. Ulsoy, Y. Koren and F. Rasmussen, "Principal developments in the adaptive control of machine tools", *Trans. ASME, J. Dyn. Syst., Meas., Control*, **105**, pp. 107–112 (1983).

34. Y. Mohamed, M.A. Elbestawi and L. Liu, "Application of Some Parameter Adaptive Control Algorithms in Machining", *Symposium on Control Methods for Manufacturing Processes*, ASME, **DSC-9**, pp. 63–70 (1988).

35. I. Yellowley and A. Kusiak, "Observations on the Use of Computers in the Process Planning of Machined Components", *Trans. CSME*, **9**(2), pp. 70–74 (1985).

36. F. Koenigsberger and J. Tlusty, *Machine Tool Structures*, v. 1, Pergamon Press (1970).

37. M. Tomizuka, J.H. Oh and D.A. Dornfield, "Model Reference Adaptive Control of the Milling Process", *Control of Manufacturing Processes and Robotics Systems*, D.E. Hardt and W.J. Book, ed., pp. 55–63, ASME Press, New York, NY (1983).

38. K.J. Åström and B. Wittenmark, *Computer-Controlled Systems*, Prentice-Hall, Englewood Cliffs, NJ (1984).

39. G.C. Goodwin and K.S. Sin, *Adaptive Filtering Prediction and Control*, Prentice-Hall, Englewood Cliffs, NJ (1984).

40. I.D. Faux and M.J. Pratt, *Computational Geometry for Design and Manufacture*, Ellis Horwood, Chichester, UK (1979).

41. T. Várady and M.J. Pratt, "Design Techniques for the Definition of Solid Objects with Free-form Geometry", *Computer-Aided Geometric Design*, **1**, pp. 207–225 (1984).

42. E.J.A. Armarego and N.P. Deshpande, "Computerized Predictive Cutting Models for Forces in End-Milling Including Eccentricity Effects", *CIRP Ann.*, **38**(1), pp. 45–49 (1989).

43. M. Yang and H. Park, "The Prediction of Cutting Force in Ball End Milling", *Int. J. of Machine Tools and Manufacturing*, **31**, pp. 45–54 (1991).

44. D. Montgomery and Y. Altintas, "Mechanism of Cutting Force and Surface Generation in Dynamic Milling", *Trans. ASME, J. Eng. Ind.*, **113**, pp. 160–168 (1991).

45. I. Greenfeld, "Open-System Machine Controllers–The MOSAIC Concept and Implementation", *Symposium on Control Issues for Manufacturing Processes*, ASME, **DSC-18**, pp. 91–97 (1989).

46. K.J. Åström, "Toward Intelligent Control", *IEEE Control Systems Magazine*, **8**(3), pp. 60–64 (1989).

47. Y. Altintas, N. Newell and M. Ito, "A Hierarchical Open Architecture Multi-Processor CNC System for Motion and Process Control", *Manufacturing Science and Engineering Symposia*, ASME, **PED-64**, pp. 195–205 (1993).

48. Y. Altintas and W.K. Munasinghe, "A Hierarchical Open Architecture CNC System for Machine Tools", *CIRP Ann.*, **43**(1), pp. 349–354 (1994).

Techniques in Discrete–Time Position Control of Flexible One Arm Robots

Sijun Wu
MagneTek
Drives and Systems
New Berlin, WI 53151

Sabri Cetinkunt
Department of Mechanical Engineering
University of Illinois at Chicago
Chicago, IL 60680

I. INTRODUCTION

In the last decade, the control problem of mechanically flexible systems has been intensively studied as a result of growing needs for fast, precise manipulators in space applications and industry.

An example of a flexible robot in space applications is the 15–meter long space shuttle remote manipulator system (SRMS) of NASA. It is usually used in assembly of space platforms and of large communication systems, etc. [1]. When all its links are extended and locked, its natural frequency is as low as 0.3 Hz without payload. As a result, it can only move slowly (0.1 m/sec). In future space applications, the manipulators will need to be made lighter and to move faster with higher accuracy, and work independently.

In industry, the automation of assembly tasks would be highly enhanced if robots could operate at high speed with high accuracy. However, these goals can not be achieved with the existing robot designs because many of them are made massive in order to increase rigidity. For high speed operation, robotic manipulators will be made lighter to reduce the driving torque requirement and to enable the robot arm to respond faster. On the other hand, lighter members of the robot are more likely to become flexible, thus making it necessary to take into consideration the dynamic effects of the distributed link flexibility.

The above mentioned applications make it necessary and challenging to investigate the dynamics and control problems for manipulators with structural flexibility. For this purpose, many one–arm flexible robot arms have been built in laboratories [2, 3, 4, 5, 6, 7, 8]. The goal is to achieve high speed and precise tip positioning of the robot. Moreover, the designed controller shall also be able to overcome the payload variations, environmental disturbances, and system parameter changes, as well as unmodeled system dynamics.

In today's industrial robots, the position of the end–effector is controlled by driving the joint angles to the appropriate positions. The specific joint angles can be derived from the analysis of the inverse kinematics of the robot. The robot links are assumed to be rigid enough so that the end–effector will be in the intended positions. This control technique is usually referred to as *colocated control*, due to the colocation of the actuator and hub angle sensors at the joints. The colocated control schemes prove to be robust against control/observation spillover, require fewer sensors, and have no stability problems [9, 10]. However, the possible closed–loop bandwidth of the system is small [11]. In this case, this bandwidth can only reach as high as two thirds of the lowest clamped–free frequency of the robot.

For high precision tip position control, it is necessary to sense the tip position directly. This type of direct tip sensing provide more reliable output measurements. The non–colocated sensor and actuator based control algorithms could also increase the bandwidth level of the overall closed–loop system. It is reported that a closed–loop bandwidth up to 2 to 3 times faster than the first clamped–free natural frequency could be reached [2]. On the other hand, stability becomes a concern in non–colocated systems when the intervening structure is flexible as the system becomes non–minimum phase [12, 13]. Therefore, the closed loop system in this case is only conditionally stable, and special attentions should be given to the robustness of the stability margin under payload variations and modeling errors.

The objective of this chapter is to design an effective adaptive controller for the tip position control of the flexible robot arm which aims for high speed and high precision manipulations. The problems in modeling and parameter identification are also addressed.

A. LITERATURE REVIEW

Most of the research in the field of tip position control has been focused on the dynamic modeling and controller designs.

1. Dynamic Modeling

A flexible robot arm can be considered as a mechanical system with an infinite number of degrees of freedom. However, a controller can only be designed based on a finite dimensional dynamics model. Using an Euler–Bernoulli beam, Balas [9, 10] examined the control and observation spillover which was shown to lead to potential instability in the closed–loop system. Some approaches, such as a phase–locked loop prefilter, were suggested to improve the closed–loop stability.

Book et al. [14] applied the transfer matrix method to describe the elastic bending motion of a two–link planar flexible robot system for small angular velocities. The investigation was conducted in the frequency domain. The elastic deflections were expanded in terms of two clamped–free modes for each link. On the other hand, Maatuk [15] derived the equation of motion for a multi–link flexible manipulator using the pinned–pinned modes. Both bending along two perpendicular axes and twist deformation were considered. The method was then applied to a six degree–of–freedom robot similar to the Space–Shuttle Remote Manipulator System (SRMS).

In most studies, finite–dimensional models were derived using model truncation where only two or three flexible modes were includes [2, 4, 8, 14, 16, 17]. Mode shapes were chosen according to boundary conditions of either pinned–free or clamped–free.

In a comparative study conducted by Cetinkunt and Yu [13], it was pointed out that the clamped–free mode shapes described the closed–loop system dynamics much more accurately than pinned–free mode shapes under feedback control. However, the difference caused by these two boundary conditions became less significant as the hub inertia increased relative to the flexible beam inertia [13]. In practice, the mode shapes were actually neither pinned–free nor clamped–free, but function of the feedback controls. Using the above observation, Book and Majette [18] designed a combined state–space and frequency domain iterative controller. But the controller could handle only colocated controllers in interconnected flexible linear systems.

Many researchers preferred the compact structure of the dynamic model under pinned–free boundary condition assumptions. In this case the dynamic model become uncoupled. Works that assumed pinned–free boundary conditions could be found in [2, 8, 12, 16, 17, 19, 20, 21, 22, 23].

2. Controller Designs

Many controllers have been designed to provide stable, high closed–loop bandwidth and high precision tip position control of a flexible robot arm. These controllers can be classified into two categories in terms of sensor and actuator locations, namely, the colocated and non–colocated controllers.

a. Colocated Joint Variable Feedback Control

The colocated control is the one that is traditionally used in motion control in industrial applications. In the design of fine and gross motion controller for a two–link flexible robot system, Cetinkunt and Book [11] developed a finite dimensional integral type adaptive model following controller. The coupling of multi–link manipulators were also investigated. Performance limitation of joint variable controls due to manipulator structural flexibility were quantified. Maizza–Neto [24] discussed the use of a pole placement algorithm to obtain full state feedback gains for a 12th order linear model of the same two–link flexible manipulator.

Liegois [6] discussed the modeling and control of a six degree–of–freedom light weight manipulator MA–23 developed by the French Atomic Energy Commission. The joint was modeled as a fourth–order linear system made of two masses free in translation and connected by a spring. A colocated PD feedback controller was presented and shown to provide inadequate damping for the dominant closed loop poles. With colocated position and rate feedback, better performance was obtained by a full state feedback controller.

Truckenbrodt [25] designed several output feedback laws and a state–feedback controller with a reduced order estimator. The sensors included joint angle and joint rate sensors. A strain gauge was also attached at a distance of $0.1L$ from the actuator. For all these colocated designs, the closed–loop bandwidth is less than the first clamped–free natural frequency of the beam.

Park and Asada [7] used a transmission mechanism to relocate the torque actuation point on the flexible arm robot studied. By analyzing the locations of the poles and zeros of the closed–loop system as the actuation point moved, it could be shown that the closed–loop system turned out to be a minimum phase system as the actuation point reached the near neighborhood of the robot tip. In that case, the controller was actually a colocated controller at the robot tip (with tip position feedback). As a result, stability is not a problem in their controller designs.

b. Non–Colocated Tip Position Feedback Control

The majority of high performance tip position controllers use non–colocated control. Cannon and Schmitz [2] designed an eighth order linear quadratic regulator (LQR) with hub rate and tip position feedbacks based on a pinned–free dynamics model. For system parameter identification in frequency domain, some experimental procedures were presented for the off–line estimation of robot resonance frequency, zeros, and DC gains. This controller is not an adaptive controller, and is also very sensitive to modeling errors. More experimental work had been conducted at Stanford University to address the on–line system identifications in discrete time domain [20, 21].

The stable factorization technique was studied by Wang and Vidyasagar [8] and Shung and Vidyasagar [22]. Wang and Vidyasagar [8] used an eighth order stable factorization controller to obtain an optimal step response in tip position control. The controller was designed based on one or two parameter compensator designs. Shung and Vidyasagar [22] extended the previous controller by adding constraints on the input energy level. They demonstrated that, with or without constraints, the SF approach could be able to provide faster transient response than the LQR in [2]. However, actuation signal level required by the SF controller is much higher than that of LQR controller.

Tip acceleration feedback was first utilized by Kotnik and Ozguner [5] with a dynamics model involving two flexible modes of the flexible arm in the real–time control algorithm. The authors concluded that the tip acceleration feedback controller could provide comparable transient response as a tip position feedback controller did. But the primary advantage of using acceleration feedback was that all the sensors (tachometer and accelerometer) could be mounted on the robot arm. Expensive sensors, such as cameras, were not needed.

In the non–colocated control, other control techniques, such as optimal state feedback [4], sliding mode control [23], singular perturbation approach [26], and direct adaptive model following [19], could also be found among the literatures.

C. Outline

In Section II, we derive a dynamic model for a single link flexible robot arm with a payload. In particular, a non–dimensionalized dynamic model for a robot arm is developed which is applicable for a wide variety of robot arms. For a discrete–time controller design, the derived continuous–time system model are sampled by means of a zero–order–hold circuit. To examine the effect of the sampling frequency on the overall system stability, we investigate the mapping of the poles and zeros of a continuous–time system to the discrete time domain. The fundamental non–minimum phase property of the open–loop transfer function from joint torque to tip position is identified in the z–plane.

Section III describes the derivation of the lattice filter parameter identifier which, for the first time, is used in the real–time parameter identification and control of a flexible robot arms. A straight forward procedure is proposed to detect the payload variations through a set of strain gauges such that a proper forgetting factor could be selected accordingly. Although the lattice filter is usually utilized in signal processing for Auto–Regressive (AR) or Moving Average (MA) processes, it is expanded to embed an Auto–Regressive and Moving Average (ARMA) process such as a flexible robot arm.

The proposed predictive adaptive controller is introduced in Section IV. In order to find an appropriate trade–off between the tip position and hub rate feedback gains, a root locus analysis approach is suggested. Also considered is the dynamic model of the actuator system including an amplifier, a permanent magnet (PM) direct current (DC) motor, and a harmonic drive. With the help of the non–dimensional conversion factors, we compare the performances of the predictive adaptive controller with that of the SF controller by Wang and Vidyasagar [8] and the LQR by Cannon and Schmitz [2].

The conclusions of this study is presented in the last section.

II. DYNAMICS MODEL OF A FLEXIBLE ONE ARM ROBOT

In this section, a simplified dynamics model for a flexible one arm robot is developed with payload considerations. Non-dimensionalization of the dynamics model is also introduced. The effect of sampling frequency on the system stability is discussed in discrete time domain.

A. DYNAMICS MODEL OF A FLEXIBLE ONE ARM ROBOT

As shown in Fig. 1, the Euler–Bernoulli model is employed for the tip position control. It is assumed that this robot arm is made of a uniformly distributed pinned–free beam of length L in the horizontal plane. The beam has a modulus of elasticity E. The moment of inertia of its cross–sectional area about the z–axis is $I(x)$, and the mass density per unit length is $m(x)$. m_p and J_p are, respectively, the payload mass and the mass moment of inertia about a z–axis passing through the center of mass of the payload. The hub is modeled by a single mass moment of inertia I_{H2}. The beam has a moment of inertia I_{H1} about the hub. The sum of I_{H1} and I_{H2} is designated by I_H. Shown also in Fig. 1, $\theta(t)$ designates the hub rotation angle corresponding to the rigid body rotation of the beam, and $\psi(t)$ represents the total hub rotation angle. The total displacement $y(x,t)$ of a beam element is defined in the global coordinate system X–Y, while x–y represents the local coordinate attached to the base of the arm. It has been assumed that the flexible deflection $u(x,t)$ is small enough to stay in the elastic range of the beam. Effects of beam rotation due to bending and distortion due to shear (Timoshenko beam theory), and friction at the joint are assumed to be negligible.

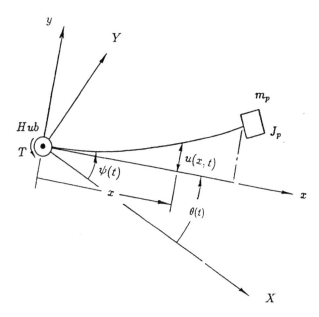

Figure 1: Schematic Diagram of a Flexible One Arm Robot

The total displacement of any point along the beam in X–Y is defined by:

$$y(x,t) = u(x,t) + x\,\theta(t) \tag{1}$$

Considering also the effect of the payloads m_p and J_p, we could write the system kinetic energy $T_k(t)$ as:

$$T_k(t) = \frac{1}{2}I_{H2}\dot{\theta}^2(t) + \frac{1}{2}\int_0^L \left(\frac{\partial y}{\partial t}\right)^2 m(x)dx + \frac{1}{2}J_p\left[\frac{\partial^2 y}{\partial t\partial x}\bigg|_{x=L}\right]^2$$

$$+ \frac{1}{2}m_p\left[\frac{\partial y(L,t)}{\partial t}\right]^2 \tag{2}$$

Only the strain energy due to bending is considered to contribute to system potential energy. Therefore, we have the system potential energy $V(t)$ as:

$$V(t) = \frac{1}{2}\int_0^L EI(x)\left(\frac{\partial^2 y}{\partial x^2}\right)^2 dx \tag{3}$$

Total work $W_{nc}(t)$ done by non–conservative forces is:

$$W_{nc}(t) = T\left(\frac{\partial y(0,t)}{\partial x}\right) \tag{4}$$

Here T is the input torque exerted at the hub. Note that $\partial y(0,t)/\partial x = \psi(t)$ is the total hub rotation angle.

Applying Hamilton's principle to Eqs. (2) to (4), one has

$$\int_{t_1}^{t_2} \delta(T_k - V + W_{nc})dt = 0 \tag{5}$$

The above equation yields:

$$\int_{t_1}^{t_2}\left\{-\int_0^L\left[m(x)\frac{\partial^2 y}{\partial t^2} + \frac{\partial^2}{\partial x^2}\left(EI(x)\frac{\partial^2 y}{\partial x^2}\right)\right]\delta y\,dx\right.$$

$$+ \left[EI(x)\frac{\partial^2 y}{\partial x^2} + T - I_{H_2}\ddot{\theta}\right]\bigg|_{x=0}\delta\theta$$

$$+ \frac{\partial}{\partial x}\left(EI(x)\frac{\partial^2 y}{\partial x^2}\right)\bigg|_{x=0}\delta y$$

$$- \left[m_p\frac{\partial^2 y}{\partial t^2} - \frac{\partial}{\partial x}\left(EI(x)\frac{\partial^2 y}{\partial x^2}\right)\right]\bigg|_{x=L}\delta y$$

$$\left.- \left[J_p\frac{\partial^3 y}{\partial x\partial t^2} + EI(x)\frac{\partial^2 y}{\partial x^2}\right]\bigg|_{x=L}\delta\theta\right\}dt = 0 \tag{6}$$

The virtual displacement $\delta\theta$ and δy are arbitrary and independent for $0 < x < L$. At $x = 0$, we have $\delta y = 0$ and $\delta\theta$ arbitrary. At $x = L$, however, they are both considered as arbitrary (pinned–free mode). As a result, we have the following partial differential equation and its corresponding boundary conditions that describe the flexible robot dynamics.

$$m(x)\frac{\partial^2 y(x,t)}{\partial t^2} + \frac{\partial^2}{\partial x^2}\left[EI(x)\frac{\partial^2 y(x,t)}{\partial x^2}\right] = 0 \tag{7}$$

$$y(0,t) = 0 \tag{8}$$

$$EI(x)\left.\frac{\partial^2 y(x,t)}{\partial t^2}\right|_{x=0} + T - I_{H2}\ddot{\theta}(t) = 0 \tag{9}$$

$$m_p\frac{\partial^2 y(L,t)}{\partial t^2} - \frac{\partial}{\partial x}\left[EI(x)\frac{\partial^2 y(x,t)}{\partial x^2}\right]\Bigg|_{x=L} = 0 \tag{10}$$

$$\left[J_p\frac{\partial}{\partial x}\frac{\partial^2 y(x,t)}{\partial t^2} + EI(x)\frac{\partial^2 y(x,t)}{\partial x^2}\right]\Bigg|_{x=L} = 0 \tag{11}$$

Equations. (10) and (11) that relate the payload tip mass m_p and the mass moment inertia J_p to the derivatives of the tip position $y(t)$ can be rewritten as:

$$m_p\frac{\partial y(L,t)}{\partial t} = \frac{\partial}{\partial x}\left[EI(x)\frac{\partial^2 y(x,t)}{\partial x^2}\right]\Bigg|_{x=L} \tag{12}$$

$$J_p\frac{\partial^4 y(L,t)}{\partial x^4} = -m_p\frac{\partial^2 y(L,t)}{\partial x^2} \tag{13}$$

Here, it has been assumed that the material and cross sectional properties $m(x)$ and $EI(x)$ are constant. Preference of Eq. (13) to Eq. (11) is due to the fact that one could identify the payloads on–line according to Eqs. (12) and (13). This can be achieved by attaching a set of strain gauges to the proximity of the arm tip, and applying the relationships between the strain and the curvature of the robot beam at the end–point (described by $\partial^2 y(L,t)/\partial x^2$). Higher position derivative of $y(L,t)$ can be obtained by numerical differentiation. Obviously, the accuracy of this identification scheme depends upon the location of those strain gauges. When identification of the overall system parameters are involved, therefore, this technique can help to identify the variation of payloads and then adjust the forgetting factors in the identification scheme accordingly.

For normal modal analysis, we assume that the solution to the partial differential equation (7) can be expressed as

$$y(x,t) = Y(x)f(t)$$

where $Y(x)$ is usually referred to as the normal mode. $f(t)$ is the generalized coordinate. Furthermore, when payloads are considered, we have the following orthogonal relations between the mode shapes:

$$\int_0^L mY_i(x)Y_j(x)dx + m_pY_i(L)Y_j(L) + J_pY_i'(L)Y_j'(L)$$

$$+I_{H2}Y_i'(0)Y_j'(0) \;=\; \left\{ \begin{array}{ll} 0 & \forall i \neq j, \quad i,j = 0,1,2,\cdots \\[2mm] I_H & \forall i = j, \quad i,j = 0,1,2,\cdots \end{array} \right. \tag{14}$$

Mode shapes are normalized such that for every $i = j$, orthogonality relation is equal to the moment of inertia of the hub plus the moment of inertia of the undeformed arm. If we apply the above relations to the dynamic model (7) to (11) with n flexible modes, the corresponding state space description in continuous time domain then becomes a set of $(n+1)$ decoupled second order systems as follows.

$$\left. \begin{array}{l} \dfrac{d\mathbf{x}}{dt} = \mathbf{A}\,\mathbf{x} + \mathbf{b}\,T \\[4mm] \mathbf{z} = \mathbf{C}\,\mathbf{x} \end{array} \right\} \tag{15}$$

Here,

$$\mathbf{x} \;=\; \left[f_0 \;\; \dot{f}_0 \;\; f_1 \;\; \dot{f}_1 \;\; \cdots\cdots \;\; f_n \;\; \dot{f}_n \right]^T \tag{16}$$

$$
\mathbf{A} \;=\; \begin{bmatrix}
0 & 1 & & & & & \\
0 & 0 & & & & & \\
& & 0 & 1 & & & \\
& & -\omega_1^2 & -2\xi_1\omega_1 & & & \\
& & & & \ddots & & \\
& & & & & 0 & 1 \\
& & & & & -\omega_n^2 & -2\xi_n\omega_n
\end{bmatrix} \tag{17}
$$

$$
\mathbf{b} \;=\; \frac{1}{I_H}\begin{bmatrix} 0 & 1 & 0 & \dfrac{dY_1(0)}{dx} & \cdots & 0 & \dfrac{dY_n(0)}{dx} \end{bmatrix}^T \tag{18}
$$

Output measurement vector z consists of hub–angle $z_1 = \psi(t)$, hub rate $z_2 = \dot{\psi}(t)$, and tip position $z_3 = y(L,t)$. As a result, the output matrix $\mathbf{C}$ has the following form:

$$
\mathbf{C} = \begin{bmatrix}
1 & 0 & \dfrac{dY_1(0)}{dx} & 0 & \cdots & \dfrac{dY_n(0)}{dx} & 0 \\[2ex]
0 & 1 & 0 & \dfrac{dY_1(0)}{dx} & \cdots & 0 & \dfrac{dY_n(0)}{dx} \\[2ex]
L & 0 & Y_1(L) & 0 & \cdots & Y_n(0) & 0
\end{bmatrix} \tag{19}
$$

The transfer functions from torque $T(t)$ to the output vector $z(t)$ could be obtained from the above equations by:

$$
\frac{\mathbf{z}(s)}{T(s)} = \mathbf{G}(s) = \begin{bmatrix}
\dfrac{\psi(s)}{T(s)} \\[3ex]
\dfrac{\dot{\psi}(s)}{T(s)} \\[3ex]
\dfrac{y(L,s)}{T(s)}
\end{bmatrix} = \mathbf{C}(s\mathbf{I} - \mathbf{A})^{-1}\mathbf{b}
$$

Table I: Non–dimensionalized variables and conversion factors

Original Variable	ND Variables	Conversion Factors
t	$\tilde{t}$	$1/\mu$
x	$\tilde{x}$	$1/\gamma$
y	$\tilde{y}$	$1/\gamma$
T	$\tilde{T}$	$\mu^2/\delta\gamma$
m_p	$\tilde{m}_p$	$1/\delta$
J_p	$\tilde{J}_p$	$1/\delta\gamma^2$
I_{H_2}	$\tilde{I}_{H_2}$	$1/\delta\gamma^2$
ω	$\tilde{\omega}$	μ

$$= \begin{bmatrix} \dfrac{1}{I_H s^2} + \dfrac{1}{I_H} \sum_{i=1}^{n} \left[\dfrac{dY_i(0)}{dx}\right]^2 \dfrac{1}{s^2 + 2\xi_i\omega_i s + \omega_i^2} \\[3ex] \dfrac{1}{I_H s} + \dfrac{1}{I_H} \sum_{i=1}^{n} \left[\dfrac{dY_i(0)}{dx}\right]^2 \dfrac{s}{s^2 + 2\xi_i\omega_i s + \omega_i^2} \\[3ex] \dfrac{L}{I_H s^2} + \dfrac{1}{I_H} \sum_{i=1}^{n} \left[Y(L)\dfrac{dY_i(0)}{dx}\right] \dfrac{s}{s^2 + 2\xi_i\omega_i s + \omega_i^2} \end{bmatrix} \quad (20)$$

Here, s is the Laplace transform variable, ω_i and ξ_i are, respectively, the natural frequency and damping ratio corresponding to the ith flexible mode of the robot.

B. NON–DIMENSIONALIZATION OF THE DYNAMIC MODEL

We can shown that parameters $\mu = \sqrt{mL^4/EI}$, $\delta = mL$ and $\gamma = L$ have units of time (sec.), mass (kg), and length (m), respectively. Therefore, a combination of them could non–dimensionalize all the variables in Eqs. (7) to (11). These variables and their non–dimensionalized counterparts, together with the corresponding conversion factors are given in TABLE I. After non–dimensionalization, Eqs.(7) to (11) have the following forms:

SIJUN WU AND SABRI CETINKUNT

$$\frac{\partial^2 \tilde{y}(\tilde{x}, \tilde{t})}{\partial \tilde{t}^2} + \frac{\partial^4 \tilde{y}(\tilde{x}, \tilde{t})}{\partial \tilde{x}^4} = 0 \qquad (21)$$

$$\tilde{y}(0, \tilde{t}) = 0 \qquad (22)$$

$$\left. \frac{\partial^2 \tilde{y}(\tilde{x}, \tilde{t})}{\partial \tilde{t}^2} \right|_{\tilde{x}=0} + \tilde{T} - \tilde{I}_{H_2} \ddot{\theta}(\tilde{t}) = 0 \qquad (23)$$

$$\tilde{m}_p \frac{\partial^2 \tilde{y}(1, \tilde{t})}{\partial \tilde{t}^2} - \left. \frac{\partial^3 \tilde{y}(\tilde{x}, \tilde{t})}{\partial \tilde{x}^3} \right|_{\tilde{x}=1} = 0 \qquad (24)$$

$$\left[\tilde{J}_p \frac{\partial}{\partial \tilde{x}} \frac{\partial^2 \tilde{y}(\tilde{x}, \tilde{t})}{\partial \tilde{t}^2} + \frac{\partial^2 \tilde{y}(\tilde{x}, \tilde{t})}{\partial \tilde{x}^2} \right]\Bigg|_{\tilde{x}=1} = 0 \qquad (25)$$

Still we assume:

$$\tilde{y}(\tilde{x}, \tilde{t}) = Y(\tilde{x}) f(\tilde{t}) \qquad (26)$$

The orthogonality relation between different mode shapes (Eq. (14)) can be rewritten as:

$$\int_0^1 Y_i(\tilde{x}) Y_j(\tilde{x}) d\tilde{x} + \tilde{m}_p Y_i(1) Y_j(1) + \tilde{J}_p Y_i'(1) Y_j'(1) + \tilde{I}_{H_2} Y_i'(0) Y_j'(0)$$

$$= \begin{cases} 0 & \forall i \neq j, \quad i, j = 0, 1, 2, \cdots \\ \tilde{I}_H & \forall i = j, \quad i, j = 0, 1, 2, \cdots \end{cases} \qquad (27)$$

where $Y(\tilde{x})$ is called the non–dimensionalized (ND) normal mode, $f(\tilde{t})$ is referred to as the non–dimensionalized (ND) generalized coordinate and $\tilde{I}_H$ the non–dimensionalized (ND) moment of inertia of the hub plus that of the undeformed beam.

Approximating the system model (21) to (25) by n–flexible modes, the continuous time state space representation is rewritten as:

$$\left. \begin{aligned} \frac{d\mathbf{x}}{d\tilde{t}} &= \tilde{\mathbf{A}}\,\mathbf{x} + \tilde{\mathbf{b}}\,\tilde{T} \\[2mm] \mathbf{z} &= \tilde{\mathbf{C}}\,\mathbf{x} \end{aligned} \right\} \qquad (28)$$

Here,

$$\mathbf{x} = \begin{bmatrix} f_0 & \dot{f}_0 & f_1 & \dot{f}_1 & \cdots\cdots & f_n & \dot{f}_n \end{bmatrix}^T \qquad (29)$$

$$
\tilde{A} \; = \; \begin{bmatrix} 0 & 1 & & & & & \\ 0 & 0 & & & & & \\ & & 0 & 1 & & & \\ & & -\tilde{\omega}_1^2 & -2\xi_1\tilde{\omega}_1 & & & \\ & & & & \ddots & & \\ & & & & & \ddots & \\ & & & & & 0 & 1 \\ & & & & & -\tilde{\omega}_n^2 & -2\xi_n\tilde{\omega}_n \end{bmatrix} \tag{30}
$$

$$
\tilde{b} \; = \; \frac{1}{\tilde{I}_H}\begin{bmatrix} 0 & 1 & 0 & \dfrac{dY_1(0)}{d\tilde{x}} & \cdots\cdots & 0 & \dfrac{dY_n(0)}{d\tilde{x}} \end{bmatrix}^T \tag{31}
$$

$$
\tilde{C} \; = \; \begin{bmatrix} 1 & 0 & \dfrac{dY_1(0)}{d\tilde{x}} & 0 & \cdots & \dfrac{dY_n(0)}{d\tilde{x}} & 0 \\[2mm] 0 & 1 & 0 & \dfrac{dY_1(0)}{d\tilde{x}} & \cdots & 0 & \dfrac{dY_n(0)}{d\tilde{x}} \\[2mm] 1 & 0 & Y_1(1) & 0 & \cdots & Y_n(1) & 0 \end{bmatrix} \tag{32}
$$

Output z consists of the ND hub angle $z_1 = \psi(\tilde{t})$, ND hub rate measurement $z_2 = \dot{\psi}(\tilde{t})$, and ND tip position measurement $z_3 = \tilde{y}(1, \tilde{t})$. Furthermore, the natural frequencies $\tilde{\omega}_i$ of the mode shapes are also non-dimensionalized.

The transfer functions from torque input $\tilde{T}(t)$ to $\psi(\tilde{t})$, $\dot{\psi}(\tilde{t})$ and $\tilde{y}(1, \tilde{t})$ could be also be obtained as in Eq. (20). Note that, it is advantageous to use the ND model (28) through (32) since matrices $\tilde{A}$, $\tilde{b}$, and $\tilde{C}$ are independent of the dimensions of the robot arm, and are only functions of the boundary conditions. Therefore, any pinned–free beam, for example, will have the same $\tilde{A}$, $\tilde{b}$, and $\tilde{C}$ matrices. Even when payloads are considered ($m_p \neq 0, J_p \neq 0$ which results in coupling between rigid and flexible modes), their effect on system behavior could be more easily investigated since only their relative magnitudes with respect to $\delta = mL$ (the beam mass) and $\delta\gamma^2 = mL^3$ (a measure of rigid beam moment of inertia) are employed. As a result, conclusions obtained this way can be applicable to a wide class of flexible beams which share the same m_p/mL and J_p/mL^3 ratios. A unified approach toward the dynamic characteristics analysis can thus be developed based on that ND model.

A system model in discrete time domain is preferable for microprocessor–

based controller designs. In practice, zero–order–hold is the most frequently used hold circuit for analog to digital (A/D) and digital to analog (D/A) conversions, hence it is included in the discrete time modeling of the flexible beam. Here, we present the equivalent discrete time transfer function including a zero–order–hold. When sampled, this continuous time system will have the following discrete time transfer function from input torque to hub angle, hub rate and tip position:

$$
\frac{z(k)}{\tilde{T}(k)} =
\begin{bmatrix}
\dfrac{\tilde{h}(q^{-1}+q^{-2})}{2\tilde{I}_H(1-q^{-1})^2} + \displaystyle\sum_{i=1}^{n} \dfrac{d_{i1}q^{-1}+e_{i1}q^{-2}}{1+b_iq^{-1}+c_iq^{-2}} \\[3ex]
\dfrac{\tilde{h}}{\tilde{I}_H(1-q^{-1})} + \displaystyle\sum_{i=1}^{n} \dfrac{d_{i2}(q^{-1}-q^{-2})}{1+b_iq^{-1}+c_iq^{-2}} \\[3ex]
\dfrac{\tilde{h}(q^{-1}+q^{-2})}{2\tilde{I}_H(1-q^{-1})^2} + \displaystyle\sum_{i=1}^{n} \dfrac{d_{i3}q^{-1}+e_{i3}q^{-2}}{1+b_iq^{-1}+c_iq^{-2}}
\end{bmatrix}
$$

$$
=
\begin{bmatrix}
\dfrac{B_1(q^{-1})}{A_1(q^{-1})} \\[3ex]
\dfrac{B_2(q^{-1})}{A_2(q^{-1})} \\[3ex]
\dfrac{B_3(q^{-1})}{A_3(q^{-1})}
\end{bmatrix}
\tag{33}
$$

where q^{-1} is the time delay operator, $\tilde{h}$ is the non–dimensionalized sampling period. Other coefficients are defined as ($i = 1, 2, \cdots, n$):

$$
b_i = -2e^{-\xi_i\tilde{\omega}_i\tilde{h}}\cos(\tilde{\omega}_i\tilde{h}) \qquad\qquad c_i = e^{-2\xi_i\tilde{\omega}_i\tilde{h}}
$$

$$
d_{i1} = \left[\frac{dY_i(0)}{dx}\right]^2 \frac{1}{\tilde{\omega}_i^2}\frac{1}{\tilde{I}_H}\left[1 - e^{-\xi_i\tilde{\omega}_i\tilde{h}}(\cos(\tilde{\omega}_i\tilde{h}) + \xi_i\sin(\tilde{\omega}_i\tilde{h}))\right]
$$

$$
e_{i1} = \left[\frac{dY_i(0)}{dx}\right]^2 \frac{1}{\tilde{\omega}_i^2}\frac{1}{\tilde{I}_H}\left[c_i + e^{-\xi_i\tilde{\omega}_i\tilde{h}}(\xi_i\sin(\tilde{\omega}_i\tilde{h}) - \cos(\tilde{\omega}_i\tilde{h}))\right]
$$

$$
d_{i2} = \frac{1}{\tilde{I}_H}\left[\frac{dY_i(0)}{dx}\right]^2 \frac{e^{-\xi_i\tilde{\omega}_i\tilde{h}}\sin(\tilde{\omega}_i\tilde{h})}{\tilde{\omega}_i}
\tag{34}
$$

$$d_{i3} = Y_i(1)\frac{dY_i(0)}{dx}\frac{1}{\tilde{\omega}_i^2}\frac{1}{\tilde{I}_H}\left[1 - e^{-\xi_i\tilde{\omega}_i\tilde{h}}(\cos(\tilde{\omega}_i\tilde{h}) + \xi_i\sin(\tilde{\omega}_i\tilde{h}))\right]$$

$$e_{i3} = Y_i(1)\frac{dY_i(0)}{dx}\frac{1}{\tilde{\omega}_i^2}\frac{1}{\tilde{I}_H}\left[c_i + e^{-\xi_i\tilde{\omega}_i\tilde{h}}(\xi_i\sin(\tilde{\omega}_i\tilde{h}) - \cos(\tilde{\omega}_i\tilde{h}))\right]$$

Note that $A_1(q^{-1})$, $A_2(q^{-1})$, $B_1(q^{-1})$, $B_2(q^{-1})$, $A_3(q^{-1})$, and $B_3(q^{-1})$, defined in Eq.(33) are polynomials in terms of the delay operator q^{-1}. $A_i(q^{-1})$ and $B_i(q^{-1})$ ($i = 1, 2, 3$) are called the AR and MA part of process (33), respectively. The damping ratios ξ_i in (34) are considered to be very small compared to unity (usually 0.01—0.03 [2, 27, 4, 5]). The parameters of discrete time model will be identified in real–time by a lattice filter as presented in Section III, and will be used in discrete–time controller design. Therefore, the whole control algorithm (identifier + controller) is based on the discrete time model.

C. SAMPLING FREQUENCY AND STABILITY

It is known that, in continuous time domain, non–colocation of sensor and actuator will yield a non–minimum phase transfer function [2, 4, 8, 11, 16, 20, 28]. whereas colocated control will give a minimum phase transfer function, and closed loop stability is robust against modeling errors and loop gain variations. The price paid for the stability robustness is that the maximum possible closed loop bandwidth with colocated output feedback control which is limited by the lowest cantilever frequency of the arm [29]. Since our goal is to achieve high bandwidth operation, non–colocated sensor–actuator configuration is inevitable. As a result, we will investigate the stability problem here. The effect of the sampling rate on system stability will also discussed.

When sampled, a continuous time system has the following discrete time z–transfer function [30]:

$$H(z) = (1 - z^{-1})\frac{1}{2\pi j}\int_{\gamma-j\infty}^{\gamma+j\infty}\frac{e^{sh}}{z - e^{sh}}\frac{G(s)}{s}\,ds \qquad (35)$$

where, $j = \sqrt{-1}$, $G(s)$ is the transfer function (can be a matrix as $\mathbf{G}(s)$ in Eq. (20)), h is the sampling interval (can be non–dimensionalized as $\tilde{h}$), and

γ is a real number such that all poles of $G(s)/s$ have real parts less than γ. Equation (35) is obtained by first using the inverse Laplace transform to determine the z – transform of the step response, and then dividing it by the z – transform of a step function.

When $G(s)$ is a proper function in s $\left(\text{i.e., } \lim_{s \to \infty} G(s) = 0\right)$, which is true for the transfer functions of flexible one arm robots, we could apply the methods in residue calculus to evaluate the integral (35) as

$$H(z) = (1 - z^{-1}) \sum_i \frac{e^{p_i h}}{z - e^{p_i h}} \text{Res}_{s=p_i} \left(\frac{G(s)}{s}\right) \qquad (36)$$

The above equation indicates that when $G(s)$ is sampled, its poles p_i, ($i = 1, 2, \cdots$) are mapped to the z–plane as:

$$z_i = e^{p_i h} \qquad (37)$$

There are, however, no explicit expressions for the zeros in the discrete time domain. The locations of the zeros depend on the residues of $G(s)/s$, sampling interval h, as well as the poles z_i [30].

If $G(s)$ is rational, i.e., $G(s) = N(s)/D(s)$ where $N(s)$ and $D(s)$ are polynomials in s, further, if $s = p_i$ is a multiple zero of $sD(s)$ of order m, we then have

$$\text{Res}_{s=p_i} \left[\frac{G(s)}{s}\right] = \frac{1}{(m - 1)!} \lim_{s \to p_i} \frac{d^{m-1}}{ds^{m-1}} \left[(s - p_i)^m \frac{N(s)}{sD(s)}\right] \qquad (38)$$

It is well known that a non–minimum phase continuous time system will not always become a discrete time system with an unstable inverse, whereas a minimum phase continuous–time system may become a discrete time non–invertible system when sampled. The sampling interval h plays an important role. Astrom [30] pointed out that when a sufficiently small sampling interval h is chosen, the discretized system will have zeros at:

$$z_i = e^{s_i h} \qquad (39)$$

where s_i's are the zeros of the original continuous system. The rest $d = r - 1$ zeros (r is the pole excess of $G(s)$) which correspond to the zeros at infinity of the continuous time system, will converge to the zeros of the polynomial

Table II: Pole excess, polynomial $P(z)$ and its unstable zeros

Pole Excess	$P(z)$	Unstable Zeros
1	1	
2	$z + 1$	-1
3	$z^2 + 4z + 1$	-3.742
4	$z^3 + 11z^2 + 11z + 1$	$-1, -9.899$
5	$z^4 + 26z^3 + 66z^2 + 26z + 1$	$-2.322, -23.20$
6	$z^5 + 57z^4 + 30z^3 + 30z^2 + 57z + 1$	$-1, -4.542, -51.22$

$P(z)$ given in TABLE II. It becomes obvious that a continuous time system with pole excess larger than 2 will surely have at least one unstable zero on or outside the unit circle of the z–plane if sampled with "sufficiently" small sampling period.

On the other hand, when sampling interval increases, the zeros of $H(z)$ in Eq. (35) will move toward the origin of the z–plane. It was shown that when a stable continuous time system is sampled and a zero–order–hold is applied, its zeros will go to the origin of the z – plane as the sampling period approaches infinity [30]. We can thus conclude that the zeros of the sampled system will lie between the origin and the locations determined by Eq. (39) and $P(z)$ for an arbitrary sampling rate. Theoretically, we could arbitrarily place the zeros by selecting suitable sampling rate, if the hold circuit is of zero order. This idea is not frequently used in practice, however, due to the fact that the sampling frequency must be greater than the Nyquist frequency of the original system according to Shannon's sampling theorem [31].

When the linearized model for a flexible one arm robot is investigated, one has the open loop system poles as:

$$p_i = -\xi_i \tilde{\omega}_i \pm j \tilde{\omega}_i \sqrt{1 - \xi_i^2} \; ; \qquad i = 0, 1, 2, \cdots, \quad j = \sqrt{-1}$$

In discrete time domain, p_i is mapped to

$$z_i = e^{-\xi_i \tilde{\omega}_i \tilde{h}} \left[\cos(\tilde{\omega}_i \tilde{h} \sqrt{1 - \xi_i^2}) + j \sin(\tilde{\omega}_i \tilde{h} \sqrt{1 - \xi_i^2}) \right] \tag{40}$$

which lies in a very close neighborhood of the unit disc, since ξ_i is usually small (between 0.01 to 0.03).

When a colocated control scheme is adopted, the continuous time transfer function is minimum phase. Moreover, the pole excess is one ($r = 1$). It can be concluded, therefore, that the discrete time transfer function has all poles and zeros inside the unit circle. In this case, stability is very easy to establish with colocated feedback. One could visualize this conclusion by plotting the root locus in the z–plane when various feedback gains are employed.

On the other hand, the non–colocated control strategy is frequntly used for precision high bandwidth tip position control of the flexible robot. In this case, the open loop transfer function is non–minimum phase and has a pole excess r equal to 2. Consequently, the sampled system will have at least one zeros on the unit circle, and a number of zeros possibly outside the unit circle, depending on the number of modes that we are investigating.

III. IMPLEMENTATION OF THE LATTICE FILTER PARAMETER IDENTIFIER

A. BACKGROUND

The lattice filters are widely used in signal processing and time series. Due to the availability of advanced computer technology, the demand for high–performance adaptive systems motivated the search for more sophisticated estimation and control algorithms that are robust against environmental changes, system dynamics variations, and uncertainties. As a result, the application of lattice filters in the area of adaptive control have grown fast during the past decade. In signal processing, a significant part of research work is based on modeling the process of interest as an output of a finite order discrete time linear system driven by noise. Such representations are called auto–regressive (AR) or auto–regressive and moving average (ARMA).

The two main filter structures that have been studied are the transversal and lattice filters. The classic sequential least square (SLS) algorithm is a good example of the transversal filter. The lattice filters became popular in the mid–1970's. Since then, they have been proven to be more robust to quantization and round–off noise than the transversal filters do. This means that under similar hardware constraints, the performance of a lattice filter will generally be better than that of a SLS filter. The lattice filters are far faster than SLS in terms of computation time per iteration when the model order N (or number of parameters to be identified) is high (in practice, $N > 6$ [32]). Still they preserve the convergence rate properties of SLS, which were shown to be of near maximum likelihood rate, i.e., the fastest possible. Graupe [32] also demonstrated that, while computational time of sequential least square increases proportionally to N^2, the computation effort for lattice filters increases proportionally only to N. Furthermore, lattice filters have another advantage that, in computing the Nth–order lattice filter, all lattice filters of order $i < N$ are also computed and are optimal. Consequently, lattice filters are a convenient hardware and/or software realization to use when the order of controlled plant is not known *a priori*.

Most of the work on lattice filter has been limited to AR (all pole) and MA (all zero) lattices, with little on ARMA. One of the difficulties with the ARMA lattice is that no general form is available. Lee, Friedlander, and Morf [33] developed an ARMA lattice filter for the case in which the order of AR terms equals to that of the MA terms. This structure also applies here.

In this chapter, we address the problem of least square AR and ARMA lattice filter formulations. The lattice filter is constructed by performing Gram–Schmidt orthogonalization on the incoming data. Each lattice section can be viewed as carrying out a one–step orthogonalization procedure. The lattice filter will be utilized in the tip position control of a flexible one–arm robot.

B. IMPLEMENTATION OF THE LATTICS FILTER PARAMETER ESTIMATOR

For the purpose of identifying the parameters of the discrete–time transfer function, we consider first the sampling of a process with M–channel measurement of the form:

$$z_k = [\, z_{1k} \quad z_{2k} \quad \cdots \quad z_{Mk} \,]^T$$

where subscript k denotes the current time instance. $\mathbf{z}_k$ is the vector of output measurements. We are interested in predicting the current output vector from past measurements. A linear auto–regressive (AR) predictor of order N is assumed and will have the form:

$$\hat{\mathbf{z}}_k = -\sum_{i=1}^{N} A_{N,i} \mathbf{z}_{k-i} \tag{41}$$

Here, $\hat{\mathbf{z}}_k$ is the predictive value of $\mathbf{z}_k$ based on the data up to time $k-1$, and $A_{N,i}$, a $M \times M$ matrix, is usually referred to as the forward prediction filter. N is the order of the filter. Similarly, we could also build a backward prediction filter:

$$\hat{\mathbf{z}}_{k-N-1} = -\sum_{i=1}^{N} B_{N,N+1-i} \mathbf{z}_{k-i} \tag{42}$$

As a result, we could define the forward estimator error vector $\epsilon_{N,k}$ and a backward estimator error vector $\mathbf{r}_{N,k}$, respectively, as:

$$\left. \begin{aligned} \epsilon_{N,k} &= \mathbf{z}_k - \hat{\mathbf{z}}_k = \mathbf{z}_k + \sum_{i=1}^{N} A_{N,i} \mathbf{z}_{k-i} \\[2em] \mathbf{r}_{N,k} &= \mathbf{z}_k - \hat{\mathbf{z}}_{k-N-1} = \mathbf{z}_{k-p-1} + \sum_{i=1}^{N} B_{N,N+1-i} \mathbf{z}_{k-i} \end{aligned} \right\} \tag{43}$$

The estimated process has the autocorrelation of the output vector as:

$$R_{i-j} = E[\mathbf{z}_i\, \mathbf{z}_j^T] \tag{44}$$

where E represents expectation.

The basic idea of a lattice filter is as follows: for an AR process, we assume that we have all the information about the pth–order lattice filter (in terms of $A_{p,i}$ and $B_{p,p+1-i}$). Then the order p is advanced to $p+1$ and a new filter is built based on the past data. The above procedure is repeated until a satisfactory filter order is achieved (meaning p could go beyond N). It is required that the $(p+1)$th order filter also outputs a set of white prediction errors when the process under estimation is fed into it. The recursive formulation for filter from order p to $p+1$ is given below.

Since the lattice filter originates from least square predictor, its coefficients can be computed by applying the orthogonality property of the least

square prediction error and the output vector:

$$E[\epsilon_{p,k} z_{k-j}^T] = 0 \qquad \text{for all } j = 1, 2, \ldots, p \qquad (45)$$

The mean square error of the forward predictor is given by:

$$R_p^\epsilon = E[\epsilon_{p,k} \ \epsilon_{p,k}^T] = R_0 + \sum_{i=1}^{p} A_{p,i} R_{-i} \qquad (46)$$

Also, we have according to Eq. (44) that $R_{-j} = R_j^T$. Applying Eqs. (44), (45), and (46), we may derive the so-called Yule–Walker equation:

$$[I \ A_{p,1} \ \ldots \ A_{p,p}] \begin{bmatrix} R_0 & R_1 & \ldots & R_p \\ R_{-1} & R_0 & \ldots & R_{p-1} \\ \vdots & \vdots & \ddots & \vdots \\ R_{-p} & R_{p+1} & \ldots & R_0 \end{bmatrix} = [R_p^\epsilon \ 0 \ \ldots \ 0] \quad (47)$$

or

$$[I \ A_{p,1} \ \ldots \ A_{p,p}] \ \mathbf{R}_p = [R_p^\epsilon \ 0 \ \ldots \ 0] \qquad (48)$$

where

$$\mathbf{R}_p = \begin{bmatrix} R_0 & R_1 & \ldots & R_p \\ R_{-1} & R_0 & \ldots & R_{p-1} \\ \vdots & \vdots & \ddots & \vdots \\ R_{-p} & R_{p+1} & \ldots & R_0 \end{bmatrix} \qquad (49)$$

Note that $\mathbf{R}_p$ is a Toeplitz and symmetric matrix.

Similarly, we have for the backward predictor:

$$[B_{p,p} \ B_{p,p-1} \ \ldots \ B_{p,1} \ I] \ \mathbf{R}_p = [0 \ 0 \ \ldots \ 0 \ R_p^r] \qquad (50)$$

where R_p^r is named as the mean square error of the backward predictor given by:

$$R_p^r = E[\mathbf{r}_{p,k}\ \mathbf{r}_{p,k}^T] = R_0 + \sum_{i=1}^{p} B_{p,p+1-i} R_{p+1-i} \qquad (51)$$

Assume that we have estimated the coefficients of a pth–order lattice filter in terms of $A_{p,i}$ and $B_{p,p+1-i}$, $(i = 1,\ldots,p)$. If we use these coefficients to build a $(p+1)$th order lattice filter, a set of errors will be resulted from a $(p+1)$th–order Yule–Walker equation:

$$\begin{bmatrix} I & A_{p,1} & \cdots & A_{p,p} & 0 \\ 0 & B_{p,p} & \cdots & B_{p,1} & I \end{bmatrix} \mathbf{R}_{p+1} = \begin{bmatrix} R_p^\epsilon & 0 & 0 & \cdots & 0 & \Delta_{p+1}^\epsilon \\ (\Delta_{p+1}^r)^T & 0 & 0 & \cdots & 0 & R_p^r \end{bmatrix}$$
$$(52)$$

where Δ_{p+1}^ϵ and Δ_{p+1}^r are the errors which can be represented by:

$$\Delta_{p+1}^\epsilon = R_{p+1} + \sum_{i=1}^{p} A_{p,i} R_{p-i+1}$$
$$(53)$$
$$(\Delta_{p+1}^r)^T = R_{-p-1} + \sum_{i=1}^{p} B_{p,i} R_{-p-1+i}$$

In order to eliminate those errors, we must choose a linear combination of the forward and the backward pth–order lattice filter such that:

$$\begin{bmatrix} K_{11} & K_{12} \\ K_{21} & K_{22} \end{bmatrix} \begin{bmatrix} I & A_{p,1} & \cdots & A_{p,p} & 0 \\ 0 & B_{p,p} & \cdots & B_{p,1} & I \end{bmatrix} \mathbf{R}_{p+1}$$

$$= \begin{bmatrix} K_{11} & K_{12} \\ K_{21} & K_{22} \end{bmatrix} \begin{bmatrix} R_p^\epsilon & 0 & \cdots & 0 & \Delta_{p+1}^\epsilon \\ (\Delta_{p+1}^r)^T & 0 & \cdots & 0 & R_{p+1}^r \end{bmatrix}$$

$$= \begin{bmatrix} I & A_{p+1,1} & A_{p+1,2} & \cdots & A_{p+1,p+1} & 0 \\ 0 & B_{p+1,p+1} & B_{p+1,p} & \cdots & B_{p+1,1} & I \end{bmatrix} \mathbf{R}_{p+1}$$

$$
= \begin{bmatrix} R_{p+1}^\epsilon & 0 & \cdots & 0 & 0 \\ \\ 0 & 0 & \cdots & 0 & R_{p+1}^r \end{bmatrix} \tag{54}
$$

Therefor, we have for K_{11}, K_{12} and R_{p+1}^ϵ the following relation:

$$
\left. \begin{aligned} K_{11}R_p^\epsilon + K_{12}(\Delta_{p+1})^T &= R_{p+1}^\epsilon \\ \\ K_{11}\Delta_{p+1}^\epsilon + K_{12}R_p^r &= 0 \end{aligned} \right\} \tag{55}
$$

For a particular solution, choose $K_{11} = I$. Then K_{12} and the mean square error of the $(p+1)$th–order forward predictor R_{p+1}^ϵ are given by:

$$
\begin{aligned} K_{12} &= -\Delta_{p+1}^\epsilon \left(R_p^r \right)^{-1} \\ \\ R_{p+1}^\epsilon &= R_p^\epsilon - \Delta_{p+1}^\epsilon \left(R_p^r \right)^{-1} \left(\Delta_{p+1}^r \right)^T \end{aligned} \tag{56}
$$

Similarly, for the backward predictor, one has $K_{22} = I$ and

$$
\begin{aligned} K_{21} &= - \left(\Delta_{p+1}^r \right)^T \left(R_p^\epsilon \right)^{-1} \\ \\ R_{p+1}^r &= R_p^r - \left(\Delta_{p+1}^r \right)^T \left(R_p^\epsilon \right)^{-1} \Delta_{p+1}^\epsilon \end{aligned} \tag{57}
$$

Substituting Eqs. (56) and (57) into Eq. (54) could yield the coefficients of a $(p+1)$th–order lattice filter. However for the sake of compactness, we define for a pth–order lattice filter:

$$
\left. \begin{aligned} A(q^{-1}) &= I + A_{p,1}q^{-1} + \cdots + A_{p,p}q^{-p} \\ \\ B(q^{-1}) &= B_{p,p} + B_{p,p-1}q^{-1} + \cdots + Iq^{-p} \end{aligned} \right\} \tag{58}
$$

where $A(q^{-1})$ and $B(q^{-1})$ are matrix polynomials which describe the forward and backward prediction filters, respectively, in terms of the delay operator q^{-1}. Then Eqs. (54), (56), and (57) give the following iteration relation:

$$
\begin{bmatrix} A_{p+1}(q^{-1}) \\ \\ B_{p+1}(q^{-1}) \end{bmatrix} = \begin{bmatrix} I & -K_{p+1}^r \\ \\ -\left(K_{p+1}^\epsilon \right)^T & I \end{bmatrix} \begin{bmatrix} A_p(q^{-1}) \\ \\ q^{-1}B_p(q^{-1}) \end{bmatrix} \tag{59}
$$

Here, K_{p+1}^{ϵ} and K_{p+1}^{r} are, respectively, the forward and backward reflection coefficient matrices given by:

$$K_{p+1}^{\epsilon} = \left(R_p^{\epsilon}\right)^{-1} \Delta_{p+1} \qquad K_{p+1}^{r} = \Delta_{p+1} \left(R_p^{r}\right)^{-1} \qquad (60)$$

Δ_{p+1} can be interpreted as the correlation between the forward and backward prediction errors which are one time step apart, and it can be shown that:

$$\Delta_{p+1} = E\left[\epsilon_{p,k}\ \mathbf{r}_{p,k-1}^{T}\right] = \Delta_{p+1}^{\epsilon} = \Delta_{p+1}^{r} \qquad (61)$$

Applying Eq. (54) and some algebraic manipulations, one can show that:

$$K_{p+1}^{\epsilon} = -A_{p+1,p+1} \qquad K_{p+1}^{r} = -B_{p+1,p+1}^{T} \qquad (62)$$

The recursive relations for forward and backward prediction errors $\epsilon_{p,k}$ and $\mathbf{r}_{p,k}$ are obtained by post–multiplying Eq. (59) by $\mathbf{z}_k$:

$$\left.\begin{array}{l} \epsilon_{p+1,k} = \epsilon_{p,k} - K_{p+1}^{r}\mathbf{r}_{p,k-1} \\[2mm] \mathbf{r}_{p+1,k} = \mathbf{r}_{p,k-1} - \left(K_{p+1}^{\epsilon}\right)^{T}\epsilon_{p,k} \end{array}\right\} \qquad (63)$$

Once a set of a properly chosen initial conditions are provided, we could recursively advance the filter order from p, until p reaches the desired order. The signal flow in such a structured filter is shown in Fig. 2.

The aforementioned estimation scheme works under the assumption that the process is wide–sense stationary such that the expectations and Toeplitz matrix can be utilized for derivation. In many cases, however, the stationarity of the process is not known *a–priori* to us, or the process is slowly time–varying. Consequently, the above algorithm may be slightly revised if the process parameters do not change too quickly compared with the convergence rate of the identification algorithm. However, for slight non–stationarities, such as piece–wise constant parameters in a process, the

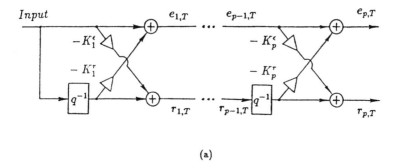

(a)

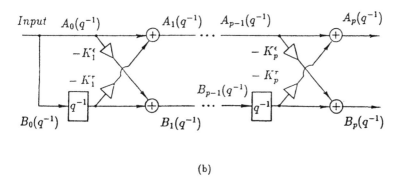

(b)

Figure 2: Structure of the Lattice Filter: (a) Error Propagation in the Lattice Filters; (b) Coefficient Propagation in the Lattice Filters

modified lattice filter still reserves the convergence rate of a least–square filter and its time saving property. This implies that the convergence property is still kept as long as the parameters have hardly changed over a time interval that is longer than that within which the identifier has adequately converged. In such cases, the terms expressed by the expectations as shown above have their equivalent forms for adaptive processes as follows:

(1) The correlation between forward and backward prediction errors is represented by:

$$\Delta_{p+1,k} = \Delta_{p+1,k-1} + \epsilon_{p,k}\, \mathbf{r}_{p,k-1}^T / \gamma_{p-1,k-1} \tag{64}$$

where

$$\gamma_{p,k} = \gamma_{p-1,k} - \mathbf{r}_{p,k}^T \left(R_{p,k}^r \right)^{-1} \mathbf{r}_{p,k}$$

(2) R_{p+1}^ϵ and R_{p+1}^r in Eqs. (56) and (57) are revised to the following form:

$$R_{p+1,k}^r = R_{p,k-1}^r - \Delta_{p+1,k}^T K_{p+1,k}^\epsilon \tag{65}$$

$$R_{p+1,k}^\epsilon = R_{p,k}^\epsilon - \Delta_{p+1,k}^T K_{p+1,k}^r \tag{66}$$

(3) Other expressions do not change for slowly time–varying systems.

Parameter $\gamma_{p,k}$ in the above algorithm has an important interpretation as an approximate log-likelihood variable related to the process z_k. Friedlander [34] showed that $\gamma_{p,k}$ can be used as a good detection statistic for the unexpectness of the recent output measurements. If the measurements are from a Gaussian distribution with second–order statistics $\mathbf{R}_p$, the variable $\gamma_{p,k}$ will be small. If the recent measurements come from a different distribution other than a Gaussian distribution, $\gamma_{p,k}$ will tend to be unity. Friedlander also recommended that this variable be used as an adaptive gain in order to achieve high speed tracking of the change in the statistics of the current output measurements.

Once the above revisions are made, the coefficients $A_{p,i}$ of a forward prediction filter could be determined and the identification results may be employed in the design of an adaptive controller. Even though the above derivation is made for AR processes, it could also be applied to the Auto–Regressive and Moving Average (ARMA) processes by transferring an n–output, m–input ARMA model into an equivalent (n+m)–dimensional AR model (Lee et al. 1982 [33]).

Assume the input/output relation of an ARMA process is represented as:

$$z = -\sum_{i=1}^{N} A_{N,i} z_{k-i} + \sum_{i=1}^{N} D_{N,i} \mathbf{T}_{k-i} \qquad (67)$$

where z_k is the M–dimensional output vector, $\mathbf{T}_k$ the n–dimensional input vector. Moreover, N is the order of the process corresponding to its AR and MA parts. $A_{N,i}$ and $D_{N,i}$ are $M \times M$ and $M \times n$ matrices. The above ARMA process can be embedded in an AR process of order $n + m$ as:

$$\begin{bmatrix} z_k \\ \mathbf{T}_k \end{bmatrix} = \sum_{i=1}^{N} \begin{bmatrix} -A_{N,i} & D_{N,i} \\ F_{N,i} & G_{N,i} \end{bmatrix} \begin{bmatrix} z_{k-i} \\ \mathbf{T}_{k-i} \end{bmatrix} \qquad (68)$$

Here, $F_{N,i}$ and $G_{N,i}$ are two auxiliary matrices. Therefore, the algorithm developed in this section can be applied to estimate the coefficients of (67).

When identifying the parameters of the discrete–time transfer function for a flexible robot arm, we could transfer this one–input, three–output model into a four–channel AR process similar to Eq.(33).

$$[z_{1k} \quad z_{2k} \quad z_{3k} \quad T_k]^T = H(q^{-1})[z_{1k} \quad z_{2k} \quad z_{3k} \quad T_k]^T \qquad (69)$$

Here, z_{1k}, z_{2k}, z_{3k}, and T_k are, respectively, the measurements of hub–angle, hub–rate, tip position, and input torque. Although we did not plan to use a torque measurement sensor at the joint, it can be indirectly estimated from the servo motor current measurement. Moreover, $H(q^{-1})$ is a matrix of polynomials in terms of the delay operator q^{-1}. Comparing with Eq.(33), we can write $H(q^{-1})$ as:

$$H(q^{-1}) = \begin{bmatrix} -A_1(q^{-1}) & 0 & 0 & B_1(q^{-1}) \\ 0 & -A_2(q^{-1}) & 0 & B_2(q^{-1}) \\ 0 & 0 & -A_3(q^{-1}) & B_3(q^{-1}) \\ F_1(q^{-1}) & F_2(q^{-1}) & F_3(q^{-1}) & F_4(q^{-1}) \end{bmatrix}$$

$A_1(q^{-1})$, $A_2(q^{-1})$, $A_3(q^{-1})$, $B_1(q^{-1})$, $B_2(q^{-1})$, and $B_3(q^{-1})$ have been de-fined in Eq. (33), while $F_1(q^{-1})$, $F_2(q^{-1})$ $F_3(q^{-1})$, and $F_4(q^{-1})$ are auxil-iary polynomials of compatible order. Once $H(q^{-1})$ is estimated, parame-ters $b_i, c_i, d_{i1}, e_{i1}, d_{i2}, d_{i3}$, and e_{i3} in Eq.(34) can be extracted from $H(q^{-1})$ based on Eq.(33).

When identifying processes that are not stationary, that is, their pa-rameters are varying slowly with time, one may use a pre–defined data window such that the identification is restarted every so many data points. Piece wise stationarity is assumed in every data window. This scheme will work if the process is stationary (more or less) in the window which is big enough for adequate convergence of the filter. If, on the other hand, the parameter variations are fast, then this scheme will not work well. In this case, one needs to employ a forgetting factor in the estimation algorithm, such that data points further in the past are made of less importance . It is obvious that the incorporation of forgetting factor will make the filter biased. The lower the forgetting factor (away from unity), the higher the resultant bias in identification. However, only a lower forgetting factor can deal with fast parameter variations of a process under estimation.

In our control of a flexible robot, since payload variations are very common, and in order to reduce bias in the identification algorithm, we choose the forgetting factor based on estimated payload change. If this change is small, the forgetting factor is set closer to unity. However, if large payload variation is detected, a lower value is assigned to the forgetting factor.

When a varying forgetting factor λ $(0 < \lambda \leq 1)$ is implemented in the lattice filter estimator, the complete identification scheme is summarized in Table III with its initial conditions.

IV. THE PREDICTIVE ADAPTIVE CONTROL ALGORITHM

A. INTRODUCTION

It has been demonstrated in Section II that the control action applied to a process can only affect the future behavior of that process in discrete time

Table III: The least square lattice filter parameter identifier

---

Input Parameter :

N : Order of the Lattice Filter

z_k : Data Sequence at Time k

λ : Forgetting Factor

Initialization :

$\epsilon_{0,k} = \mathbf{r}_{0,k} = \mathbf{z}_k$

$R_{0,k}^{\epsilon} = R_{0,k}^{r} = \lambda R_{0,k-1}^{\epsilon} + \mathbf{z}_k \mathbf{z}_k^T$

$\gamma_{-1,k} = 0$

Do For $p = 0$ to $\text{Min}(N,k) - 1$

$\Delta_{p+1,k} = \lambda \Delta_{p+1,k-1} + \epsilon_{p,k} \ \mathbf{r}_{p,k-1}^T / \gamma_{p-1,k-1}$

$\gamma_{p,k} = \gamma_{p-1,k} - \mathbf{r}_{p,k}^T \left(R_{p,k}^r \right)^{-1} \mathbf{r}_{p,k}$

$K_{p+1}^r = \Delta_{p+1,k} \left(R_{p,k-1}^r \right)^{-1}$

$\epsilon_{p+1,k} = \epsilon_{p,k} - K_{p+1}^r \mathbf{r}_{p,k-1}$

$R_{p+1,k}^{\epsilon} = R_{p,k}^{\epsilon} - \Delta_{p+1,k}^{\epsilon} \left(K_{p+1,k}^r \right)^T$

$K_{p+1,k}^{\epsilon} = \left(R_{p,k}^{\epsilon} \right)^{-1} \Delta_{p+1,k}$

$\mathbf{r}_{p+1,k} = \mathbf{r}_{p,k-1} - \left(K_{p+1,k}^{\epsilon} \right)^T \epsilon_{p,k}$

$R_{p+1,k}^{r} = R_{p,k-1}^{r} - \left(\Delta_{p+1,k} \right)^T K_{p+1,k}^{\epsilon}$

---

domain due to time delay from sampling as well as the inherent time delay in the process. As a result, for the design of effective control schemes, it is important to be able to predict not only which disturbance will influence the process in the future, but also how the process will respond to different control variables. The control strategy presented in this section, namely the *predictive adaptive control* (PAC), can be used to serve for the above purposes.

The predictive adaptive control method adds an alternative to the self-tuning adaptive control techniques. It is capable of stable control of systems with time-varying parameters, time delays, and variable model orders provided that the input/output data are rich enough to allow reasonable system identification [32, 35]. It is also effective with non-minimum phase, and lightly damped systems whose model might be overparameterized by the estimation scheme. Hence, it is suitable for high-performance control of flexible robot arms.

The origin of the predictive adaptive control can be traced back to a set of papers which appeared in the late 1970s. Richalet et al. [35] in 1978 developed and applied a "Model Predictive Heuristic Control" method to some industrial processes. This method was also called Model Algorithmic Control (MAC). The Dynamic Matrix Control (DMC) scheme was proposed by Cutler and Ramaker [36] and was utilized to control a fluid catalytic cracker. Other predictive adaptive control algorithms have also been developed (See Garcia at al. [37] for a complete review on this subject). The methods in [12, 16] used an explicit dynamic model to predict the effect of future actions of the manipulation variables on the output. The predictive identification, in essence, is an on-line estimation of the process pulse response [32], and is based upon the input/output data sampled. The control action is determined by optimization with the objective of minimizing the predicted output error subjected to operating constraints. The optimization is repeated at each sampling instance based on updated information from the system output measurements. The predictive control scheme is valid for linear and non-linear systems if perturbation is small over a control interval. The present approach, however, does not necessarily assumes the presence of system noise and measurement noise. Such noises, when present, may deteriorate the performance of the predictive adaptive controller. A Kalman filter or a Wiener filter may be applied to attenuate the noise level. On the other hand, the use of the lattice filter in the identification will help to reduce the effect of noise due to its robustness and fast convergence speed.

This section is devoted to the development of the predictive adaptive control algorithm for single–input–multi–output (SIMO) discrete–time systems. Together with a parameter identifier, the predictive adaptive controller gives a special self–tuning controller. This controller is then applied to the tip position control of the one arm flexible robot where tip position and hub–rate feedbacks are utilized. Model of actuator dynamics (such as a moving coil, permanent magnet direct current motor, a servo amplifier, and a harmonic drive speed reducer) has also been included in the analysis. The effect of the actuator dynamics on the overall system response and stability are investigated. Actuator parameters are chosen from the actual specifications provided by manufacturers.

B. THE PREDICTIVE ADAPTIVE CONTROL SCHEME

It has been mentioned earlier that precision high bandwidth control could be achieved using non–colocated tip position feedback, whereas the colocated hub rate feedback tends to stabilize the overall control system. Therefore, by choosing a proper trade–off between these two output feedback gains, one could obtained a desirable stable control, yet provide a satisfactory transient characteristics. The predictive adaptive controller is formulated based on the above observations.

The control signals are determined at each sampling instance under the assumption that parameters of the process under control remain constant (and equal to their estimates) during the current control sampling interval. The control is determined so as to minimize a quadratic cost function of the predicted output error and control increment. Note that the following formulation is directly derived for single input, multiple output (SIMO) processes.

Consider a SIMO process modeling a flexible robot arm by a reduced order ARMA process given in Eq. (33). We can rewrite that equation in a state space representation as:

$$\left. \begin{array}{l} \mathbf{x}(k+1) = \mathbf{A}\,\mathbf{x}(k) + \mathbf{b}\,\tilde{T}(k) \\[2mm] \mathbf{z}(k) = \mathbf{C}\,\mathbf{x}(k) \end{array} \right\} \qquad (70)$$

Here elements of $\mathbf{A}$, $\mathbf{b}$ and $\mathbf{C}$ depend on the coefficients of the polynomials $A_2(q^{-1}), B_2(q^{-1}), A_3(q^{-1})$, and $B_3(q^{-1})$ given in Eq. (33). Since only the hub–rate and tip position are utilized in the feedbacks, the output vector here will contain the non–dimensionalized hub–rate and tip position measurements.

The main steps in the proposed predictive adaptive control algorithm are shown in Fig. 3 where the ith element $z_i(k)$ of the output vector $\mathbf{z}(k)$ is shown. At time instant k, we may predict output $z_i(k)$ at time $k+1$ according to information from the past upto k, assuming that the input remains unchanged after time instant $k-1$. However, the predicted output may be different from the desired output $z_d(k)$. The goal of the predictive adaptive controller is to bring the predicted system output to the desired position as close as possible. Hence, let us define $z_0(k+1)$ as the value of output $\mathbf{z}(k)$ when assume:

$$\Delta \tilde{T}(k) = \tilde{T}(k) - \tilde{T}(k-1) = 0$$

If instead a nonzero $\Delta \tilde{T}(k)$ is applied, the corresponding system output $\mathbf{z}(k+1)$ renders the predicted system response to $\Delta \tilde{T}(k)$ as follows [32]:

$$\hat{\mathbf{u}}(k) = \frac{\mathbf{z}(k+1) - \hat{\mathbf{z}}_0(k+1)}{\Delta \tilde{T}(k)} \qquad (71)$$

where $\hat{\mathbf{z}}_0(k+1)$ denotes the estimated value of $\mathbf{z}_0(k+1)$. Furthermore, in Eq. (71), $\hat{\mathbf{u}}(k)$ explicitly gives the estimated system response to a unit step pulse. It has been pointed out [32, 38] that because of the application of zero–order hold (every control signal is in the form of a step pulse), $\hat{\mathbf{u}}(k)$ will provide enough information for the purpose of control. Consequently, control increment $\Delta \tilde{T}(k)$ is determined to minimize the following cost-functional J_{k+1} such that, at sampling interval $k+1$, system output $\mathbf{z}(k+1)$ is brought to its desired position $z_d(k+1)$ as close as possible, with a penalty on the magnitude of the control signal $\tilde{T}(k)$.

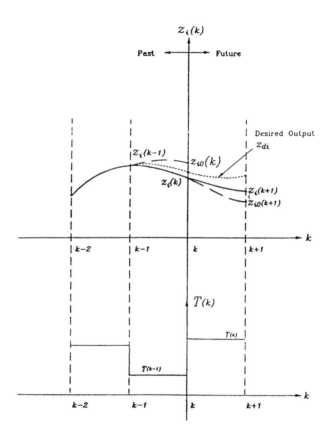

Figure 3: Prediction of System Output

$$J_{k+1} = \left[\hat{e}^T(k+1) \mathbf{P} \hat{e}(k+1) + \rho \tilde{T}^2(k) \right] \tag{72}$$

where $\hat{e}(k+1)$ designates the estimated output error vector at time instance $k+1$, and ρ is a weighting factor on the control cost. $\mathbf{P}$ is a chosen matrix which determines the relative weighting on the predicted tip position error and hub rate error. Definitions of $\hat{e}(k+1)$ and $\hat{e}_0(k+1)$ are given below.

$$\hat{e}(k+1) = \hat{z}_0(k+1) + \Delta \tilde{T}(k)\hat{u}(k) - z_d(k+1)$$

$$\hat{e}_0(k+1) = \hat{z}_0(k+1) - z_d(k+1) \tag{73}$$

Here, $\hat{e}_0(k+1)$ is the estimated output error assuming $\Delta \tilde{T}(k) = 0$.

The optimal solution to Eq. (72), subjected to constraints (70) and (73) can be obtained by choosing $\Delta \tilde{T}(k)$ such that:

$$\frac{\partial J_{k+1}}{\partial \Delta \tilde{T}(k)} = 0$$

which yields

$$\Delta \tilde{T}(k) = -\frac{\rho \tilde{T}(k-1) + \hat{e}_0^T(k+1) \mathbf{P} \hat{u}(k)}{\rho + \hat{u}^T(k) \mathbf{P} \hat{u}(k)} \tag{74}$$

This algorithm can be applied directly to the control of the flexible one arm robot. The control calculation procedure is as follows: $\hat{e}_0(k+1)$ is derived by prediction (Eq. (70) with $\Delta \tilde{T}(k) = 0$), and $\hat{u}(k)$ is approximated by applying Eq. (71) at the $k-1$ interval. One assumption has been made in the above procedure that $\hat{u}(k-1)$, estimate at the previous interval, also applies to the current interval. Consequently, it is required that the system parameters remain constant or change slowly relative to sampling frequency (which is not a conservative assumption considering the availability of micro–computer technology for fast sampling rates). Furthermore, if adequate predictions $\hat{z}_0(k+1)$ and $\hat{z}(k+1)$ can be achieved, and if $z(k)$ changes slowly from one control interval to the next, the above control scheme will provide an adequate control sequence.

The stability of closed loop system is guaranteed if $V(k+1) = J_{k+1}$ is a Lyapunov function of system (70) as stated in the following theorem.

Theorem:　　*The discrete time system (70) is asymptotically stable in* *the sense of Lyapunov if given any symmetric, positive definite matrix* $\mathbf{Q}$, *there exists a symmetric, positive definite matrix* $\mathbf{P}$, *such that the following equality holds:*

$$\mathbf{A}^T\mathbf{C}^T\mathbf{P}\,\mathbf{C}\,\mathbf{A} - \mathbf{C}^T\mathbf{P}\,\mathbf{C} - \frac{\mathbf{A}^T\mathbf{C}^T\mathbf{P}\mathbf{C}\mathbf{b}\mathbf{b}^T\mathbf{C}^T\mathbf{P}\mathbf{C}\mathbf{A}}{\rho + \mathbf{b}^T\mathbf{C}^T\mathbf{P}\mathbf{C}\mathbf{b}} = -\mathbf{Q} \quad (75)$$

and ρ is chosen non–zero.

In such a case, the PAC algorithm shown above is also optimal.

Or in a design problem, for a chosen weighting matrix $\mathbf{P}$ (symmetric and positive definite), if

$$\mathbf{Q} = -(\mathbf{A}^T\mathbf{C}^T\mathbf{P}\,\mathbf{C}\,\mathbf{A} - \mathbf{C}^T\mathbf{P}\,\mathbf{C} - \frac{\mathbf{A}^T\mathbf{C}^T\mathbf{P}\mathbf{C}\mathbf{b}\mathbf{b}^T\mathbf{C}^T\mathbf{P}\mathbf{C}\mathbf{A}}{\rho + \mathbf{b}^T\mathbf{C}^T\mathbf{P}\mathbf{C}\mathbf{b}}\,) \quad (76)$$

is symmetric and positive definite, then the PAC control strategy is optimal and asymptotically stable in the sense of Lyapunov.

Proof. In an optimal control problem, even though the quadratic performance index is minimized by a physically realizable system trajectory, the stability is not guaranteed [39, 40]. The system is asymptotically stable, however, if the cost function is also a suitable Lyapunov function. Hence, we chose a Lyapunov function for system (70) as:

$$V(k+1) = \hat{\mathbf{e}}^T(k+1)\,\mathbf{P}\,\hat{\mathbf{e}}(k+1) + \rho\tilde{T}^2(k) \quad (77)$$

Obviously, $V(k+1)$ is continuous in the state variable $\mathbf{x}(k)$ and the driving torque $\tilde{T}(k)$, and is positive definite. Without loss of generality, it is assumed that $\mathbf{z}_d = 0$. Therefore, $V(k+1)$ has the following increment.

$$\begin{aligned} \Delta V(k+1) &= V(k+1) - V(k) \\ &= \mathbf{x}^T(k)\left\{\mathbf{A}^T\mathbf{C}^T\mathbf{P}\,\mathbf{C}\,\mathbf{A} - \mathbf{C}^T\mathbf{P}\,\mathbf{C}\right\}\mathbf{x}(k) + 2\mathbf{x}^T(k)\mathbf{A}^T\mathbf{C}^T\mathbf{P}\mathbf{C}\mathbf{b}\tilde{T}(k) \\ &\quad + \tilde{T}^T(k)\mathbf{b}^T\mathbf{C}^T\mathbf{P}\mathbf{C}\mathbf{b}\tilde{T}(k) + \rho\tilde{T}^2(k) - \rho\tilde{T}^2(k-1) \end{aligned} \quad (78)$$

Furthermore, $\Delta V(k+1)$ has a minimum value with respect to $\tilde{T}(k)$ because it is in a quadratic form of $\tilde{T}(k)$. $\Delta V(k+1)$ is minimized if we choose $\tilde{T}(k)$ in such a way that

$$\frac{\partial \Delta V(k+1)}{\partial \tilde{T}(k)} = 0$$

which renders the following solution:

$$\tilde{T}(k) = -\frac{\mathbf{x}^T(k)\mathbf{A}^T\mathbf{C}^T\mathbf{P}\mathbf{C}\mathbf{b}}{\rho + \mathbf{b}^T\mathbf{C}^T\mathbf{P}\mathbf{C}\mathbf{b}} \tag{79}$$

Substituting Eq. (79) into Eq. (78) gives:

$$\Delta V(k+1) = -\mathbf{x}^T(k)\mathbf{Q}\mathbf{x} - \rho\tilde{T}^2(k-1) \tag{80}$$

where $\mathbf{Q}$ is given in Eq. (75). As a result, $\Delta V(k+1)$ is negative if $\mathbf{Q}$ is positive definite, guaranteeing an asymptotically stable control. Note that, for positive ρ ($\rho > 0$), $\mathbf{Q}$ can be a symmetric positive semidefinite matrix for asymptotic stability.

To illustrate that the control sequence from Eq. (79) is truly the PAC given by Eq. (74), keep in mind that:

$$\tilde{T}(k) = \tilde{T}(k-1) + \Delta\tilde{T}(k) \,, \quad \mathbf{e}_0(k+1) = \mathbf{C}[\mathbf{A}\mathbf{x}(k) + \mathbf{b}\tilde{T}^T(k-1)]$$

$$\hat{\mathbf{u}}(k) = \frac{\mathbf{z}(k+1) - \mathbf{z}_0(k+1)}{\Delta\tilde{T}(k)} = \mathbf{C}\mathbf{b}$$

Then we have Eq. (79) as:

$$\Delta\tilde{T}(k) = -\frac{\rho\tilde{T}(k-1) + \hat{\mathbf{e}}_0^T(k+1)\,\mathbf{P}\,\hat{\mathbf{u}}(k)}{\rho + \hat{\mathbf{u}}^T(k)\,\mathbf{P}\,\hat{\mathbf{u}}(k)} \tag{81}$$

And it agrees with Eq.(74).

To summarize, the above shows that, if Eq.(75) holds and the control increment is determined according to the PAC algorithm Eq.(74), then the increment of the Lyapunov function Eq. (78) is negative definite. In other words, the control system is asymptotically stable in the sense of Lyapunov, and the control is also optimal in the sense that cost function (72) is minimized.

Furthermore, the increment of the Lyapunov function can be utilized as a measure of the transient system behavior, because it can be interpreted as a "distance" from an arbitrary state to the origin of the state plane. If this increment is minimized, the system has fast speed of response. We may therefore select proper value for ρ for the above argument. However, ρ must remain non-negative. ·

The weighting matrix P plays also a important role in the system transient behavior through Eq. (80). This equation reveals that, if Eq. (75)

holds, the PAC will be stable. Moreover, the speed of response, measured by $\Delta V(k+1)$, depends on the eigenvalues λ_i of Q. The bigger λ_i, the faster the response. The physical interpretation of the above argument in the control of a flexible robot is that increasing the weighting on the hub–rate, while decreasing the penalty on the tip position error, yields a stable control but with a slower transient response. If P is changed in the opposite way, one may obtain a higher closed–loop bandwidth (faster transient response), whereas the stability margin will be reduced. A good relative weighting between hub–rate and tip position error feedbacks should be such that all eigenvalues of Q in Eq. (75) are positive with bigger magnitude. In the following sections, the predictive adaptive output feedback control combined with lattice filter parameter estimator is applied to a specific one–arm flexible robot.

C. DYNAMICS MODEL OF ACTUATOR SYSTEMS

The actuator in our study consists of a voltage servo amplifier, a PM DC motor, and a harmonic drive speed reducer. In practice, an amplifier has a limited bandwidth which can be characterized by its time constant. But this time constant is usually small, and thus is assumed negligible. As a result, the idealized amplifier model becomes a constant gain for all frequencies, and can be represented as:

$$\frac{e_a}{e_{in}} = K_a \tag{82}$$

where e_{in} and e_a are, respectively, input and output voltage signals to the amplifier. e_a is directly fed into the DC motor. K_a is the amplifier gain.

In modeling a DC motor, the following observations are made: the armature of the PM DC motor is modeled as a circuit with a resistance R, and an inductance L_a in parallel with another resistance R_a as shown in Fig. 4(a). In this figure, e_a is the applied motor input voltage, i_a the motor current, and e_b the induced back emf across the armature coil. Furthermore, T_m denotes the output torque from the motor, ω_m and θ_m are the motor angular speed and displacement. T_L designates the payload torque exerted on the motor shaft. If we denote motor moment of inertia by J_m, the rotational friction coefficient by μ_m, the we could have:

1. The Electrical Equation of the Motor:

$$\frac{di_a}{dt} = \frac{1}{L_a}e_a - \frac{R}{L_a}i_a - \frac{1}{L_a}e_b \tag{83}$$

Here, we neglect the effect of R_a, since in practice R_a is usually 5 to 10 times larger than R.

2. The Dynamic Equation of the Motor

$$T_m = J\frac{d\omega_m}{dt} + \mu_m\omega_m + T_L \tag{84}$$

Here,

$$J = J_m + J_h + I_H/n_h^2 \tag{85}$$

where J_h is the moment of inertia of the harmonic drive, and J is the effective total moment of inertia of the whole system on the motor axis. n_h is the gear ratio of the harmonic drive ($n_h > 1$).

3. Motor Characteristics

$$T_m = K_i i_a \tag{86}$$

where K_i is the motor torque constant in N-m/amp.

Furthermore, the back emf e_b of the motor is proportional to the angular velocity of the motor

$$e_b = K_b\omega_m \tag{87}$$

where K_b is called the back emf constant. It can be shown that K_i and K_b are equal in magnitude when the SI units are utilized.

The transfer function from motor input voltage e_a to motor output torque T_m in the absence of payload torque T_L is given by:

$$\frac{T_m(s)}{e_a(s)} = \frac{K_i(Js + \mu_m)}{(L_aJ)s^2 + (L_a\mu_m + RJ)s + (K_iK_b + R\mu_m)} \tag{88}$$

The back emf, characterized by coefficient K_b, is a stabilizing effect for the motor. e_b could also improve the stability of systems where the motor is used as an actuator.

If we assume that the motor damping factor μ_m is negligible ($\mu_m = 0$), and inductance is small ($L_a \leq 10^{-4}$ Henry for our system), then we can simplify the motor transfer function as

$$\frac{T_m(s)}{e_a(s)} = \frac{K_m s}{(s\tau_m + 1)(s\tau_e + 1)} \tag{89}$$

with

$$K_m = \frac{J}{K_b} \qquad \tau_m = \frac{RJ}{K_i K_b} \qquad \tau_e = L_a/R$$

Here, K_m is the motor gain, τ_m and τ_e are named as the motor mechanical and electrical time constants, respectively. It is usually true that $\tau_m > 10\tau_e$.

The harmonic drive, in our study, is represented as a gear train with gear ratio n_h as shown by the schematic diagram in Fig. 4(b). On its input is a inertia J_h named the effective input inertia of the harmonic drive. T_m is the motor output torque. μ_h is the damping coefficient of this component. ω_h is the angular velocity ($\omega_h = \omega_m$). As a result, we have for this component

$$\left.\begin{aligned} T_m &= T_1 + J_h\frac{d\omega_h}{dt} + \mu_h\omega_h \\[2mm] T &= n_h T_1 \end{aligned}\right\} \tag{90}$$

Hence, the relationship between torques T and T_m, when the motor dynamics equations are considered, is given by:

$$T = n_h\frac{(J - J_h)s + \mu_h}{Js + \mu_h}T_m = n_h\frac{(J_m + I_H/n_h^2)s + \mu_h}{(J_m + J_h + I_H/n_h^2)s + \mu_h}T_m \tag{91}$$

In cases that μ_h is negligible, we have

$$T = n_h\frac{J_m + I_H/n_h^2}{J_m + J_h + I_H/n_h^2}T_m = K_h T_m \tag{92}$$

where K_h denotes the gain of the harmonic drive. For the control of the flexible robot, T is the controlling torque exerting on the robot joint.

The simplest way with DC motor control is the open–loop actuator control system where no inner loop feedback is adopted within the actuator. In this case, the amplifier is a voltage amplifier. The input to the actuator is a control voltage e_{in} to the amplifier. The amplified voltage signal is thus fed into the DC motor to produce a output torque T_m. This torque is then increased by the harmonic drive (speed reducer) which applies a

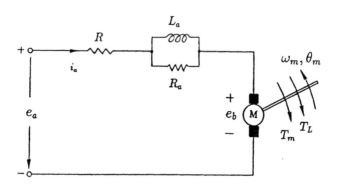

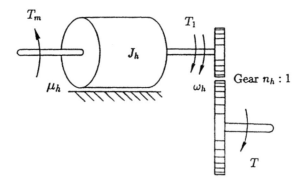

Figure 4: (a) Circuit Representation of s PM DC motor. (b) Schematic Diagram of the Harmonic Speed Reducer

control torque T to the robot joint. The input e_{in} to the actuator system is determined by a predictive adaptive control algorithm, and is generated by a controlling computer. Therefore, the transfer function of this actuator system can be obtained as:

$$\frac{T(s)}{e_{in}(s)} = K_a K_m K_h \frac{s}{(s\tau_m + 1)(s\tau_e + 1)} \qquad (93)$$

This transfer function will be used in the simulation examples presented in the next section.

D. PREDICTIVE ADAPTIVE CONTROL OF THE FLEXIBLE ROBOT ARM

Presented here are a number of simulated examples which demonstrate the properties of the proposed predictive adaptive controller. Comparisons have been made between systems with and without actuator dynamics consideration. The PAC scheme is also compared with LQG [2] and stable factorization (SF) [8, 22] type control methods.

Presented in Table IV are the dimensions and material properties of the flexible one arm robot studied. Its (non–dimensionalized) natural frequencies are given in Table V. Two sensors are necessary for controller implementation. One is a tachometer measuring the hub rate $\dot{\psi}(t)$, while the other measures the tip position of the beam. Information from these two sensors (plus the torque information), after non–dimensionalization, are fed into the parameter estimator and controller.

In the simulation study, we use a system model of the flexible arm with one rigid and three flexible modes. With the choice of sampling interval $h = 1$ millisecond (or $\tilde{h} = 0.0047$), the discrete time transfer function from ND driving torque to the ND hub rate and tip position will be in the form of:

$$\frac{z(k)}{\tilde{T}(k)} = \begin{bmatrix} \dfrac{B_2(q^{-1})}{A_2(q^{-1})} \\[2ex] \dfrac{B_3(q^{-1})}{A_3(q^{-1})} \end{bmatrix} = \begin{bmatrix} \dfrac{b_{21}q^{-1} + b_{22}q^{-2} + \cdots + b_{27}q^{-7}}{1 + a_{21}q^{-1} + \cdots + a_{27}q^{-7}} \\[2ex] \dfrac{b_{31}q^{-1} + b_{32}q^{-2} + \cdots + b_{38}q^{-8}}{1 + a_{31}q^{-1} + \cdots + a_{38}q^{-8}} \end{bmatrix} \qquad (94)$$

The values of coefficients a_{ij}, b_{ij}, $(i = 2, 3; \ j = 1, 2, \cdots, 8)$ are listed in Table VI.

Table IV: Physical property and dimensions of the flexible arm

Material		Aluminum
Modulus of Elasticity E (N/m^2)		70×10^9
Volume Density ρ (kg/m^3)		2710.0
Length L (m)		1.21×10^0
Height h (m)		1.90×10^{-2}
Width t (m)		4.76×10^{-3}
Total Mass (kg)		2.99×10^{-2}
Parameters for Non–Dimensionalization μ (sec.)		2.12×10^{-1}
mL (kg)		2.99×10^{-1}
L (m)		1.21×10^0
ND Rigid Beam Moment of Inertia $\bar{I}_{H1}$		3.32×10^{-1}
ND Hub Moment of Inertia $\bar{I}_{H2}$		1.98×10^{-2}
ND Beam and Hub Inertia $\bar{I}_H$		3.52×10^{-1}
ND Mass of Payload $\tilde{m}_p$	(a)	1.51×10^{-1}
	(b)	0.0
ND Moment of Inertia of Payload $\tilde{J}_p$		0.0

Table V: ND natural frequencies and damping ratios of the flexible beam

Mode		0	1	2	3
	Pinned–Free Beam	0	15.43	49.89	104.22
Non–Dimen.	Pinned–Mass Beam	0	9.27	24.38	40.32
Frequency	Clamped–Free Beam	0	3.52	22.08	61.80
	Clamped–Mass Beam	0	2.29	16.26	46.54
Damping Ratio		0	0.02	0.01	0.01

Table VI: Coefficients for ARMA model

j	a_{2j}	b_{2j}	a_{3j}	b_{3j}
1	-0.667×10^1	0.154×10^1	-0.767×10^1	-0.177×10^{-3}
2	0.193×10^2	-0.904×10^1	0.260×10^2	0.895×10^{-3}
3	-0.316×10^2	0.222×10^2	-0.510×10^2	-0.162×10^{-2}
4	0.315×10^2	-0.294×10^2	0.632×10^2	0.920×10^{-3}
5	-0.191×10^2	0.221×10^2	-0.506×10^2	0.876×10^{-3}
6	0.653×10^1	-0.895×10^1	0.256×10^2	-0.159×10^{-2}
7	-0.970×10^0	0.152×10^1	-0.750×10^1	0.882×10^{-3}
8			0.970×10^0	-0.176×10^{-3}

1. Predictive Adaptive Control without Actuator Consideration

The control parameters in the predictive adaptive output feedback control are $\mathbf{P}$ and ρ. They can be chosen in such a way that Eq.(75) has a feasible solution for $\mathbf{Q}$ whose positiveness could be guaranteed. However, searching for suitable $\mathbf{P}$ and ρ is not an easy task due to the complexity of Eq.(75). Since simple feedback control with tip position and hub rate feedback is a very special case of PAC, where no restriction has been placed on the input energy, one can visualize the motion of the closed–loop poles in the z–plane as the feedback gains change, then apply the result from this root locus as a reference for determining the elements of $\mathbf{P}$. An example of the root locus for the flexible robot is presented in Fig. 5 (the actuator system is not included here). Part (a) of this figure shows a complete sketch of the loci as K_y (tip position feedback gain) varies from zero to infinity and $K_{\dot\psi}$ (hub rate feedback gain) is unity. The loci start from the poles of the closed loop system with unity hub rate feedback (marked by ×) inside the unit circle. As K_y increases, two pairs of poles corresponding to the first and third flexible modes will move toward outside the unit disc. Enlarged view of part A of Fig. 5(a) demonstrates that this pair of poles always tends to unstabilize the system. As K_y increases approximately to 5, this pair

crosses the unit circle. The ratio between $K_{\dot\psi}$ and K_y which results in good
closed loop pole locations can be used as a guide in selecting the matrix **P**
in the PAC design. The appropriate selection of scalar penalty on control
magnitude, ρ, can easily be studied by direct computer simulations.

In the example shown in Fig. 6, the weightings on the predicted hub
rate and tip position errors are 1.0 and 3.0, respectively, and the penalty
on input magnitude ρ is chosen to be 0.1. The robot arm is commanded to
follow a prescribed trajectory with ω_d approximately equal to $\tilde\omega_1 = 3.5286$.
Damping ratio ξ_d for this desired trajectory is selected as 0.707. This
trajectory has a non–dimensional settling time of 1.175. When tracking this
trajectory, the flexible arm tip does not move for a little while (about 0.094
non–dimensional time, that is interpreted as the time delay in the response
of tip position due to the time needed for the bending wave to travel from
base to tip), then it goes in a wrong direction. This behavior is totally
due to the non–minimum phase characteristics of the transfer function from
driving torque to tip position. However, the controller corrects the response
after around 0.122 non–dimensional time, and begins tracking in the desired
direction. As the non–dimensional time reaches around 1.265, close tracking
is obtained with very small tracking error of less than 5%. The system
settles down thereafter.

In this study, we observed that the dominant dynamics of a flexible arm
is observed to be sufficiently represented by its rigid body mode plus one
flexible mode. The error in identifying the system parameters during typical
moves is in the range of 3%, and the identification scheme converges within
50 steps. For the sake of completeness, an example for AR parameters
identification is presented in Fig. 7. The true values of these parameters
are presented in Table VII. The higher modes usually do not get excited
enough to be detected by the identifier in discrete time domain. On the
other hand, in cases where higher modes are excited, such as by an impact
force or by collision with obstacles, the order of the identifier must be
increased to accommodate such changes. Under these circumstances, lattice
filter has shown great advantages over other ordinary least square parameter
identifiers since the order of the filter can be changed in real time without
causing any computational difficulty.

It is worth to note that the non–dimensionalized model proposed in
this study makes it easier to compare different control schemes in the tip
position control of flexible robot arm. Here, comparison will be made be-

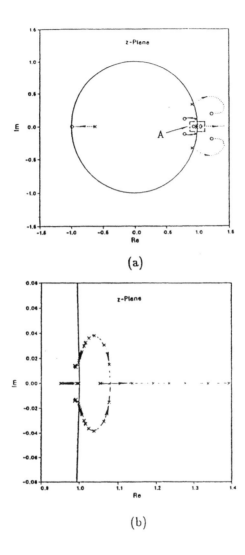

Figure 5: Root Locus when Hub Rate Feedback Gain is Unity while Tip Position Feedback Gain Changes from Zero to Infinity: (a) A Whole View; (b) An Enlarged View of Area A.

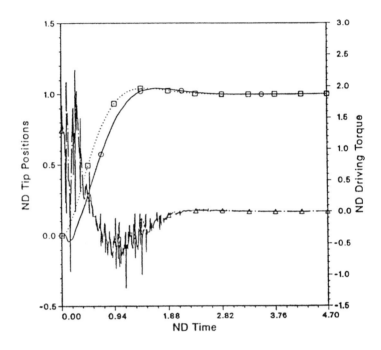

Figure 6: Predictive Adaptive Control without Actuator Consideration ($\rho = 0.1$, $P = \text{diag}(1.0,\ 3.0)$, $\omega_d = \tilde{\omega}_1 = 3.5286$, and $\xi_d = 0.707$)

    o: ND Tip Position
    □:   ND Desired Tip Position
    △:   ND Driving Torque

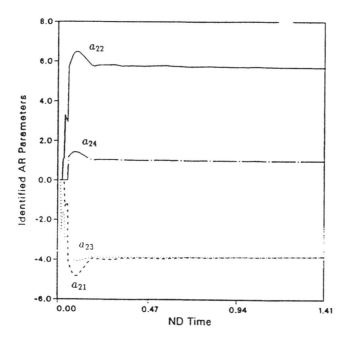

Figure 7: Identification of AR Parameters when the Rigid and One Flexible Modes is Identified

Table VII: Coefficients for ARMA model when one flexible mode is used

j	a_{2j}	b_{2j}	a_{3j}	b_{3j}
1	-0.299×10^1	0.150×10^0	-0.399×10^1	-0.860×10^{-4}
2	0.299×10^1	-0.301×10^0	0.598×10^1	0.862×10^{-4}
3	-0.998×10^0	0.150×10^2	-0.399×10^1	0.860×10^{-4}
4			0.998×10^0	-0.859×10^{-4}

tween the PAC algorithm and the LQG [2], and the stable factorization
(SF) method [8, 22]. In [22], it is shown that SF controller is superior
to the LQG in that faster response and shorter settling time can be obtained (either with or without input constraints). On the other hand, SF
controller requires very large control torque levels which is over ten times
higher than that required by LQG. Therefore, we first compare PAC with
SF, using the example presented by Wang and Vidyasagar [8]. Based on
the dimensions and materials of the flexible robot of that study, the robot
has the following time and torque conversion factor μ and ν, respectively,
for non–dimensionalization as:

$$\mu = \sqrt{mL^4/EI} = 1.3632 \times 10^{-1} \text{ (sec.)} \quad \nu = EI/L = 1.8229 \times 10^{+1} \text{ (N} \cdot \text{m)}$$

As a result, our results can be easily compared with the example in [8]
by simply scaling our non–dimensionalized tip position by μ and the input
torque by ν. Fig. 8 illustrates the scaled results by PAC as well as the
step response under SF controller presented in [8]. From this figure, the
predictive adaptive controller provides shorter rising and settling times. A
5% overshoot is observed in PAC. Moreover, the maximum torque by PAC
is about 20 N-m while the SF method requires a maximum torque up to
9.5×10^3 N-m. If compared with the LQG, we deduce from the above that
the PAC scheme has promise of providing faster response.

The examples show that accurate, fast motions can be achieved by
proper choice of design parameters in PAC. However, the required control
signal of PAC algorithm has high frequency components, especially at the
beginning stage of motion (Fig. 6). The fast changing input torque, in
practice, may not be physically realizable by DC motors. To study the
more realistic situation, we next include the dynamics of actuator system

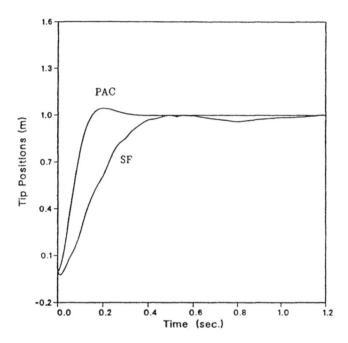

Figure 8: Comparison between PAC and SF methods

in the control algorithm and simulation.

2. Predictive Adaptive Control with Actuator Consideration

When considering the effect of the actuator dynamics, we have in our study a PM DC motor manufactured by PMI Motion Technologies. The output shaft of the DC motor is connected to a harmonic drive from Emhart Machinery Corporation with a gear ratio of $n_h = 100$, and an effective inertia of $J_h = 1.4807 \times 10^{-3}$ kg-m^2.

In the tip position control with a real actuator system described by the transfer function (93), the input to the whole system (actuator plus flexible arm) is a control voltage sent from the control computer to the power amplifier. This control signal is required by the lattice filter for the purpose of parameter identification. Furthermore, due to the inclusion of the actuator control system, the order of the lattice filter is increased by 2.

Shown in Fig. 9 is a simulation result of tip position response of the flexible arm under the PAC algorithm where actuator dynamics is included. ω_d and ξ_d are respectively 3.5286 and 0.707 which are the same as in the example shown in Fig. 6. Since the actuator system acts like a low pass filter for high frequency components in the input signal, the input torque is much smoother and physically realizable by DC motors compared with Fig. 6.

When the restriction on input energy is relaxed by reducing the value of ρ, faster tracking speed can be reached as shown in Fig. 10, where $\rho = 0.01$. The robot tip moves faster as expected in this case, but an 8% overshoot relative to the steady state tip position is observed. In the mean time bigger input energy is also required to control the system. On the other hand if the value of ρ is increased, the response will be much slower, and the overshoot will be reduced.

The noise rejection characteristics of the control algorithm can be investigated by contaminating the input signal to the actuator by a white noise and the studying the system response under this environment. Shown in Fig. 11 is such an example. Here, the disturbance sequence is generated by IMSL routine and added to the input sequence (the noise has a variance of about 2.35). Furthermore, the weightings on the predicted hub–rate and tip position errors are again chosen as 1.0 and 3.0. ρ is equal to 0.1 as in

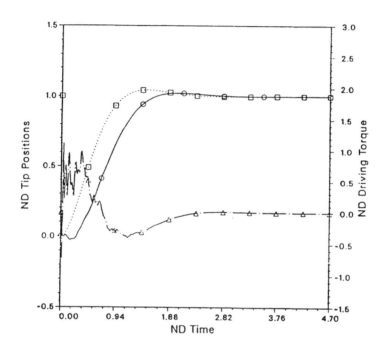

Figure 9: Predictive Adaptive Control with Actuator Consideration ($P =$ diag(1.0, 3.0), $\rho = 0.1$, $\omega_d = \omega_1$, and $\xi_d = 0.707$)

o: ND Tip Position
□:   ND Desired Tip Position
△:   ND Driving Torque

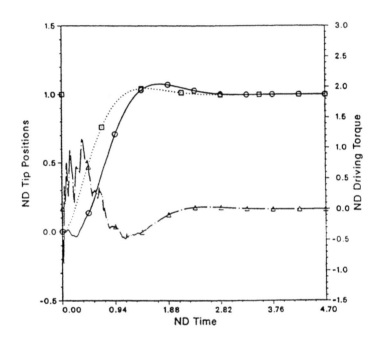

Figure 10: Predictive Adaptive Control with $P = \text{diag}(1.0,\ 3.0)$, $\rho = 0.01$, $\omega_d = \tilde{\omega}_1 = 3.5286$, and $\xi_d = 0.707$

   o: ND Tip Position
   □: ND Desired Tip Position
   △: ND Driving Torque

the case shown in Fig. 9. The noise slows down the system slightly and the system has a more difficult time to settle down. However, the control algorithm is able to overcome the disturbance and bring the robot tip to the desired position.

In order to study the behavior of the predictive adaptive control algorithm when fast response is desired, we choose ω_d to be three times as fast as the ND first natural frequency $\tilde{\omega}_1$ of the robot. $\rho = 0.1$ in this case. Although input energy needs to be large and big overshoot (about 20%) has been observed, the system is still stable under PAC (Fig. 12).

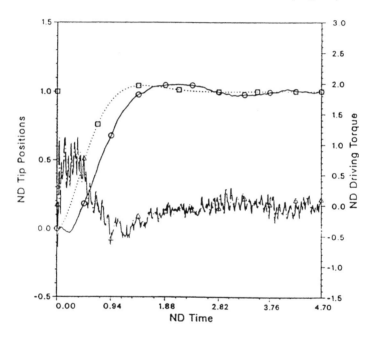

Figure 11: Noise Rejection Characteristics of Predictive Adaptive Control with $\rho = 0.1$, $P = \text{diag}(1.0,\ 3.0)$, $\omega_d = \tilde{\omega}_1$, and $\xi_d = 0.707$

    o: ND Tip Position
    □:   ND Desired Tip Position
    △:   ND Driving Torque

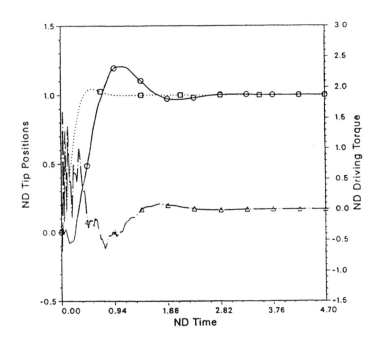

Figure 12: Predictive Adaptive Control with $P = \mathrm{diag}(1.0, \; 3.0)$, $\rho = 0.1$, $\omega_d = 3\tilde{\omega}_1$, and $\xi_d = 0.707$

o: ND Tip Position
□:   ND Desired Tip Position
△:   ND Driving Torque

V. CONCLUSIONS

A non–dimensionalized dynamics model for the flexible robot arm is developed. Payload mass and moment of inertia are also considered in the modeling. It is demonstrated in discrete time domain that the controlled system tends to become non–minimum phase when sampling frequency is high enough.

The lattice filter proves to be a good parameter identifier in the on–line identification of flexible robot due to its high convergence rate and noise rejection capability. Investigation shows that, in discrete time domain, the dynamics of the flexible robot arm could be sufficiently described by one rigid and one flexible mode. It has been shown in simulations that the lattice filter could estimate the parameters of the robot with 5% estimation errors or less (whether or not input noise is present). Furthermore, it converges within 50 time steps, and no special input excitation is required to drive the filter during identification.

The predictive adaptive control algorithm could provide wider closed-loop bandwidth control. If the dynamics of the actuator systems are not included, however, the control action would contain very high frequency signals. When the actuator system are put into consideration, the closed loop system bandwidth could reach as high as three times the first clamped free natural frequency of the flexible robot. In order to choose control parameters properly, we suggested the use of a root locus as an alternative. Compared with other control techniques such as stable factorization and linear quadratic Gaussian, the predictive adaptive controller could provide faster control at a reasonably low input energy level.

VI. REFERENCES

1 Henrichfreise, H., "The Control of an Elastic Manipulation Device Using Digitial Signal Processors", *Proc. American Control Conference*, Atlanta, GA, pp. 1029-1035, 1988.

2 Cannon, R.H.Jr., and Schmitz, E., "Initial Experiments on the End-Point Control of a Flexible One-Link Robot", *The International Journal of Robotics Research*, Vol. 3, No. 3, pp. 62–75, Fall 1984.

3 Chalhoub, N.G., and Ulsoy, A.G.: "Control of a Flexible Robot Arm: Experimental and Theoretical Results", *ASME Journal of Dynamic Systems, Measurement, and Control*, Vol. 109, No. 4, pp. 299–309, 1987.

4 Hastings, G.G., and Book, W.J., "Verification of a Linear Dynamic Model for Flexible Robotic Manipulators," *IEEE Control Systems Magazine*, IEEE Control Systems Society, April 1987.

5 Kotnik, P.T., Yurkovich, S., and Ozguner, U. "Acceleration Feedback for Control of a Flexible Manipulator Arm," *Journal of Robotic Systems*, 5(3), pp. 181–196, 1988.

6 Liegois, A., Dombre, E., and Borrel, P., "Learning and Control for a Compliant Computer–Controlled Manipulator," *IEEE Trans. on Autimatic Control*, Vol. 25, No. 6, pp. 724–738, Dec. 1980.

7 Park, J.-H., and Asada, H.: "Design and Control of Minimum Phase flexible Arms with Torque Transmission Mechanism," *IEEE Proc. Robotics and Automation*, Vol. 3, pp. 1790–1795, 1990.

8 Wang, D., and Vidyasagar, M., "Control of a Flexible Beam from Optimum Step Response," *IEEE Int. Conf. on Robotics and Automation*, Raleigh, NC., pp.1567–1572, 1987.

9 Balas, M.J., "Active Control of Flexible Systems", *Journal of Optimization Theory and Applications*, Vol.25, No. 3, pp. 415-436, July 1978.

10 Balas, M.J., "Feedback Control of Flexible Systems," *IEEE Trans. Automatic Control*, Vol.23, No.4, pp.673–679, 1978.

11 Cetinkunt, S., and Book, W.J., "Performance Limitations of Joint Variable–Feedback Controllers Due to Manipulator Structural Flexibility," *IEEE Trans. Robotics and Automation*, Vol. 6, No. 2, pp. 219–231, 1990.

12 Cetinkunt, S., and Wu, S., "Tip Position Control of a Flexible One Arm Robot with Predictive Adaptive Output Feedback Implemented with Lattice Filter Parameter Identifier," *International Journal of Computers & Structures*, Vol. 36, No. 3, pp. 429–441, 1990.

13 Cetinkunt, S., and Yu, W.L., "Closed–Loop Behavior of a Feedback Controlled Flexible Arm: A Comparative Study", *International Journal of Robotics Research*, Vol. 10, No. 3, June, 1991.

14 Book, W.J., Maizza-Netto, O., and Whitney, D.E., "Feedback Control of Two Beam, Two Joint Systems with Distributed Flexibility", *ASME Journal of Dynamic Systems, Measurement, and Control*, 97G, pp.424–431, Dec. 1975.

15 Maatuk, J., "A Study of the Dynamics and Control of Flexible Spatial Manipulators," *Ph.D. Thesis*, Department of Applied Mechanics, University of California, Los Angeles, CA, 1976.

16 Cetinkunt, S., and Wu, S., "Output Perdictive Adaptive Control of a Single Link Flexible Arm," *International Journal of Control*, Vol. 53, No. 2, pp. 311–333, 1991.

17 Wu, S., and Cetinkunt, S., "Model Reference Adaptive Inverse Control of a Single Link Flexible Robot", *International Journal of Robotics Research*, 1993 (to appear).

18 Book, W.J., and Majette, M., "Controller Design for Flexible, Distributed Parameter Mechanical Arms Via Combined State Space and Frequency Domain Techniques", *Journal of Dynamic Systems, Measurement, and Control*, Vol. 105, pp. 245-254, Dec. 1985.

19 Meldrum, D.R., and Balas, M.J., "Direct Adaptive Control of a Flexible Remote Manipulator Arm," in *Robotics and Manufacturing Automation*, eds. M. Donath and Ming Leu, ASME, New York, pp.115–119.

20 Rovner, D.M., and Cannon, R.H. Jr., "Experiments Toward On-Line Identification and Control of a Very Flexible One-Link Manipulator", *Int. Journal of Robotics Research*, Vol. 6, No. 4, pp. 3–19, Winter 1987.

21 Rovner, D.M., and Franklin, G.F., "Experiments In Load–Adaptive Control of a Very Flexible One–Link Manipulator," *Automatica*, Vol. 24, No.4, pp.541–548, 1988.

22 Shung, I.Y., and Vidyasagar, M., "Control of a Flexible Robot Arm with Bounded Input: Optimum Step responses", *IEEE Int. Conf. on Robotics and Automation*, Raleigh, NC., pp. 916-922, 1987.

23 Yeung, K.S., and Chen, Y.P., "Regulation of One–Link Flxible Robot Arm Using Sliding–Mode Tchnique," *Int. J. Control*, Vol. 49, No. 6, pp. 126–145, 1989.

24 Maizza–Neto, O., "Modal Analysis and Control of Flexible Manipulator Arms," *Ph.D. Thesis*, Department of Mechanical Engineering, Mass. Institute of Technology, Cambridge, Mass., 1974.

25 Truckenbrodt, A.: "Modeling and Control of Flexible Manipulator Structure," *Proc. 4th CISM–IFTOMM Symp. on Theory and Practice of Robots and Manipulators*, Warsar, pp. 1125–1131, 1981.

26 Siciliano, B., and Book, W.J.: "A Singular Perturbation Approach to Control of Lightweight Flexible Manipulator," *International Journal of Robotics Research*, Vol. 7, No. 4, pp.79–90, 1988.

27 Gevarter, W.B., "Basic Relations for Control of Flexible Vehicles", *AIAA Journal*, Vol. 8, No. 4, pp. 666–672, April 1970.

28 Siciliano, B., Book, W.J., and Yuan, B., "Model Reference Control of a One–Link Flexible Arm," Proc. 25th IEEE Conf. on Decision and Control, Athens, Greece, pp. 1379–1384, Dec. 1986.

29 Book, W.J., "Recursive Lagrangian Dynamics of Flexible Manipulator Arms", *International Journal of Robotic Research*, Vol. 3, No. 3, pp.87-101, Fall 1984.

30 Åström, K.J., Hagander, P., and Sternby, J., "Zeros of Sampled Systems," *Automatica*, Vol. 20, No. 1, pp. 31–38, 1984.

31 Åström, K.J., and Wittenmark, B., *Computer Controlled Systems: Theory and Design*, Englewood Cliffs, N.J., Prentice–Hall, Inc.,1984.

32 Graupe, D., *Time Series Analysis, Identification and Adaptive Filtering*, Malabar, Fla., R.E. Krieger Pub., 1984.

33 Lee, D. T. L., Friedlander, B., and Morf, M., "Recursive Ladder Algorithms for ARMA Modeling," *IEEE Trans. on Automatic Control*, Vol. 27, No. 4, pp. 753–764, August 1982.

34 Friedlander, B., "Lattice Filters for Adaptive Processing," *IEE Proceedings*, Vol. 70, No. 8, pp.829–867, August 1982.

35 Richalet, J.A., Rault, A., and Testud, J.L., and Papon, J., "Model Predictive Heuristic Control: Applications to a Industrial Precess", *Automatica*, Vol. 14, pp. 413–428, 1978.

36 Cutler, C.R., and Ramaker, B.L., "Dynamic Matrix Control–A Computer Control Algorithm", *Proc. Joint Automatic Control Conf.*, San Fransisco, CA, pp. 210–223, 1980.

37 Garcia, C.E., Prett, D.M., and Morari, M., "Model Predictive Control: Theory and Practice—A Survey," *Automatica*, Vol. 25, No. 3, pp. 335–348, 1989.

38 Graupe, D., and Cassir, G.R., "Adaptive Control by Predictive Identification and Optimization," *IEEE Transaction on Automatic Control*, Vol. 12, pp.191–196, 1967.

39 Kalman, R.E., and Bertram, J.E., "Control System Analysis and Design Via the 'Second Method' of Lyapunov; I. Continuous–Time Systems," *Journal of Basic Engineering*, pp. 371–393, June 1960.

40 Kalman, R.E., and Bertram, J.E., "Control System Analysis and Design Via the 'Second Method' of Lyapunov; II. Discrete–Time Systems," *Journal of Basic Engineering*, pp. 394–400, June 1960.

Pole assignment by memoryless periodic output feedback

Dirk Aeyels   Jacques L. Willems

Faculty of Engineering
Universiteit Gent
Technologiepark-Zwijnaarde, 9
9052 GENT (Zwijnaarde)
BELGIUM
e-mail:aeyels@gpx.autoctrl.rug.ac.be
e-mail:willems@gpx.autoctrl.rug.ac.be

Contents

CONTROL AND DYNAMIC SYSTEMS, VOL. 70

1   Introduction

1.1   Stabilization and pole assignment in linear time invariant systems

The purpose of closed loop control is to monitor the behavior of a dynamical system by means of appropriate feedback in order to achieve a prescribed goal. An important characteristic is the stability, or for linear systems the location of the closed loop poles. These poles determine the qualitative behavior of the system variables. The design of appropriate feedback laws to obtain stability or a preassigned location of the poles, as well as the study of the limitations imposed by constraints on the possible feedback strategies constitute a major research area in control theory. A complete theory [18] exists for

- pole assignment for linear time invariant systems by means of linear time invariant state feedback;

- pole assignment for linear time invariant systems by means of linear time invariant dynamic output feedback.

In the former case the closed loop poles can be assigned arbitrarily[1], provided the system is controllable. In the latter case controllability and observability imply the existence of a state reconstructor and a controller with feedback of the reconstructed state, such that the poles of the closed loop system can be assigned arbitrarily. The disadvantage of a state reconstructor is that a dynamic feedback component is needed, which means that the output measurements have to be memorized for some time.

Two interesting areas of further research extending the above results are related to the following questions.

- Is pole assignment or stabilization realizable by means of static (i.e. memoryless) output feedback?

- Are extensions of the results to linear time varying systems possible?

In the present contribution some results in both these areas are discussed.

As to the first question, it is well known that in general pole assignment is not possible by means of linear time invariant memoryless output feedback. If A, B and C denote the plant, input and output matrices, then the pole placement problem corresponds to assigning the eigenvalues of $A + BKC$ by a suitable choice of the matrix K of the feedback gains. Kimura [12] and other authors derived sufficient conditions for arbitrary pole assignment by means of time invariant static output feedback control.

---

[1] Obviously the complex poles must occur in complex conjugate pairs.

The criteria involve a relation between the number m of inputs, the number p of outputs and the order of the system n (i.e. the number of state variables). The result derived by Kimura states that (almost) pole assignment is possible if the system is controllable and observable and

$$m + p > n$$

Rosenthal [13] proves that this condition can be relaxed except for m or p equal to 1, or for both m and p equal to 2. For a single input single output system arbitrary pole assignment cannot be achieved if the order of the system is two or more. This is almost obvious, since it is clear that the eigenvalues of $A + kbc$ cannot arbitrarily be located by choosing the scalar gain k, if A is a square matrix, b a column vector and c a row vector. To overcome these constraints another idea has recently been put forward: can the use of linear *time varying* memoryless output feedback lead to pole assignment and/or stabilization? This idea comes about because it has been shown, as discussed in the next section, that for some particular properties of time invariant systems, time varying feedback control can achieve performances which a time invariant feedback cannot. An important part of the present contribution deals with time varying feedback control for achieving pole assignment in time invariant linear systems. In particular it is shown that for almost all controllable and observable linear time invariant systems pole assignment is possible by means of memoryless linear time varying output feedback.

Another problem discussed in the present contribution is related to the second question formulated above, namely the possibility of extending these results to linear periodically time varying systems. For such systems it is shown that under broad conditions pole assignment is possible by means of memoryless periodic output feedback.

In the present contribution only discrete time systems are dealt with. The analysis of continuous time systems with time varying elements is technically more complicated. Many of the results obtained for discrete time systems can however also be formulated as results for continuous time systems with a sample and hold in the feedback loop.

1.2   Periodic feedback strategies for time invariant systems

In recent years a number of results have been obtained showing that when controlling time invariant systems, time varying feedback may yield results superior to those obtained by means of time invariant feedback.

First we want to point out that there are situations where time varying feedback cannot relax the inherent limitations typical for time invariant feedback [16]:

- When a time invariant linear system has uncontrollable modes, these modes are fixed as well for any time invariant state feedback as for any time varying state feedback.

- When a time invariant linear system has modes which are uncontrollable and/or unobservable, these modes are fixed with respect to any linear dynamic output feedback, be it time invariant or time varying.

On the other hand in the following applications the introduction of time varying feedback improves the performance of time invariant systems.

- For decentralized control systems the decentralization constraint implies that some modes which are controllable and observable, can nevertheless not be moved by time invariant feedback. It has been shown that in most cases time varying feedback overcomes these structural constraints

imposed by decentralization. [4, 15, 17].

- For time invariant controllable systems arbitrary pole assignment can be achieved by means of time invariant state feedback. However if the desired closed loop pole pattern contains multiple poles, there are limitations on the minimal polynomial that can be realized. These restrictions can be removed by the use of time varying state feedback [16].

- Khargonekar *et al.* [11] have shown that for linear time invariant plants better stability robustness and higher stability margins can be obtained using periodic feedback strategies rather than time invariant feedback.

One wonders whether the limitations of time invariant memoryless output feedback with respect to stabilization and pole assignment cannot be overcome by the use of time varying output control strategies. A number of results in this vein were obtained by, among others, Chammas and Leondes [6], Kabamba [8], and Kaczorek [9]. The problem considered by these authors is as follows. Let u_i and y_i denote the input and the output at time i for a time invariant discrete time system. The output is measured at times $0, T, 2T, \ldots$. The question is whether there exists a control strategy

$$u_{rT+i} = k_i y_{rT}$$

to stabilize the system. The controller measures the output periodically. Between two measurements the latest available measurement is used to generate the control input. The feedback gain depends on the difference between the time at which the measurement is taken and the time at which the control is applied. The control strategy is nondynamic in the sense that every control input only depends on a single measurement, namely the latest one available, and not on previous measurements; the control does not require state reconstruction. However the feedback law cannot be considered as static or memoryless since the measurements are used for generating later inputs. The authors referred to above show that under weak conditions a periodic closed loop system can be obtained with preassigned modes.

An alternative approach is to consider the potential of time varying memoryless output feedback where the input at time i depends only on the output at the *same time i*:

$$u_i = k_i y_i$$

The controller is thus constrained to have no memory at all. This problem was discussed by Greschak and Verghese [7] for second order systems. They showed that pole assignment for second order systems is possible by memoryless periodic output feedback control, except in some particular cases. This result was rederived by the authors of the present contribution [2] using a different methodology. Subsequently this led to a general result for systems of arbitrary order [3]. These results are discussed in depth in Section 2 of the present contribution. The extension to periodic linear systems is the subject of Section 3.

This contribution is written with the technical details and proofs collected in the appendices of Section 4. The first three sections contain the main results and can be read independently.

2   Memoryless periodic output feedback control for time invariant systems

2.1   Problem statement

Consider a linear time invariant discrete time system with scalar input and scalar output

$$
\begin{aligned}
x_{i+1} &= Ax_i + bu_i \\
y_i &= cx_i
\end{aligned}
\tag{1}
$$

where $x_i \in \mathbf{R}^n$, $u_i \in \mathbf{R}$, $y_i \in \mathbf{R}$, $i \in \mathbf{Z}$. The matrices A, b, c are constant and of appropriate dimension. Let the system be controllable and observable. It is well known that arbitrary pole assignment can be realized by dynamic output feedback, but not by memoryless output feedback (i.e. u_i depending only on the instantaneous output y_i). The fundamental question that arises is to what extent the pole assignment problem can be solved by introducing time varying (but memoryless) output feedback. In particular we assume that the time dependent feedback gain is periodic. Consider the system (1) with periodic feedback

$$
u_{rT+i} = k_i y_{rT+i}
$$

where T is a positive integer denoting the period of the feedback strategy, $r \in \mathbf{Z}$ and $i \in \{0, 1, ..., T-1\}$. With this feedback the closed loop system is a periodic linear system of period T. This system can be considered as a time invariant system over a time interval equal to the period T:

$$
x_{(r+1)T} = (A + k_{T-1}bc)(A + k_{T-2}bc)...(A + k_0bc)x_{rT}
\tag{2}
$$

The eigenvalues of the closed loop matrix

$$
A_{cl} = (A + k_{T-1}bc)...(A + k_0bc)
$$

of (2) – from now on called the poles of the periodic system – determine the dynamics of the periodic system. In particular the Hurwitz character of A_{cl} implies asymptotic stability of the closed loop system. The problem of pole assignment can then be restated as selecting feedback gains $k_0, ..., k_{T-1}$ such that the eigenvalues of the closed loop system matrix A_{cl} are the roots of a given polynomial

$$
\alpha(z) = z^n + \alpha_{n-1}z^{n-1} + ... + \alpha_1 z + \alpha_0
\tag{3}
$$

2.2 Reformulation of the problem

First the system equation (2) is rewritten as

$$
\begin{aligned}
x_{(r+1)T} &= A(A + k_{T-2}bc)...(A + k_0bc)x_{rT} \\
&\quad + k_{T-1}bc(A + k_{T-2}bc)...(A + k_0bc)x_{rT}
\end{aligned}
$$

After introducing the notation

$$
\begin{aligned}
\Pi &= (A + k_{T-2}bc)...(A + k_0bc) \\
A_{eq} &= A\Pi \\
c_{eq} &= c\Pi \\
b_{eq} &= b
\end{aligned}
$$

the system equation (2) becomes

$$
\begin{aligned}
\xi_{r+1} &= A_{eq}\xi_r + b_{eq}v_r \\
z_r &= c_{eq}\xi_r
\end{aligned}
\tag{4}
$$

with $\xi_r = x_{rT}$ and $v_r = k_{T-1}z_r$. The question is whether the eigenvalues of

$$
A_{cl} = A_{eq} + k_{T-1}b_{eq}c_{eq}
\tag{5}
$$

can be arbitrarily assigned by means of the feedback gains.

For the analysis of this question a result by van der Woude [14] on time invariant memoryless output feedback is used. It is formulated in Appendix 4.1, applied to the system equation (4).

2.3 Main result

The analysis in the appendices leads to the following result [3]:

Theorem 1 *Consider the single input single output system (1), where the pair A, b is controllable and the pair A, c observable. Assume that all numerator coefficients of the transfer function (23) are non-zero. Suppose moreover that all ratios p_i/q_i for $i = 1, ..., n - 1$ are different from p_0/q_0, and that*

$$
rank\begin{bmatrix} b & A\Pi^\circ b & ... & (A\Pi^\circ)^{n-1}b \end{bmatrix} = n
\tag{6}
$$

where

$$
\Pi^\circ = (A + \frac{p_{n-1}}{q_{n-1}}bc)...(A + \frac{p_0}{q_0}bc)
$$

Then the poles of the system can arbitrarily be assigned by a periodic output feedback strategy with period equal to $n+1$, provided that there are no closed loop poles at the origin.

2.4 Discussion

- Some of the assumptions featuring in Theorem 1 are necessary for the pole assignment property to hold, others are implied by the technique used to derive the results.

  - The conditions on observability and controllability of the system are clearly necessary, as discussed in the Introduction.

  - In the analysis of the appendices it is assumed that the matrix A is nonsingular. Because of controllability and observability, this can always be achieved by a preliminary time invariant output feedback, which is then combined with the periodic feedback obtained in the analysis to realize pole assignment. The combination also yields a periodic feedback. The nonsingularity assumption on A is therefore not included in the formulation of the theorem.

  - That all q_i/p_i $(i = 1, ..., n-1)$ are assumed to be unequal to q_0/p_0 is probably not necessary. It is a consequence of the approach, especially the part based on the implicit function theorem. In any case, for second order systems the condition has been proved to be not necessary [2]. Note that the assumption on the ratios q_i/p_i is invariant with respect to a preliminary time invariant output feedback.

  - Condition (6) is also a consequence of the procedure used to derive the result. However it is a very weak condition as discussed in a previous paper [3].

  - The assumption that the coefficients of the numerator of the transfer function are all nonzero is a consequence of our approach, but is not necessary in general. The condition that q_0 does not vanish, is necessary: otherwise the determinant of the closed loop matrix A_{cl}, and hence the product of the eigenvalues, would be independent of the gains [3]. On the other hand, from the analysis of the second order system [2] it is known that $q_{n-1} = cb \neq 0$ is not necessary. Indeed it was shown that for a second order system a zero value of cb can be allowed, but then the trace of the matrix A has to be assumed nonzero. Generalization of this type of conditions to higher order systems (and thus relaxing the conditions $q_i \neq 0$) should be subject to further research.

  - The period of the output feedback is equal to $n + 1$. This is in line with the paper by Greschak and Verghese [7], where it has been shown that for second order systems periodic feedback of period 2 cannot solve the pole assignment problem.

- In Theorem 1 it is assumed that the desired pole pattern does not contain one or more closed loop poles at the origin. This restriction comes about because of the method used to derive the general result, and can most probably be waived, as is illustrated by an example discussed in Section 2.5.

- The result of Theorem 1 can readily be generalized to MIMO systems. Consider the system

$$
\begin{aligned}
x_{i+1} &= Ax_i + Bu_i \qquad (7)\\
y_i &= Cx_i
\end{aligned}
$$

where $x_i \in \mathbf{R}^n$, $u_i \in \mathbf{R}^m$, $y_i \in \mathbf{R}^p$, $i \in \mathbf{Z}$. The matrices A, B, C have appropriate dimensions. Let the system be controllable and observable. Then it can readily be shown [3] that the problem can be transformed to a single input single output problem by means of a preliminary output feedback, by linearly combining the inputs into a scalar input and considering a linear combination of the outputs as a scalar output. A relevant question is to investigate whether the degrees of freedom involved in these transformations can help to relax some of the assumptions in Theorem 1.

- It is important to stress that the result of Theorem 1 is more than a statement on the genericity of pole assignability by periodic memoryless output feedback. The theorem provides explicit conditions under which the pole assignability problem is solvable. The main part of the proof of the theorem consists of a procedure to calculate the feedback gains.

• It is interesting to discuss the relationship between the periodic control strategy used in the above result and the periodic output feedback as proposed by Kaczorek or others [6, 8, 9]. As explained in 1.2 the feedback control in their approach

$$
u_{rT+i} = k_i y_{rT}
$$

depends on the output at the beginning of the period, where T can be assumed equal to the order of the system. If this control is applied to (1), the closed loop system equation is

$$
x_{(r+1)T} = A^T x_{rT} + A^{T-1} b k_0 c x_{rT} + A^{T-2} b k_1 c x_{rT} + \dots + b k_{T-1} c x_{rT}
$$

The closed loop system modes are the eigenvalues of the matrix

$$
A^T + A^{T-1} b k_0 c + \dots + A b k_{T-2} c + b k_{T-1} c
$$

or of its transpose

$$(A')^T + k_0 c' b' (A')^{T-1} + \ldots + k_{T-2} c' b' A + k_{T-1} c' b'$$

where A', b' and c' are respectively the transposes of A, b and c. If the pair A', b' is observable, or equivalently the pair A, b controllable, then the information contained in $b'(A')^{T-1} x_{mT}, \ldots, b' A' x_{mT}, b' x_{mT}$ is sufficient to derive x_{mT}. Hence the above feedback strategy can be associated with a *state feedback*. Pole assignment is thus possible if moreover the pair $(A')^T, c'$ is controllable, or equivalently the pair A^T, c is observable. This corresponds to the result derived by Kaczorek [9]. The above analysis shows that these results are essentially different from the results of the present paper. The control can actually not be seen as a memoryless feedback control, since the measurement has to be memorized over a period.

2.5   Examples

- The first example concerns a second order system. In this example [2] the gains are determined following the demonstration of Theorem 1. There are of course other, perhaps numerically better ways to obtain the required gains. The theorem serves then as a guarantee that the pole assignment problem is solvable. Consider the system (1) with

$$A = \begin{bmatrix} 0 & 1 \\ 2 & 1 \end{bmatrix}, \ b = \begin{bmatrix} 0 \\ 1 \end{bmatrix}, \ c = \begin{bmatrix} -1 & 1 \end{bmatrix},$$

The system is controllable and observable. The transfer function

$$H(z) = \frac{z-1}{z^2 - z - 2}$$

satisfies the condition of Theorem 1. Hence the poles of this system can be arbitrarily assigned by 3-periodic memoryless output feedback. Notice also that the open loop system is unstable (poles -1 and 2) and cannot be stabilized by time invariant output feedback (it follows from the root locus that the poles remain real valued, with at least one unstable). Suppose the characteristic equation to be realized is

$$z^2 + z + 0.5 = 0$$

i.e. the closed loop system is required to have two complex conjugate poles $-0.5 + 0.5j$ and $-0.5 - 0.5j$. We introduce the time varying memoryless output feedback

$$
\begin{aligned}
u_{3r} &= y_{3r} \\
u_{3r+1} &= 2.0915 \, y_{3r+1} \\
u_{3r+2} &= -3.4665 \, y_{3r+2}
\end{aligned}
$$

with $r \in \mathbf{Z}$. The resulting closed loop matrix is

$$A_{cl} = \begin{bmatrix} 10.933 & -3.933 \\ 33.299 & -11.933 \end{bmatrix}$$

This matrix has indeed the desired characteristic equation.

- As a second example a third order system [3] is considered. As mentioned in the previous section there are strong indications that pole assignment is also possible when some closed loop poles are required to lie at the origin ($\alpha_0 = 0$). The following example supports this conjecture. Consider the discrete-time third order system with

$$A = \begin{bmatrix} 0 & 1 & 0 \\ 0 & 0 & 1 \\ -p_1 & -p_2 & -p_3 \end{bmatrix}, \ b = \begin{bmatrix} 0 \\ 0 \\ 1 \end{bmatrix}, \ c = \begin{bmatrix} q_0 & q_1 & q_2 \end{bmatrix},$$

and the numerical data

$$p_0 = 1.875, \ p_1 = 5.75, \ p_2 = 4.5, \ q_0 = 2, \ q_1 = 3, \ q_2 = 1$$

The system has a transfer function with three real poles (located at -0.5, -1.5 and -2.5), and two real zeros (at -1 and -2). By the root locus method it is readily seen that this system cannot be stabilized by constant memoryless output feedback. Suppose one wants to locate all poles at the origin by means of linear periodic output feedback. This corresponds to a zero value of α_0.

A numerical analysis yields the following feedback gains for a periodic 4-step output feedback strategy

$$k_0 = 0.9375, \ k_1 = 2.528322, \ k_2 = -8.928145, \ k_3 = 10$$

The characteristic equation turns out to be $z^3 = 0$, corresponding to dead beat behavior.

3   Memoryless periodic output feedback control for periodic linear systems

3.1   Preliminaries

In the previous section it was shown that it is possible to obtain arbitrary assignment of the poles of a linear time invariant system by means of memoryless output feedback, provided the feedback gains are not chosen to be constant, but periodically varying. It is natural to examine the the possibilities of periodic feedback control when applied to *periodic* systems. This is the subject of the present section. This problem has only received limited attention in the literature. However we would like to mention a recent paper that addresses the same question [19]. We describe briefly this approach. Basically the original open loop periodic system (of period T) is in a natural way "lifted" to a linear time invariant system of the same order, with the open loop monodromy matrix as drift term and with input and output spaces of higher dimension than for the open loop system. This system is then identified as a multichannel system with T channels. The authors proceed to establish that pole assignment of the original periodic system by means of periodic output feedback of the same period is equivalent to pole assignment by decentralized static output feedback of the associated T-channel system. This relation introduces the notions of "fixed mode" and "strongly connectedness" into the field of periodic systems. Finally invoking some well-known results on pole assignment of linear time invariant systems by means of static output feedback, a number of results on pole assignability are obtained. In this way Yan and Bitmead establish the following result, which is of particular relevance with respect to the subject of this chapter: a periodic system which is strongly connected in the above sense is almost pole assignable by means of periodic output feedback with the same period if there are no fixed modes and the order of the system is less than the sum of the input dimension and output dimension. Note the following aspects of this result which are relevant with respect to the further analysis:

- it is clear that because of the condition on the dimensions of the system the theorem does not cover periodic SISO systems;

- the sufficient condition is independent of the period T of the open loop system;

- the theorem is concerned with *almost* pole assignability.

3.2 Problem statement

Consider a linear periodic discrete time system with scalar input and scalar output

$$x_{i+1} = A_i x_i + b_i u_i \qquad (8)$$
$$y_i = c_i x_i$$

where $x_i \in \mathbf{R}^n$, $u_i \in \mathbf{R}$, $y_i \in \mathbf{R}$, $i \in \mathbf{Z}$. The matrices A_i, b_i, c_i have appropriate dimensions. The system is assumed to be periodic with period T:

$$A_{T+i} = A_i$$
$$b_{T+i} = b_i$$
$$c_{T+i} = c_i$$

The fundamental question that arises is to what extent the pole assignment problem can be solved by introducing time varying (but memoryless) output feedback. In particular we assume that the time dependent feedback gain is periodic with the same period as the system. Consider the system (8) with periodic feedback

$$u_{rT+i} = k_i y_{rT+i}$$

with $r \in \mathbf{Z}$ and $i \in \{0, 1, ..., T-1\}$. With this periodic feedback also the closed loop system is a periodic linear system of period T. This system can be considered as a time invariant system over a time interval equal to the period T:

$$x_{(r+1)T} = (A_{T-1} + k_{T-1} b_{T-1} c_{T-1})(A_{T-2} + k_{T-2} b_{T-2} c_{T-2})... $$
$$(A_0 + k_0 b_0 c_0) x_{rT} \qquad (9)$$

The eigenvalues of the system matrix

$$A_{cl} = (A_{T-1} + k_{T-1} b_{T-1} c_{T-1})(A_{T-2} + k_{T-2} b_{T-2} c_{T-2})...(A_0 + k_0 b_0 c_0)$$

of (9) – from now on also called the poles of the periodic system – determine the dynamics of the periodic system. The original problem of pole assignment can then be restated as selecting feedback gains $k_{T-1}, ..., k_0$ such that the eigenvalues of the closed loop system matrix A_{cl} are the roots of a given polynomial given by (3).

3.3 Reformulation of the problem

The system equation (9) are first rewritten as

$$x_{(r+1)T} = A_{T-1}(A_{T-2} + k_{T-2} b_{T-2} c_{T-2})...(A_0 + k_0 b_0 c_0) x_{rT}$$
$$+ k_{T-1} b_{T-1} c_{T-1}(A_{T-2} + k_{T-2} b_{T-2} c_{T-2})...(A_0 + k_0 b_0 c_0) x_{rT}$$

After introducing the notation[2]

$$\begin{aligned}
\Pi &= (A_{T-2} + k_{T-2}b_{T-2}c_{T-2})...(A_0 + k_0 b_0 c_0) \\
A_{eq} &= A_{T-1}\Pi \\
c_{eq} &= c_{T-1}\Pi \\
b_{eq} &= b_{T-1}
\end{aligned}$$

the system equation (9) is represented in exactly the same form as (4). The same is true for the closed loop system matrix defined in (5). Here also use can be made of the result on time invariant memoryless output feedback derived by van der Woude [14] which was referred to in Section 2.2 and discussed in Appendix 4.1.

3.4   Main result

The analysis of the appendices yields the pole assignment property for systems with period T equal to $n + 1$. Only this case is considered in the present section. The general case is discussed in the next section. For the sake of the presentation a number of assumptions are formulated before the theorem statement is given.

Assumption 1 *The matrices A_i for $i = 0, ..., n$ are nonsingular.*

If necessary an a priori periodic feedback should be introduced to achieve nonsingularity. If this is not possible, then the system (9) has a zero eigenvalue for all feedback coefficients and pole assignment is not possible.

Assumption 2 *The n vectors*

$$\begin{aligned}
&c_1 \\
&c_2 A_1 \\
&... \\
&c_{n-1}A_{n-2}A_{n-3}...A_1 \\
&c_n A_{n-1}A_{n-2}...A_1
\end{aligned}$$

are linearly independent.

Assumption 3 *The n vectors*

$$\begin{aligned}
&c_0 A_0^{-1} \\
&c_1 \\
&c_2 A_1 \\
&... \\
&c_{n-1}A_{n-2}A_{n-3}...A_1
\end{aligned}$$

are linearly independent.

---

[2] The same symbols are used here as in Section 2.2. There should be no confusion.

Assumption 4

$$rank \begin{bmatrix} b_{eq} & A_{eq}b_{eq} & \cdots & A_{eq}^{n-1}b_{eq} \end{bmatrix} = n \qquad (10)$$

with the values of the gains k_i , for $i = 1, ..., n - 1$, given by (22), and with k_0 equal to $k_0^o = -1/c_0 A_0^{-1} b_0$.

Assumption 5

$$rank \begin{bmatrix} b_{eq} & A_{eq}b_{eq} & \cdots & A_{eq}^{n-2}b_{eq} & A_0^{-1}b_0 \end{bmatrix} = n \qquad (11)$$

with the values of the gains k_i, for $i = 1, ..., n - 1$, given by (22), and with k_0 equal to $k_0^o = -1/c_0 A_0^{-1} b_0$.

Assumption 6 *The parameters of system (8) satisfy:*

-
$$c_0 A_0^{-1} b_0 \neq 0$$

- *The denominators of the gains k_i^o, for $i = 1, ..., n-1$, defined by (22), are nonzero.*

-
$$1 + k_i^o c_i A_i^{-1} b_i \neq 0 \text{ for } i = 1, ..., n - 1$$

Theorem 2 *Consider the periodic single input single output system (8) with period $n + 1$. Then under Assumptions 1 to 6, the poles of the system can arbitrarily be assigned by a periodic output feedback strategy with period equal to $n + 1$, provided that there are no closed loop poles at the origin,*

3.5 Discussion

Several of the comments made in Section 2.4 for time invariant systems are also relevant here. Therefore they are
   not repeated. There are also some specific comments to make for the case of periodic systems.

- The result for periodic systems is in a sense more interesting than the result obtained for time invariant systems. Indeed, an interesting aspect of Theorem 2 is that the feedback strategy has exactly the same periodic structure as the system itself. Hence the system structure is not altered by the feedback, in contrast to Theorem 1 where the time invariant system is transformed to a periodic system by the feedback.

- It is instructive to see what Assumptions 1 to 6 imply for time invariant systems and how they relate to the conditions required in the formulation of Theorem 1. Assumption 1 on the nonsingularity of the system matrix has been discussed in Section 2.4. Assumptions 2 and 3 are equivalent for time invariant systems and correspond to observability of the system. Assumptions 10 and 5 coincide for time invariant systems

  [3] and correspond to condition 6 in the statement of Theorem 1. The conditions of Assumption 6 correspond respectively to the condition that q_0 is not zero, that the other coefficients of the numerator polynomial of the transfer function do not vanish, and that the ratios p_i/q_i are different from p_0/q_0.

- The result in Theorem 2 requires the period T of the system to be exactly equal to $n + 1$, where n is the order of the system. Other periods can also readily be dealt with as shown below.

  - If the period T is larger than $n + 1$ then the result can still be used. Indeed if e.g. at $rT + 1, r \in \mathbf{Z}$ the input is set equal to zero, then

    $$x_{rT+2} = A_1 A_0 x_{rT} + A_1 b_0 u_{rT}$$

    This reduces the period by one, considering rT, $rT+2$, $rT+3$, ..., as sampling instants, and $A_1 A_0$ as the new plant matrix and $A_1 b_0$ as the new input matrix at rT. Repeating this procedure (if necessary), one finally obtains a system of period $n + 1$ to which Theorem 2 can be applied. This periodic feedback of period $n+1$ corresponds to a feedback of period T in the original system.
    Another approach would be to consider more variables k_i in the analysis of the appendices, such that some of the variables can be chosen (almost) arbitrarily.

  - If the period T is less than $n + 1$, then the system can also be seen as a periodic system with period $2T$, $3T$, or in general, kT, with $k \in \mathbf{Z}$. In this way a period equal to $n + 1$ or larger can easily be obtained and the problem is reduced to either the case considered above, or the case considered in Theorem 2 itself.

- An important feature of the result of Theorem 2 distinguishing it from the result of Yan and Bitmead [19] is that no minimum number of inputs or outputs is imposed, whereas Yan and Bitmead have to relate the number of inputs and outputs to the order of the system. This is a restrictive condition similar to the requirement needed in the results obtained by Kimura [12] and Rosenthal [13] for time invariant systems.

- Some of the restrictions implied by the assumptions of Theorem 2 can be overcome by a cyclic permutation of the time indices 0, 1, 2, ..., T-1.

References

[1] J. Ackermann, *Abtastregelung*, Springer-Verlag, Berlin (1972)

[2] D. Aeyels and J.L. Willems, "Pole assignment for linear time- invariant second-order systems by static output feedback", *IMA Journal of Mathematical Control and Information*, 8, pp. 267-274 (1991).

[3] D. Aeyels and J.L. Willems, "Pole assignment for linear time-invariant systems by periodic memoryless output feedback", *Automatica*, 28, pp. 1159-1168 (1992).

[4] B.D.O. Anderson and J.B. Moore, "Time-varying feedback laws for decentralized control", *IEEE Transactions on Automatic Control*, AC-26, pp. 1133-1139 (1981).

[5] R.W. Brockett, *Finite Dimensional Linear Systems*, John Wiley and Sons, New York (1970).

[6] A.B. Chammas and C.T. Leondes, "On the design of linear time invariant systems by periodic output feedback", Part I and Part II, *International Journal of Control*, 27, pp. 885-903 (1978).

[7] J.P. Greschak and G.C. Verghese, "Periodically varying compensation of time-invariant systems", *Systems and Control Letters*, 2, pp. 88-93 (1982).

[8] P.T. Kabamba, "Control of linear system using generalized sampled-data hold functions", *IEEE Transactions on Automatic Control*, AC-32, pp. 772-783 (1987).

[9] T. Kaczorek, "Pole assignment for linear discrete-time systems by periodic output feedbacks", *Systems and Control Letters*, 6, pp. 267-269 (1985).

[10] T. Kailath, *Linear Systems*, Prentice-Hall, New York (1980).

[11] P.P. Khargonekar, K. Poolla and A. Tannenbaum, "Robust control of linear time-invariant plants using periodic compensation", *IEEE Transactions on Automatic Control*, AC-30, pp. 1088-1096 (1985).

[12] H. Kimura, "A further result on the problem of pole assignment by output feedback", *IEEE Transactions on Automatic Control*, AC-22, pp. 458-463 (1977).

[13] J. Rosenthal, "New results in pole assignment by real output feedback", *SIAM Journal on Control and Optimization*, 30, pp. 203-211 (1992).

[14] J.W. Van der Woude, "A note on pole placement by static output feedback for single-input systems", *Systems and Control Letters*, **11**, pp. 285-287 (1988).

[15] S.H. Wang, "Stabilization of decentralized control system via time-varying controllers", *IEEE Transactions on Automatic Control*, **AC-18**, pp. 741-744 (1982).

[16] J.L. Willems, V. Kucera and P. Brunovsky, "On the assignment of invariant factors by time-varying feedback strategies", *Systems and Control Letters*, **5**, pp. 75-80 (1984).

[17] J.L. Willems, "Time-varying feedback for the stabilization of fixed modes in decentralized control systems", *Automatica*, **25**, pp. 127-131 (1989).

[18] M. Wonham, *Linear Multivariable Control: a Geometric Approach*, Springer-Verlag, New York (1979).

[19] W.Y. Yan and R.R. Bitmead, "Control of linear discrete-time periodic systems: a decentralized control approach", *IEEE Transactions on Automatic Control*, **37**, pp. 1644-1648 (1992).

Acknowledgments

The authors gratefully acknowledge research support from the Belgian Program on Interuniversity Poles of Attraction initiated by the Belgian State, Prime Minister's Office, Science Policy Programming, and from the EC-Science Project SC1-0433-C(A). The scientific responsibility is assumed by the authors.

4 Appendices

4.1 Van der Woude's lemma

In the discussion of the pole assignment problem in Section 2.2 for time invariant systems as well as in Section 3.3 for periodic systems it was shown that the problem is reduced to finding the feedback gains such that the eigenvalues of (5) are the roots of a given characteristic polynomial (3). A result derived by Van der Woude [14] is well suited to analyze this condition. It is formulated below applied to the problems formulated in Sections 2.2 and 3.3.

Lemma 1 *If the pair A_{eq}, b_{eq} is controllable, there exists an output feedback*

$$v_r = k_{T-1} z_r \tag{12}$$

for the system (4) such that the polynomial $\alpha(z)$, given by (3), is the characteristic polynomial of (5) if and only if

$$\alpha(A_{eq}) ker(c_{eq}) \subset im \begin{bmatrix} b_{eq} & A_{eq} b_{eq} & \cdots & A_{eq}^{T-3} b_{eq} \end{bmatrix} \tag{13}$$

The problem is hence to ensure, if possible, that (13) can be satisfied by a suitable choice of real $k_{T-2}, ..., k_0$. Notice that both sides of (13) depend on $k_{T-2}, ..., k_0$ either through A_{eq} or c_{eq}. Once $k_{T-2}, ..., k_0$ have been computed, it must be checked that the pair A_{eq}, b_{eq} is controllable. The control, given by (12), must then be determined such that

$$\alpha(z) = \det[zI - (A_{eq} - k_{T-1} b_{eq} c_{eq})]$$

The existence of the control follows from Lemma 1. One approach for determining k_{T-1} is to use Ackermann's formula [1, 10]. The original problem is then reduced to ensure that the feedback constants can be chosen such that (13) holds and such that the controllability matrix has full rank:

$$\text{rank} \begin{bmatrix} b_{eq} & A_{eq} b_{eq} & \cdots & A_{eq}^{T-2} b_{eq} \end{bmatrix} = n \tag{14}$$

The expressions of A_{eq} and b_{eq} are different for the time invariant case of Section 2.2 and for the periodic case of Section 3.3. Therefore in the following appendices the solution is first discussed for the case of a periodic system, which of course contains the time invariant system as a particular case. Afterwards the expression of the solution for the time invariant case is derived.

In the sequel we assume that all matrices A_i are nonsingular. If necessary an a priori feedback control is introduced to achieve this. If it is not possible, then the system (8) has a zero eigenvalue for all feedback coefficients, and arbitrary pole assignment can certainly not be achieved.

From the above analysis it is clear that the pole assignment problem is solvable if $k_{T-2}, ..., k_0$ exist which satisfy (13) and (14). Indeed k_{T-1} can then readily be computed. The following appendices are dedicated to an in depth analysis of relations (13) and (14). The main line of the analysis is as follows. First (13) is reformulated as an algebraic equation in $k_{T-2}, ..., k_0$. This equation is explicitly solved in Appendix 4.2 for a special choice of k_0 which is denoted by k_0^o; the obtained solutions are denoted as $k_{T-2}^o, ..., k_1^o$. However these solutions are not acceptable since the choice of k_0 violates one of the conditions imposed. Therefore a perturbation analysis is performed in Appendix 4.3 to show that for k_0 close to k_0^o the equation can also be solved and the solutions can be accepted. The analysis requires some weak restrictions on the system data, which are introduced as the discussion proceeds.

4.2   The determination of the nominal feedback gains

In the analysis of the present and the next appendix the period T is set equal to $n + 1$. This choice is motivated

by the paper of Greschak and Verghese [7], where it is shown that for time invariant second order systems a periodic feedback of at least period 3 is needed for pole assignability. In order to solve (13) the equation is transformed into a more tractable expression under the additional assumption that the matrices $A_i + k_i b_i c_i$, for $i = 0, ..., n-1$, are nonsingular. Then

$$\ker(c_{eq}) = \ker(c_n \Pi) = \Pi^{-1} \ker(c_n) \qquad (15)$$

and (13) is equivalent with

$$\alpha(A_{eq}) \Pi^{-1} \ker(c_n) \subset \text{im} \begin{bmatrix} b_n & A_n \Pi b_n & ... & (A_n \Pi)^{n-2} b_n \end{bmatrix} \qquad (16)$$

Of course it must be checked a posteriori that the nonsingularity conditions are satisfied for the obtained solutions.

Assume that the controllability condition (14) is satisfied and let v^+ be a row vector (dependent on k_i) perpendicular to the $n-1$ independent vectors

$$b_n, \quad A_n \Pi b_n, \quad (A_n \Pi)^2 b_n, ..., (A_n \Pi)^{n-2} b_n$$

Let $c_1^+, c_2^+, ..., c_{n-1}^+$ be a basis of column vectors spanning the null space of c_n (i.e. perpendicular to c_n). Then (16) is equivalent with the set of $n-1$ equations in the n variables $k_0, k_1, ..., k_{n-1}$:

$$v^+ \alpha(A_{eq}) \Pi^{-1} c_i^+ = 0$$

for $i = 1, ..., n-1$. Explicitly these equations are

$$v^+ (A_{eq}^{n-1} A_n + ... \alpha_1 A_n) c_i^+ + \alpha_0 v^+ (A_0 + k_0 b_0 c_0)^{-1} ...$$
$$(A_{n-1} + k_{n-1} b_{n-1} c_{n-1})^{-1} c_i^+ = 0 \qquad (17)$$

The Woodbury formula [10] yields

$$(A + kbc)^{-1} = (1 + kcA^{-1}b)^{-1}[(1 + kcA^{-1}b)A^{-1} - kA^{-1}bcA^{-1}]$$

Then (17) is equivalent with

$$(1 + k_0 c_0 A_0^{-1} b_0) v^+ (A_{eq}^{n-1} A_n + \alpha_{n-1} A_{eq}^{n-2} A_n + ... + \alpha_1 A_n) c_i^+$$
$$+ \alpha_0 v^+ [(1 + k_0 c_0 A_0^{-1} b_0) A_0^{-1} - k_0 A_0^{-1} b_0 c_0 A_0^{-1}]$$
$$(A_1 + k_1 b_1 c_1)^{-1} ... (A_{n-1} + k_{n-1} b_{n-1} c_{n-1})^{-1} c_i^+ = 0 \qquad (18)$$

for $i = 1, ..., n - 1$. Recall that the desired gains k_i must be such that the matrices $A_i + k_i b_i c_i$, with $i = 1, ..., n$, are nonsingular; i.e., for each k_i,

$$1 + k_i c_i A_i^{-1} b_i \neq 0$$

Note that (18) consists of $n - 1$ equations with n variables k_i; hence in general one of the gains may be chosen and the others computed from the equations. We assume that $c_0 A_0^{-1} b_0$ does not vanish. Although we have required that the matrices $A_i + k_i b_i c_i$ are nonsingular, we first solve (18) under the assumption that k_0 is equal to $-1/c_0 A_0^{-1} b_0$, denoted as k_0^o. Appendix 4.3 is devoted to solving (18) for k_0 different from that value.

The gains k_i satisfying (18) for the particular choice of k_0 are called the nominal solutions; they are denoted by k_i^o. It is assumed that α_0 is non-zero. This implies that the desired closed loop system has no poles at the origin. Then (18) yields

$$c_0 A_0^{-1} (A_1 + k_1 b_1 c_1)^{-1} ... (A_{n-1} + k_{n-1} b_{n-1} c_{n-1})^{-1} c_i^+ = 0 \qquad (19)$$

This is equivalent with

$$c_0 A_0^{-1} \div c_n (A_{n-1} + k_{n-1} b_{n-1} c_{n-1}) ... (A_1 + k_1 b_1 c_1) \qquad (20)$$

where the symbol $\div$ stands for proportionality.

Although the above set of equations for the gains $k_1, ..., k_{n-1}$ is non-linear, it is shown below that they can brought in a triangular form and explicitly solved. Assume that the vectors

$$\begin{aligned}
v_1 &= c_1 \\
v_2 &= c_2 A_1 \\
&\quad \cdot \quad \cdot \quad \cdot \\
v_{n-1} &= c_{n-1} A_{n-2} A_{n-3} ... A_1 \\
v_n &= c_n A_{n-1} A_{n-2} ... A_1
\end{aligned}$$

are linearly independent. This is actually an observability condition, requiring that the state can be reconstructed form n successive output measurements. Then $c_0 A_0^{-1}$ can be expressed as

$$c_0 A_0^{-1} = a_1 v_1 + a_2 v_2 + ... + a_n v_n \qquad (21)$$

The constant a_n is assumed to be nonzero; it determines the proportionality factor in (20). The solutions k_i for $i = 1, ..., n-1$ can readily be calculated form (20); the solution k_i^o is expressed as a function of $k_{i+1}^o, ..., k_{n-1}^o$:

$$k_i^o = \frac{a_i}{a_n c_n (A_{n-1} + k_{n-1}^o b_{n-1} c_{n-1}) ... (A_{i+1} + k_{i+1}^o b_{i+1} c_{i+1}) b_i}$$

where the denominators are assumed to be nonzero. More explicitly

$$k_{n-1}^o = \frac{a_{n-1}}{a_n c_n b_{n-1}}$$

$$k_{n-2}^o = \frac{a_{n-2}}{a_n c_n A_{n-1} b_{n-2} + a_{n-1} c_{n-1} b_{n-2}}$$

and in general for $i = 1, ..., n-1$:

$$k_i^o = \frac{a_i}{a_n c_n A_{n-1} ... A_{i+1} b_i + ... + a_{i+1} c_{i+1} b_i} \qquad (22)$$

For the time invariant system (1) the gains $k_1^o, ..., k_n^o$ can be expressed in terms of the coefficients of the transfer function of the system (1). Indeed the Cayley-Hamilton theorem yields

$$A^n = -p_{n-1} A^{n-1} - p_{n-2} A^{n-2} - ... - p_1 A - p_0 I$$

where the constants p_i are the coefficients of the characteristic polynomial

$$\det(zI - A) = z^n + p_{n-1} z^{n-1} + ... + p_1 z + p_0$$

of the matrix A. Hence

$$p_0 = \frac{-1}{a_n}$$

and

$$p_i = \frac{a_i}{a_n}$$

for $i = 1, ..., n-1$. This yields

$$k_i^o = \frac{p_i}{c A^{n-i-1} b + ... + p_{i+2} c A b + p_{i+1} c b}$$

or [3]

$$k_i^o = \frac{p_i}{q_i}$$

This expression is also valid for $i = 0$. The constants q_i are the coefficients of the numerator polynomial of the transfer function of system (1):

$$H(z) = c(Iz - A)^{-1}b = \frac{q_{n-1}z^{n-1} + ... + q_1 z + q_0}{z^n + p_{n-1}z^{n-1} + ... + p_1 z + p_0} \qquad (23)$$

Therefore the condition for the nominal gains k_i^o to be well defined is that all coefficients of the numerator of the transfer function are nonzero.

4.3 Perturbation analysis

In Appendix 4.2 we obtained a set of expressions (22) determining the constants k_i^o $(i = 1, ..., n)$ satisfying relation (16) for a choice of k_0 equal to k_0^o. This value cannot be accepted as was explained in Appendix 4.2. Therefore relation (16) is reconsidered for $k_0 = k_0^o + \epsilon$, with ϵ small. It is shown below that for this choice of k_0 there exist solutions k_i $(i = 1, ..., n-1)$ satisfying (16) or equivalently (18).

The application of the Woodbury formula to all matrices $(A_i + k_i b_i c_i)^{-1}$ in (18) yields

$$(1 + k_0 c_0 A_0^{-1} b_0)(1 + k_1 c_1 A_1^{-1} b_1)...(1 + k_{n-1}c_{n-1}A_{n-1}^{-1}b_{n-1})$$
$$v^+(A_{eq}^{n-1}A_n + \alpha_{n-1}A_{eq}^{n-2}A_n + ... + \alpha_1 A_n)c_i^+$$
$$+\alpha_0 v^+[(1 + k_0 c_0 A_0^{-1}b_0)A_0^{-1} - k_0 A_0^{-1}b_0 c_0 A_0^{-1}]...$$
$$[(1 + k_{n-1}c_{n-1}A_{n-1}^{-1}b_{n-1})A_{n-1}^{-1} - k_{n-1}A_{n-1}^{-1}b_{n-1}c_{n-1}A_{n-1}^{-1}]c_i^+ = 0 \qquad (24)$$

for $i = 1, ..., n - 1$. The vectors c_i^+ constitute a basis of $ker(c_n)$. We take the following particular choice. Consider the set of vectors

$$w_i = v_i A_1^{-1} A_2^{-1}...A_{n-1}^{-1}$$

for $i = 1, ..., n$. These vectors are independent because of the observability assumption introduced in Appendix 4.2. Notice that

$$w_i = v_i A_1^{-1} A_2^{-1}...A_{n-1}^{-1} = c_i A_i^{-1} A_{i+1}^{-1}...A_{n-1}^{-1}$$

for $i = 1, ..., n - 1$, and

$$w_n = v_n A_1^{-1} A_2^{-1}...A_{n-1}^{-1} = c_n$$

Then the vectors c_i^+ are chosen such that c_i^+ is orthogonal to all w_j, for $j \neq i$. Each vector is unambiguously defined up to a scalar.

In an obvious notation each equation in (24) is written as

$$\Phi_i(k_0, ..., k_{n-1}) + \alpha_0 g_i(k_0, ..., k_{n-1}) = 0 \qquad (25)$$

for $i = 1, ..., n-1$. Let k^o denote the vector of the nominal gains $k_0^o, ..., k_{n-1}^o$. Below it is shown that the Jacobian matrix

$$\frac{\partial(\Phi_i + \alpha_0 g_i)}{\partial k_j} \tag{26}$$

with $i, j = 1, ..., n - 1$, evaluated at k^o is nonsingular. This establishes by the implicit function theorem that there exists a solution to

(24) also for $k_0 = k_0^o + \epsilon$ with ϵ sufficiently small. The corresponding solutions k_i are smoothly dependent on ϵ and equal to k_i^o for ϵ equal to zero. Assume

$$k_i^o c_i A_i^{-1} b_i \neq -1 \text{ for } i = 1, ..., n - 1 \tag{27}$$

Then the solutions k_i also satisfy $1 + k_i c_i A_i^{-1} b_i \neq 0$ for small ϵ.

We now prove the nonsingularity of the Jacobian matrix. First notice that evaluated at k^o, we have

$$\frac{\partial(\Phi_i + \alpha_0 g_i)}{\partial k_j} = \alpha_0 \frac{\partial g_i}{\partial k_j} \tag{28}$$

for $i, j = 1, ..., n - 1$, since each Φ_i contains the factor $1 + k_0 c_0 A_0^{-1} b_0$ which vanishes at k^o. For $k_0 = k_0^o$ the function g_i can be written as follows

$$g_i = \frac{v^+ A_o^{-1} b_0}{c_0 A_0^{-1} b_0} f_i$$

with

$$f_i = c_o A_o^{-1}[(1 + k_1 c_1 A_1^{-1} b_1) A_1^{-1} - k_1 A_1^{-1} b_1 c_1 A_1^{-1}]...$$
$$[(1 + k_{n-1} c_{n-1} A_{n-1}^{-1} b_{n-1}) A_{n-1}^{-1} - k_{n-1} A_{n-1}^{-1} b_{n-1} c_{n-1} A_{n-1}^{-1}] c_i^+$$

From (19) it follows that f_i vanishes at k^o. Therefore

$$\alpha_0 \frac{\partial g_i}{\partial k_j} = \alpha_0 \frac{v^+ A_o^{-1} b_0}{c_0 A_0^{-1} b_0} \frac{\partial f_i}{\partial k_j} \tag{29}$$

Notice that with the particular choice of the vectors c_1^+ the function f_1 has the following form

$$f_1 = (1 + k_2 c_2 A_2^{-1} b_2)...(1 + k_{n-1} c_{n-1} A_{n-1}^{-1} b_{n-1})$$
$$c_0 A_0^{-1}[A_1^{-1}(1 + k_1 c_1 A_1^{-1} b_1) - k_1 A_1^{-1} b_1 c_1 A_1^{-1}] A_2^{-1} A_3^{-1}...A_{n-2}^{-1} A_{n-1}^{-1} c_1^+$$

With the particular choice of c_1^+ and invoking (21) one obtains

$$c_0 A_0^{-1} A_1^{-1}...A_{n-1}^{-1} c_1^+ = a_1 c_1 A_1^{-1} A_2^{-1}...A_{n-1}^{-1} c_1^+$$

Hence

$$f_1 = (1 + k_2 c_2 A_2^{-1} b_2)...(1 + k_{n-1} c_{n-1} A_{n-1}^{-1} b_{n-1})$$
$$[a_1(1 + k_1 c_1 A_1^{-1} b_1) - k_1 c_0 A_0^{-1} A_1^{-1} b_1] c_1 A_1^{-1}...A_{n-1}^{-1} c_1^+$$

With

$$c_0 A_0^{-1} = a_1 c_1 + a_2 c_2 A_1 + ... + a_n c_n A_{n-1}...A_1 \qquad (30)$$

one obtains

$$f_1 = (1 + k_2 c_2 A_2^{-1} b_2)...(1 + k_{n-1} c_{n-1} A_{n-1}^{-1} b_{n-1})$$
$$c_1 A_1^{-1}...A_{n-1}^{-1} c_1^+ [a_1 - k_1(a_2 c_2 + a_3 c_3 A_2 + ... + a_n c_n A_{n-1} A_{n-2}...A_2) b_1]$$

From the explicit expressions (22) of k_i^o we conclude that all elements above the diagonal in the Jacobian matrix (26) vanish when they are evaluated at k^o:

$$\frac{\partial f_i}{\partial k_j}(k^o) = 0$$

for $j > i$. As for the diagonal elements one obtains e.g.

$$\frac{\partial f_1}{\partial k_1}(k^o) = -(1 + k_2^o c_2 A_2^{-1} b_2)...(1 + k_{n-1}^o c_{n-1} A_{n-1}^{-1} b_{n-1})$$
$$c_1 A_1^{-1}...A_{n-1}^{-1} c_1^+ (a_2 c_2 + a_3 c_3 A_2 + ... + a_n c_n A_{n-1} A_{n-2}...A_2) b_1$$

This expression is different from zero since (27) holds and the nominal k_i^o exist (i. e. denominators not equal to zero).

To calculate the second diagonal element of the Jacobian matrix (26) the expression of f_2 is written as follows, using the definition of c_2^+:

$$f_2 = (1 + k_3 c_3 A_3^{-1} b_3)...(1 + k_{n-1} c_{n-1} A_{n-1}^{-1} b_{n-1})$$
$$c_0 A_0^{-1}[(1 + k_1 c_1 A_1^{-1} b_1) A_1^{-1} - k_1 A_1^{-1} b_1 c_1 A_1^{-1}]$$
$$[(1 + k_2 c_2 A_2^{-1} b_2) A_2^{-1} - k_2 A_2^{-1} b_2 c_2 A_2^{-1}] A_3^{-1}...A_{n-2}^{-1} A_{n-1}^{-1} c_2^+$$

Invoking (21) and the definition of c_2^+, one obtains:

$$f_2 = (1 + k_3 c_3 A_3^{-1} b_3)...(1 + k_{n-1} c_{n-1} A_{n-1}^{-1} b_{n-1})$$
$$[(1 + k_1 c_1 A_1^{-1} b_1)(1 + k_2 c_2 A_2^{-1} b_2) a_2 - (1 + k_1 c_1 A_1^{-1} b_1) k_2 c_0 A_0^{-1} A_1^{-1} A_2^{-1} b_2$$
$$+ k_1 k_2 c_0 A_0^{-1} A_1^{-1} b_1 c_1 A_1^{-1} A_2^{-1} b_2] c_2 A_2^{-1} A_3^{-1}...A_{n-2}^{-1} A_{n-1}^{-1} c_2^+$$

and hence

$$\frac{\partial f_2}{\partial k_2}(k^o) = -(1 + k_1^o c_1 A_1^{-1} b_1)(1 + k_3^o c_3 A_3^{-1} b_3)...(1 + k_{n-1}^o c_{n-1} A_{n-1}^{-1} b_{n-1})$$
$$c_2 A_2^{-1}...A_{n-1}^{-1} c_2^+ (a_3 c_3 + a_4 c_4 A_3 + ... + a_n c_n A_{n-1} A_{n-2}...A_3) b_2$$

This expression is different from zero since (27) holds and the nominal k_i^o exist (i. e. the denominators not equal to zero).

Similar conclusions can be obtained for the other diagonal elements of the Jacobian matrix. Therefore the Jacobian matrix, evaluated at k^o, is nonsingular if

- the nominal gains k_i^o, for $i = 1, ..., n - 1$, exist and satisfy (27),

- $v^+ A_0^{-1} b_0 \neq 0$, $c_0 A_0^{-1} b_0 \neq 0$ and $\alpha_0 \neq 0$.

Note that $v^+ A_0^{-1} b_0 \neq 0$ is equivalent to

$$\text{rank} \begin{bmatrix} b_{eq} & A_{eq} b_{eq} & \cdots & A_{eq}^{n-2} b_{eq} & A_0^{-1} b_0 \end{bmatrix} = n \qquad (31)$$

if the controllability condition (14) is satisfied at k^o.

For the time invariant case the conditions can be formulated in terms of the coefficients of the numerator of the transfer function (23). In particular the following expressions are obtained[3]:

$$\frac{\partial f_1}{\partial k_1}(k^o) = (1 + k_2^o cA^{-1}b)...(1 + k_{n-1}^o cA^{-1}b)cA^{-(n-1)}c_1^+ q_1/p_0 \qquad (32)$$

$$\frac{\partial f_2}{\partial k_2}(k^o) = (1 + k_1^o cA^{-1}b)(1 + k_3^o cA^{-1}b)...(1 + k_{n-1}^o cA^{-1}b)cA^{-(n-2)}c_2^+ q_2/p_0$$
$$(33)$$

These expressions show that the elements $\frac{\partial f_i}{\partial k_i}(k^o)$ for $i = 1, ..., n - 1$ are nonzero if the coefficients of the numerator of the transfer function (23) do not vanish and if the nominal solutions k_i^o for $i = 1, ..., n - 1$ differ from $-1/cA^{-1}b$. These conditions are included in the formulation of Theorem 1. Moreover it has been shown that for time invariant systems [3] the condition that $v^+ A_0^{-1} b_0$ does not vanish is implied by the controllability condition (14).

---

[3]Note a difference with the expressions given in our original paper [3]; this is due to a computation error. This leads to a slightly different formulation of Theorem 1.

INDEX

ISBN 0-12-012770-9

90065

9 780120 127702

Printed and bound by CPI Group (UK) Ltd, Croydon, CR0 4YY

03/10/2024

01040416-0006